SOUTH-WESTERN

PRACTICAL MATH APPLICATIONS

Sharon Burton

Brookhaven College

Dallas County Community College District

Nelda Shelton

Tarrant County Junior College

South Campus

APR

Face Value

I

$P \times R \times T$

I Quarterly

$P \times R \times T$

Annual Percentage

80%

25%

APR

60%

Editor-in-Chief: Peter McBride
Managing Editor: Eve Lewis
Production Coordinator: Patricia M. Boies
Editor: Edna Stroble
Marketing Manager: Colleen J. Thomas
Cover Design: Alan Brown, Photonics Graphics, Inc.
Electronic Prepress Production: A. W. Kingston Publishing Services

I(T)P®

International Thomson Publishing

South-Western Educational Publishing
is a division of
International Thomson Publishing Inc.

The ITP trademark is used under license.

ISBN: 0-538-70726-7

6 7 8 9 BN 04 03

Printed in the United States of America

Contents

Preface

Even with calculators and spreadsheets, there is still a need for basic math skills. When workers do not have the ability to perform the basic computations required in business, both the worker and the business suffer.

There is a growing emphasis in education at all levels to ensure that students learn the skills that are required in the workplace. The US Department of Labor's Secretary's Commission on Achieving Necessary Skills (SCANS) was set up to help identify these skills. Math skills, and the ability to apply them, are foundation skills. *PRACTICAL MATH APPLICATIONS* provides the opportunity to review basic math skills and apply them to personal and business applications.

Math is a perishable skill. Once learned, it must be used constantly to maintain a high rate of proficiency. Usually, review and drill of basic math concepts bring back a high degree of ability. *PRACTICAL MATH APPLICATIONS* contains many features that will aid your learning review:

Chapter Openers: Each chapter begins with a discussion of how the information taught in the chapter is used in business.

Terminology: Words related to the math and business concepts taught are defined before the concept is taught. Additional terminology may be presented with an individual concept where necessary.

Step-by-Step Approach: Each new concept is illustrated by an example, followed by a step-by-step explanation of how the problem is completed.

Practice Problems: Presenting practice problems early in the lesson allows immediate reinforcement of each new concept as it is covered.

Tips: Information that will aid your learning is scattered throughout the text. Tips are located near the concepts to which they are related. They provide useful shortcuts and reminders.

Math Alert: A challenging problem at the end of each chapter lets you apply the concepts taught to a real-life situation. Math Alert problems are based on a realistic or actual business source, such as a newspaper ad, an article, or an invoice.

Study Guide: Each chapter concludes with a study guide that contains a summary of the vocabulary, concepts, and formulas taught in the chapter. The guide includes a page reference, an example, and a solution.

End-of-Chapter Assignments: Each chapter includes six assignments at the end of each chapter, including three to four word problem assignments.

How to Solve a Word Problem
1. Define the problem. Identify what you are looking for (for example, rate, product, or the result of a combination of calculations).
2. Identify each number in the problem. Remember, not all numbers may be relevant to finding the solution.
3. Plan the solution. Analyze the problem by asking yourself the following questions: What key word(s) identify the math function to be performed. Here are some key words to look for: *Add*: sum, total, increase, how many *Subtract*: less, balance, remainder, difference, withdrawal, outstanding, decrease, left over, How much less? How many fewer? *Multiply*: per, at each, total, number of items in several equal groups *Divide*: per, determine the number of groups, determine the number in each group, how many
4. Decide which numbers relate to what is to be calculated. Write down the number and label it; include the mathematical symbol to be used.
5. Solve the problem. Complete the calculations.
6. Check the solution. If you are using a calculator, do the problem twice to check your answer. Make certain your solution is reasonable. For instance, ask the following questions: In addition: Is my answer greater than any one of the addends? In subtraction: Is my answer less than my subtrahend or minuend? In multiplication: Is my answer greater than my multiplicand? In division: Is my answer less than my dividend?

Acknowledgments

The authors wish to thank all those who encouraged, supported, and shared their insights and suggestions during the writing of this text. We especially want to thank the following people for their meticulous attention to detail and for their many helpful comments:

Thomas Burns Formerly of Phillips Junior College, Raleigh, NC
Sue Gilchrist Southeastern Community College, West Burlington, IA
Charlene Hoon Formerly of Harding Junior College, Struthers, IA
Joann Marler Meadows Business College, Columbus, GA
Kaye Shelton Consultant, Grand Prairie, TX
And finally to our husbands, George and Shell, who
without their love and patience we could not endure.

We would like to hear from you. Please address your comments and questions to the following addresses:

Sharon Burton
Professor, Brookhaven College
Dallas County Community College District
3939 Valley View
Farmers Branch, TX 75244

Nelda Shelton
Associate Professor
South Campus Tarrant County Junior College District
5301 Campus Drive
Fort Worth, TX 76119

CHAPTER
Numbers
1

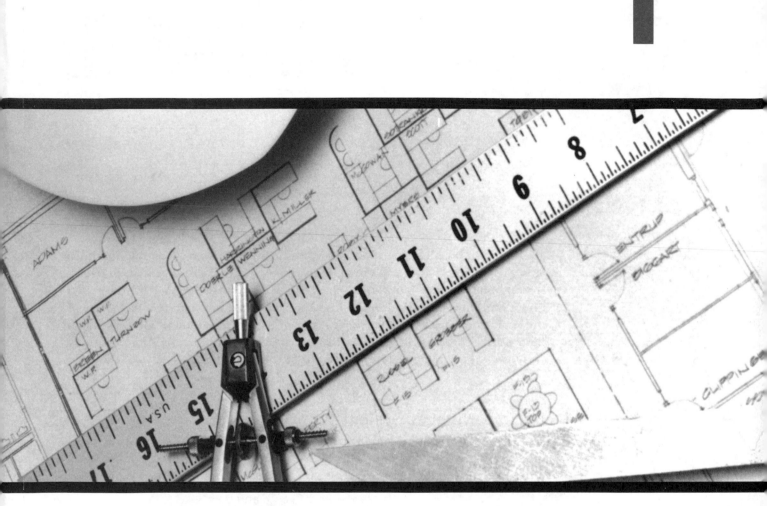

1

Numbers

OBJECTIVES
After completing this chapter, you will be able to:
1. Write numbers or amounts in word form.
2. Approximate and round numbers.

The study of business math is a very practical approach to learning math. In studying business math, you will learn basic math skills that will be useful throughout your life. Checking a sales slip, balancing a checking account, and understanding the various ways interest is charged on loans are just a few of the many practical skills you will learn in the coming chapters. Let's begin your study of business math with an explanation of the *decimal number system*—the number system most used in the United States.

1.1 Decimal Number System

The decimal number system (also called the Hindu-Arabic system) is the most universally used number system today. It is based on ten *digits*; a digit is a numeral. The ten digits used in this system are:

$$1, 2, 3, 4, 5, 6, 7, 8, 9, \text{ and } 0$$

In the decimal number system, the starting point is a dot (.), which is called the *decimal point*. Figure 1-1 shows the names of the digits to the left and right of the decimal point—sometimes called *positions* or *places*—using the number

$$4{,}512{,}367{,}498.26821$$

as an example.

Notice that commas are used to the left of the decimal point to separate each group of three digits. Digits to the left of the decimal point form what is called the *whole-number* part; those to the right of the decimal point form the *decimal* part.

To read a decimal number, begin by reading the whole-number part; then say *and* to indicate the decimal; then read the decimal part.

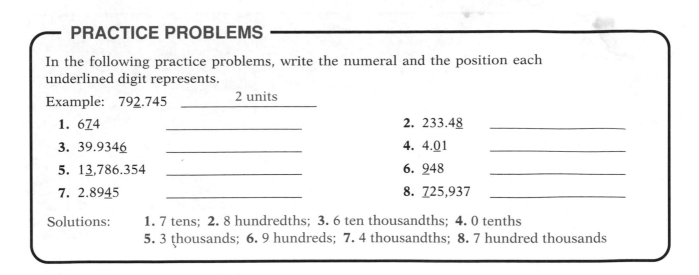

PRACTICE PROBLEMS

In the following practice problems, write the numeral and the position each underlined digit represents.

Example: 79<u>2</u>.745 _____2 units_____

1. 6<u>7</u>4 _____

2. 233.4<u>8</u> _____

3. 39.934<u>6</u> _____

4. 4.<u>0</u>1 _____

5. 1<u>3</u>,786.354 _____

6. <u>9</u>48 _____

7. 2.89<u>4</u>5 _____

8. <u>7</u>25,937 _____

Solutions: **1.** 7 tens; **2.** 8 hundredths; **3.** 6 ten thousandths; **4.** 0 tenths
5. 3 thousands; **6.** 9 hundreds; **7.** 4 thousandths; **8.** 7 hundred thousands

1.2 Numbers and Word Forms

Numbers or amounts may be written using *numerals* or spelled out in word forms. When amounts are written in word forms, the decimal point is represented by the word *and*. (15.78 is written fifteen and seventy-eight hundredths.)

Numbers	Word Forms
3	three
28	twenty-eight
39.06	thirty-nine and six hundredths
837	eight hundred thirty-seven
4,169	four thousand one hundred sixty-nine
72.1	seventy-two and one tenth
283.512	two hundred eighty-three and five hundred twelve thousandths

A common use for word forms is in writing checks. The bank requires you to write the amount in both number and word forms, as shown in the figure on the next page. Notice that the fraction of a dollar $\frac{35}{100}$ is written as a fraction along with the word forms.

FIGURE 1-1
Decimal Number
System

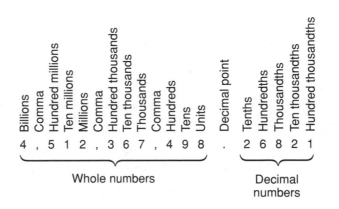

FIGURE 1-2

Check Showing
Number and Word
Forms

To say aloud a number such as 0.98765, pretend it is a whole number (98,765) and notice the final position (place) that is given (hundred thousandths). Then say "ninety-eight thousand, seven hundred sixty-five hundred thousandths."

PRACTICE PROBLEMS

In Problems 1 and 2, write the word forms as numbers:

1. four hundred twenty-five = _____

2. six million two thousand = _____

In Problems 3 and 4, write the numbers in word forms:

3. 431 = _____

4. 11,118.6 = _____

Solutions: **1.** 425; **2.** 6,002,000; **3.** four hundred thirty-one
4. eleven thousand one hundred eighteen and six tenths

1.3 Approximation

Occasionally it is helpful to come near or to *approximate* an amount rather than give an exact amount. For example, suppose you purchased a VCR and two movie tapes for $219.95. If a friend asks what you paid for them, you might say $220. You rounded up the amount to the nearest whole dollar (unit position) rather than saying the actual dollars and cents. Had you paid $219.05, you might have rounded down to the nearest whole dollar and said $219. Study the following examples.

EXAMPLE _____

VCR and two movie tapes cost $219.95 or approximately $220

VCR and two movie tapes cost $219.05 or approximately $219

1.4 Rounding Numbers

If you understand approximation and positions (places), you will more easily understand *rounding* numbers. Here are the rules for rounding numbers:

REMEMBER
Rules for Rounding

Step 1   Determine the position (place) being rounded.
Step 2   If the digit to the right of the position (place) is 5 *or more*, round up by 1. An example of rounding up to the nearest unit is:
<u>6</u>.8 rounds up to 7 (8 is more than 5)
Step 3   If the digit to the right of the position is *less than* 5, do not round up— drop all digits from that position to the right. An example of rounding down to the nearest unit is:
<u>6</u>.4 rounds to 6 (4 is less than 5)

You can easily round to any position by looking at the number to its right and applying one of the two rules just given.

EXAMPLE —————————————————————————————

5.<u>3</u>5 rounded to the nearest tenth is 5.4 (5 is equal to 5)

0.9<u>8</u>1 rounded to the nearest hundredth is 0.98 (1 is less than 5)

When a number in the round-off position is 9, as in 0.<u>98</u>, and the digit to the right is 5 or more, round up by adding 1 to the 9 even though a decimal number may change to a whole number.

EXAMPLE —————————————————————————————

0.<u>9</u>9   rounded to the nearest tenth is 1.0 (9 is more than 5)

0.<u>0</u>9   rounded to the nearest tenth is 0.1 (9 is more than 5)

358.85<u>4</u>9 rounded to the nearest thousandth is 358.855
          (9 is more than 5)

For you to gain a better understanding of rounding, all references thus far have stressed the word "position" rather than "place." However, you need to be aware that place is the most common way to refer to rounding. We might say, "Round your answer to the third place," "Round to three places," or "Carry your answer to three places."

For all problems in this text unless otherwise stated, when instructed to round answers, carry your answers one place past the desired number of places and round up if 5 or larger or drop off if less than 5. This text assumes you will complete all answers by hand. If you are allowed to use a calculator, set your calculator according to the instructions given by your instructor and turn on the round-off key if you have one. If you are allowed to use a hand-held calculator, maximum decimals will be given. To arrive at the correct rounding, look one place beyond the desired number of decimals and round up if 5 or larger or drop off if less than 5.

Round 2,732.971 to the nearest:

1. thousand   _____
2. hundred   _____
3. ten   _____
4. unit   _____
5. tenth   _____
6. hundredth   _____

Round 44,936.1384 to the nearest:

7. ten thousand   _____
8. thousand   _____
9. hundred   _____
10. unit   _____
11. hundredth   _____
12. thousandth   _____

Solutions: **1.** 3,000; **2.** 2,700; **3.** 2,730; **4.** 2,733; **5.** 2,733.0; **6.** 2,732.97; **7.** 40,000; **8.** 45,000; **9.** 44,900; **10.** 44,936; **11.** 44,936.14; **12.** 44,936.138

MATH ALERT

1. The price per gallon of gasoline varies from state to state. Here are some approximate prices per gallon in a few states. The prices on the gas pumps are shown to three decimal places. Round each amount to the nearest penny (hundredth) and write your answers in the blanks provided. Be sure to place dollar signs in all answers.

   (a) Arkansas: $1.415 per gallon   = _____

   (b) Texas: $1.195 per gallon   = _____

   (c) California: $1.455 per gallon   = _____

   (d) New York: $1.605 per gallon   = _____

2. Complete the following check, writing the amounts in numbers and word forms. Start with check number 101. Sign your name to the check.

   Write the check for $455.12.

Your Name Address City, State Zip	No. _____ $\frac{11-71}{690}$
	Nov. 1 19 __
PAY TO THE ORDER OF __*Chase Visa*_____ $ _____	
_____ DOLLARS For Classroom Use Only	
First Texas Bank San Antonio, Texas 78223-6031	
⑈069000712⑈ ⑈016172⑈	

Chapter 1   Numbers
Study Guide

I.   Terminology

page 2	decimal number system	Also called the Hindu-Arabic system. The most commonly used system today.
page 2	decimal point	The starting point (.) or dot in a number.
page 2	whole numbers	Numbers to the left of the decimal point.
page 2	decimal parts	Numbers to the right of the decimal point.
page 3	word forms	Numbers written in words.
page 4	approximation	An amount rounded up or down rather than an exact amount.

II.   Decimal Number System *page 2*

Example Number	*Position*
1. 6,543,896,63<u>2</u>.38904	Represents the <u>units</u> position.
2. 6,543,896,6<u>3</u>2.38904	Represents the <u>tens</u> position.
3. 6,543,896,<u>6</u>32.38904	Represents the <u>hundreds</u> position.
4. 6,543,89<u>6</u>,632.38904	Represents the <u>thousands</u> position.
5. 6,543,8<u>9</u>6,632.38904	Represents the <u>ten thousands</u> position.
6. 6,543,<u>8</u>96,632.38904	Represents the <u>hundred thousands</u> position.
7. 6,54<u>3</u>,896,632.38904	Represents the <u>millions</u> position.
8. 6,5<u>4</u>3,896,632.38904	Represents the <u>ten millions</u> position.
9. 6,<u>5</u>43,896,632.38904	Represents the <u>hundred millions</u> position.
10. <u>6</u>,543,896,632.38904	Represents the <u>billions</u> position.
11. 6,543,896,632.<u>3</u>8904	Represents the <u>tenths</u> position.
12. 6,543,896,632.3<u>8</u>904	Represents the <u>hundredths</u> position.
13. 6,543,896,632.38<u>9</u>04	Represents the <u>thousandths</u> position.
14. 6,543,896,632.389<u>0</u>4	Represents the <u>ten thousandths</u> position.
15. 6,543,896,632.3890<u>4</u>	Represents the <u>hundred thousandths</u> position.

III.   Writing in Word Forms *page 3*

Example Number	*Word Form*
1. 1	One
2. 11	Eleven
3. 111	One hundred eleven
4. 1,111	One thousand, one hundred eleven
5. 11,111	Eleven thousand, one hundred eleven
6. 111,111	One hundred eleven thousand, one hundred eleven
7. 1,111,111	One million one hundred eleven thousand one hundred eleven
8. 0.1	One tenth
9. 0.11	Eleven hundredths

III. Writing in Word Forms (continued) *page 3*

	Example Number	Word Form
10.	0.111	One hundred eleven thousandths
11.	0.1111	One thousand one hundred eleven ten thousandths
12.	0.11111	Eleven thousand one hundred eleven hundred thousandths

IV. Approximation *page 4*

Rounding up or down an amount to the nearest unit, ten, hundred, thousand, etc., rather than give an exact amount.

Examples:

$475.25 rounded up to the nearest hundred would become approximately $500

$1,999 rounded up to the nearest thousand would become approximately $2,000

$22.75 rounded down to the nearest unit would become approximately $22

$525 rounded down to the nearest hundred would become approximately $500

V. Rounding *page 5*

1. Determine the position being rounded.
2. If the digit to the right of the position is 5 or more, round up by 1.
3. If the digit to the right of the position is less then 5, do not round up—drop all digits from that position to the right.

Study Skills

To the Student

At the end of each chapter in Chapters 1 through 6, you will find tips to help you develop better math study habits. Whether you feel confident or nervous about your study of math, these tips can help you. You will learn a variety of practical ways to strengthen your chances of success in this math course.

Managing Your Time

Most instructors suggest the average study time is two hours outside study for every one hour in class. In other words, a three hour class requires six hours of outside study time. This, of course, is an average; some courses require more, some less.

Tips

1. Make a schedule and stick to it. Ask your instructor for a blank study schedule.
2. Set a goal for each class based on the assignments made.
3. Make a "To Do" list. List your top priorities first.
4. Anticipate the unexpected: What if I am interrupted for a long period of time? What if I am called in to work unexpectedly? What if I become ill?
5. Break major tasks into smaller, more manageable ones.
6. Build in breaks and pleasure time for yourself.

Assignment 1

Name_____ Date _____

Complete the following problems. Be sure to mark off thousands and millions with commas.

A. Write the numeral and the position each underlined digit represents.

1. 6<u>4</u>5 _____

2. 42,<u>3</u>49 _____

3. 27.0<u>2</u> _____

4. <u>6</u>7,256.1 _____

5. <u>2</u>,765,437 _____

6. 95.54<u>9</u> _____

7. 8<u>4</u> _____

8. 2,<u>4</u>25 _____

9. <u>4</u>48.394 _____

10. 232.2<u>6</u>5 _____

11. <u>5</u>67 _____

12. 1,00<u>7</u>.1 _____

13. 1.9795<u>1</u> _____

14. <u>7</u>3.50 _____

15. <u>3</u>00,511 _____

16. 0.4<u>3</u>8 _____

17. 0.0005<u>9</u> _____

18. <u>6</u>43.95 _____

19. <u>7</u>,811 _____

20. 1.<u>6</u>2 _____

B. Write numerals for the following word forms.

21. One hundred thirty _____

22. Ninety-five thousand, two hundred twenty _____

23. Seventy-three and sixty-five hundredths _____

24. Six million, three thousand, twenty-one _____

25. Five thousand, six hundred and thirty-four thousandths _____

26. Three hundred eleven _____

27. One thousand, two _____

28. Nine hundred six _____

29. Four billion _____

30. Twenty-six hundredths _____

C. Write word forms for the following numbers.

31. 595 _____

32. 3,721.2 _____

Assignment 2

Name_____ Date _____

Write word forms for the following numbers.

1. 465,602 _____

2. 3,942,001 _____

3. 2.469 _____

4. 3.24 _____

5. 3,452,000.042 _____

6. 0.94265 _____

7. 2.42 _____

8. 45.002 _____

9. 0.462 _____

10. 9,200,001 _____

11. 12,592.7 _____

12. 0.1110 _____

13. 5,429,001,019 _____

14. 66.987 _____

15. 0.889 _____

16. 1.9001 _____

17. 2,335 _____

18. 0.25 _____

19. 47.9 _____

20. 1.92245 _____

21. 255 _____

22. 9.9 _____

23. 6,881.1 _____

Assignment 3

Name_____ Date _____

Complete the following problems.

A. Write the numeral and the position each underlined digit represents.

1. 49<u>5</u> _____

2. 7,<u>4</u>54 _____

3. <u>9</u>,624,246 _____

4. 0.0<u>2</u>014 _____

5. 9.3912<u>4</u> _____

6. 884,2<u>9</u>1 _____

7. 7<u>5</u>,469 _____

8. 5.20<u>4</u> _____

9. 54,269.2<u>4</u>8 _____

10. 3,<u>9</u>12,215 _____

B. Write word forms for the following numbers.

11. 269 _____

12. 4.92 _____

13. 30.9 _____

14. 274.68 _____

15. 4,999.262 _____

C. Write numerals for the following word forms. Be sure to mark off thousands and millions with commas.

16. nine thousand eighty-one _____

17. two million and eleven hundredths _____

18. six hundred thousand, three hundred eleven and five hundredths _____

19. nine hundred sixty-four and twelve hundredths _____

Assignment 4

Name_____ Date _____

Complete the following problems.

A. Write word forms for these numbers.

1. 8.14   _____

2. 9,761   _____

3. 1,457,999   _____

4. 733,908.212   _____

5. 6.168   _____

6. 32,413.29   _____

7. 8.9047   _____

8. 0.0009   _____

9. 23.222   _____

10. 798,443.11   _____

B. Write numerals for the following word forms. Be sure to mark off the thousands, millions, and billions with commas.

11. Sixty-eight thousand twenty-two   _____

12. Four million two hundred three   _____

13. Eighty billion one thousand one hundred and one-tenth   _____

14. Fourteen thousandths   _____

15. Seven hundred eight   _____

C. Write the numeral and the position each underlined digit represents.

16. 28<u>7</u>   _____   **17.** 6.28<u>9</u>   _____

18. 13.8<u>8</u>888   _____   **19.** <u>4</u>4,962.448   _____

20. <u>8</u>,293,398,123.112   _____   **21.** 2,1<u>1</u>9,777,265.9999   _____

22. 7,390.2<u>2</u>73   _____   **23.** 10,42<u>8</u>,331   _____

24. 335,<u>2</u>57.1   _____   **25.** <u>1</u>.4789   _____

Assignment 5

Name _____ Date _____

Complete the following problems. Be sure to mark off thousands or millions with commas.

A. Round to the nearest ten.

1. 946 _____

2. 35.427 _____

3. 5,320 _____

4. $12.15 _____

5. $5,267.82 _____

B. Round to the nearest hundred.

6. 465.2749 _____

7. 642,497.4 _____

8. 269.5 _____

9. $426.48 _____

10. 249.1207 _____

C. Round to the nearest unit.

11. 422.75 _____

12. 7.495 _____

13. 42,962.854 _____

14. 26.426 _____

15. 5,472.1 _____

D. Round to the nearest thousand.

16. 9,742 _____

17. 59,012,345 _____

18. 3,062.475 _____

19. 36,469.5 _____

20. 45,999 _____

E. Round to the nearest tenth.

21. 0.974526 _____

22. 945.65 _____

23. 8,494.201 _____

24. 4.2354 _____

25. 9,999.987 _____

F. Round to the nearest hundredth.

26. 0.497 _____

27. 469.2465 _____

28. 9.4603 _____

29. 29,465.2486 _____

30. 1.769 _____

G. Round to the nearest thousandth.

31. 249.9573 _____

32. 0.56792 _____

33. 5.0679 _____

34. 30,777.7779 _____

35. 7.6490035 _____

H. Round to the nearest ten thousandth.

36. 44.24862486 _____

37. 0.3612159 _____

38. 7.2468901 _____

39. 0.0036999 _____

40. 500.1416822 _____

I. Round to the nearest unit.

41. 49.5 _____

42. 9.99 _____

43. 8.98 _____

44. 10,501.90 _____

45. 5,001.30 _____

46. 4.98 _____

J. Round to the nearest ten thousandth.

47. 41.962483 _____

48. 0.1211111 _____

49. 0.999999 _____

50. 6.42193 _____

K. Round to the nearest tenth.

51. 47.88 _____

52. 0.99 _____

53. 1.498 _____

54. 27.33 _____

55. 8.1192 _____

L. Round to the nearest hundred.

56. 5,968.50 _____

57. 138.59 _____

58. 51.60 _____

59. 809.11 _____

60. 1,509 _____

Assignment 6

Name_____ Date _____

Complete the following problems. Be sure to mark off thousands or millions with commas.

A. Round to the nearest unit.

1. 651.1 _____

2. 4.23 _____

3. 7,908.8 _____

4. 14,543.98 _____

5. 0.99885 _____

B. Round to the nearest ten.

6. 459 _____

7. 4,481 _____

8. 86.12 _____

9. 99,830 _____

10. 72 _____

C. Round to the nearest tenth.

11. 0.3498 _____

12. 7,342.66 _____

13. 237.85 _____

14. 0.875676 _____

15. 7.2312 _____

D. Round to the nearest hundred.

16. 333.86 _____

17. 621.8865 _____

18. 986,123.4 _____

19. 432.015 _____

20. 1,111.21 _____

E. Round to the nearest hundredth.

21. 0.123 _____

22. 8.78 _____

23. 35.3466 _____

24. 19,875.121 _____

25. 1.4561 _____

F. Round to the nearest thousand.

26. 8,765 _____

27. 27,656.1 _____

28. 5,555 _____

29. 23,003 _____

30. 99,999 _____

G. Round to the nearest thousandth.

31. 0.4865 _____

32. 0.2222 _____

33. 69.1784 _____

34. 258.6613 _____

35. 0.837695 _____

H. Round to the nearest ten thousandth.

36. 0.57897 _____

37. 0.12121 _____

38. 13.67986 _____

39. 258.55881 _____

40. 8.633422 _____

I. Round to the nearest ten thousand.

41. 1,798,421 _____

42. 25,595.5 _____

43. 145,659.52 _____

44. 1,234,432 _____

45. 5,001.30 _____

J. Round to the nearest hundred thousand.

46. 4,886,250 _____

47. 299,364.4 _____

48. 711,121.11 _____

49. 874,388.85 _____

50. 49,654,327 _____

K. Round to the nearest unit.

51. 408.1 _____

52. 0.9826 _____

53. 75.863 _____

54. 8,006.9 _____

55. 10,695.56 _____

L. Round to the nearest hundred.

56. 587.90 _____

57. 10,800.90 _____

58. 409.50 _____

59. 399.75 _____

60. 811.20 _____

CHAPTER  2

Operations with Whole Numbers and Decimals

2

Operations with Whole Numbers and Decimals

Addition, subtraction, multiplication, and division of whole numbers and decimals are essential math skills that are used every day by nearly everyone. Like any other skill, these require practice. Some everyday uses of these skills include balancing your checking account, keeping score in a game, or making purchases.

2.1 Terms Used in Addition

Addition is the process of combining two or more numbers and arriving at a larger number. Each number is called an *addend*. The solution is called the *sum*, *total*, or *amount*.

EXAMPLE _____

$$
\begin{array}{r}
931 \leftarrow \text{addend} \\
\text{plus sign} \rightarrow \quad + 27 \leftarrow \text{addend} \\
\hline
958 \leftarrow \text{sum, total, or amount}
\end{array}
$$

2.2 Aligning Digits in Whole Number and Decimal Addition

In addition it is necessary to align the digits in columns so that the units are above units, tens above tens, hundreds above hundreds, and so on.

EXAMPLE _____

$$
\text{Incorrectly written:} \quad
\begin{array}{r}
8,712 \\
256 \\
3,421 \\
+ \quad 951 \\
\hline
13,340
\end{array}
\qquad
\text{Correctly written:} \quad
\begin{array}{r}
8,712 \\
256 \\
3,421 \\
+ \quad 951 \\
\hline
13,340
\end{array}
$$

When adding numbers with decimals, it is important that you align the decimal points. This will automatically insure the proper placement of digits in the appropriate column. The number of decimal places in the answer is determined by the addend with the largest number of decimal places.

EXAMPLE _____

$$
\begin{array}{r}
5,146.11 \\
61.044 \\
10.07 \\
694.081 \\
+1,237.0548 \\
\hline
7,148.3598
\end{array}
$$

Addend with the largest number of
← decimal places

(Answer has 4 decimal places)

┌─ PRACTICE PROBLEMS ─

Rewrite these problems to practice aligning the addends correctly, and write your solutions beneath each problem in the blanks provided showing the maximum decimal places.

1.

$$
\begin{array}{r}
968 \\
1,433 \\
9 \\
809 \\
+ \quad 3 \\
\hline
\end{array}
$$

2.

$$
\begin{array}{r}
39 \\
2,360 \\
1,968 \\
2 \\
+ \quad 14 \\
\hline
\end{array}
$$

3.

$$
\begin{array}{r}
27.6 \\
9.083 \\
156.1 \\
0.089 \\
+ \quad 1.8 \\
\hline
\end{array}
$$

1. _____

2. _____

3. _____

_____

_____

_____

Solutions: **1.** 3,222; **2.** 4,383; **3.** 194.672

2.3 Terms Used in Subtraction

Subtraction is the process of determining the difference between two numbers. The *difference* or *remainder* is obtained by subtracting the *subtrahend* from the *minuend*, as shown:

EXAMPLE _____

$$
\begin{array}{r}
\$95.66 \quad \leftarrow \text{minuend} \\
\text{minus sign} \rightarrow \quad - \ 36.12 \quad \leftarrow \text{subtrahend} \\
\hline
\$59.54 \quad \leftarrow \text{difference or remainder}
\end{array}
$$

In subtraction, as in addition, it is important that you align numbers and decimals as shown in the preceding example. The terminology used in stating a subtraction problem varies. You might say "9 minus 5," "9 take away 5," "9 subtract 5," "5 subtracted from 9," or " the difference between 9 and 5 is."

Another term commonly used in subtraction is *negative number* or *credit balance*, sometimes abbreviated **CR**. A negative number, or credit balance, results when you subtract a larger number from a smaller number. Negative numbers are often expressed by enclosing the number within angle brackets, such as ⟨29⟩. They may also be enclosed within parentheses (29) or printed in red.

2.4 Learning to Subtract Whole Numbers and Decimals

Subtraction is one of the building blocks for other basic math skills. It is important, therefore, that you thoroughly understand subtraction before you attempt to develop other math skills. To subtract vertically, work from the right to the left, subtracting the bottom number from the top.

Because subtraction is based on addition, you can check the results of your subtraction by using addition.

EXAMPLE

$$\begin{array}{llll}
\text{subtraction:} & 36 \text{ minuend} & \text{add:} & 15 \text{ subtrahend} \\
& \underline{-15} \text{ subtrahend} & & \underline{+21} \text{ difference} \\
& 21 \text{ difference} & & 36 \text{ minuend}
\end{array}$$

STEPS

1. Align the minuend, 36, over the subtrahend, 15.

2. Working from right to left, subtract the units position (6 − 5 = 1), writing the difference, 1, beneath the 5 in the units position.

3. Subtract the tens position (3 − 1 = 2), writing the difference, 2, under the 1 in the tens position.

4. Check your answer by adding the subtrahend and the difference. (15 + 21 = 36)

    Solution: 21

— PRACTICE PROBLEMS —

Complete the following problems. Write your answers in the blanks provided. Check you answers by adding the subtrahend and the difference.

1.	947	2.	8,498	3.	$387.11	4.	$46.95
	−138		−3,747		−273.07		−13.42

Solutions: **1.** 809; **2.** 4,751; **3.** $114.04; **4.** $33.53

2.5 Regrouping in Subtraction

Regrouping, or *borrowing* as it is sometimes called, should not be difficult for you. Let's take the example shown below rewritten to show the positions or places to help you learn how to borrow.

EXAMPLE

$$\begin{array}{lll}
75 & \text{or} & 7 \text{ tens} + 5 \text{ units} \\
\underline{-39} & & \underline{-3 \text{ tens} + 9 \text{ units}}
\end{array}$$

STEPS

1. Working from right to left, subtract the units position. Because the 9 cannot be subtracted from the 5, we must borrow 10 from the tens position and add it to the 5 to make 15. In so doing, we are reducing the tens position by 1, leaving 6 tens. Draw a line through the 7 and write a 6 above it to help you remember that you have reduced the 7 to a 6 when you begin subtracting the tens position. We can now subtract the units position ($15 - 9 = 6$).

2. Subtract the tens position ($6 - 3 = 3$).

   Solution: 36

Anytime you cannot subtract a number, regroup by borrowing ten from the place value column(s) to the left of the column in which you are working. Always check your answer by adding the subtrahend and the difference.

PRACTICE PROBLEMS

Complete the following problems. Write your answers in the blanks provided.

1. 61	**2.** 48	**3.** 831	**4.** $68.21
-25	-39	-276	-52.74

Solutions: **1.** 36; **2.** 9; **3.** 555; **4.** $15.47

Difficulties with Zeros in Regrouping

Many students have difficulty with borrowing when the problems contain zeros. The following steps will be helpful to you.

EXAMPLE

$$\begin{array}{r} 900 \\ -\ 18 \\ \hline \end{array} \quad \text{or} \quad \begin{array}{r} 9\text{ hundreds} + 0\text{ tens} + 0\text{ units} \\ -\phantom{9\text{ hundreds} + {}} 1\text{ tens} + 8\text{ units} \\ \hline \end{array}$$

STEPS

1. Subtract the units position. Because 8 cannot be subtracted from 0, borrow 10 from the tens position. Since the tens position is 0, you must borrow 10 from the hundreds position, reducing it 1. Study how the numbers are now regrouped:

   8 hundreds + 10 tens + 0 units

   $$\begin{array}{r} \overset{8\ 10}{\cancel{9}\,\cancel{0}\,0} \\ -\ 1\ 8 \\ \hline \end{array}$$

   Now you can borrow 1 from the tens position as shown here:

   8 hundreds + 9 tens + 10 units

   $$\begin{array}{r} \overset{9}{\underset{}{8\ \cancel{10}\,10}} \\ \cancel{9}\,\cancel{0}\,\cancel{0} \\ -\ 1\ 8 \\ \hline 2 \end{array}$$

   Subtract the units position ($10 - 8 = 2$).

2. Subtract the tens position—remember—the tens position has been reduced 1 because you borrowed 1; it is now 9 (9 − 1 = 8).

$$
\begin{array}{r}
{}^{9}_{8}\ {}^{10}_{\cancel{1}}\ 10 \\
\cancel{9}\cancel{0}\cancel{0} \\
-\ 1\ 8 \\
\hline
8\ 2
\end{array}
$$

3. Subtract the hundreds position (8 − 0 = 8).

$$
\begin{array}{r}
{}^{9}_{8}\ {}^{10}_{\cancel{1}}\ 10 \\
\cancel{9}\cancel{0}\cancel{0} \\
-\ 1\ 8 \\
\hline
8\ 8\ 2
\end{array}
$$

Solution: 882

PRACTICE PROBLEMS

Complete the following problems. Write your answers in the blanks provided.

1. 500	**2.** 4,000	**3.** 3,080	**4.** 205
−269	− 288	− 969	− 88

Solutions: **1.** 231; **2.** 3,712; **3.** 2,111; **4.** 117

2.6 Negative Numbers: Combining Addition and Subtraction

Learning to Express Negative Numbers

A negative number results when a larger number is subtracted from a smaller number.

EXAMPLE _____

        31 minuend smaller
      −48 subtrahend larger
      −17 remainder is a negative number

Negative numbers are sometimes used to express degrees of temperature, such as −2 degrees Fahrenheit. Think of a thermometer. When the temperature drops below 0, we think in terms of negative numbers to describe in degrees the temperature, such as −8 degrees or −10 degrees. Numbers above zero are positive numbers. Notice that the minus sign signals a negative number as well as indicating subtraction.

Another example of negative numbers is when you write a check for more money than you have deposited in the bank. The bank will return the check indicating that you have a negative balance.

On paper it is possible to subtract a larger number from a smaller number.

EXAMPLE _____

          31
        −48

A simple way to determine the answer is to reverse the problem and subtract (48 − 31 = 17), but you must remember to place a minus sign before the answer (−17) or the answer will be incorrect.

Complete the following problems. Write your answers in the blanks provided.

1. 53
  −87
	2. 75
	  −93
		3. 109 − 578 = _____
		4. 413 − 732 = _____

Solutions: **1.** −34; **2.** −18; **3.** −469; **4.** −319

Combining Addition and Subtraction

You may be required to complete calculations on the job that combine addition and subtraction, resulting in a negative or a positive number.

EXAMPLE ——————————————————————————

$$+25$$
$$+13$$
$$-88$$
$$+\ 7$$
$$\underline{-26}$$

Simply total the positive numbers (+) 25, 13, 7; total the negative numbers (−) 88, 26; then subtract the two. If the total of the positive numbers is the larger of the two, the answer is a positive number; if the total of the negative numbers is the larger of the two, the answer is a negative number. You would:

$$\text{Total:} \quad +25 + 13 + 7 = +\ 45$$
$$\text{Total:} \quad -88 - 26 \qquad = -114$$
$$\text{Subtract:} \quad -114$$
$$\underline{\quad\quad\quad +45}$$
$$\text{Solution:} \quad -69$$

■ **TIP**   If all your numbers are negative numbers, it is faster to add them and then place a minus sign beside the answer.

Complete the following problems and write your answers in the blanks provided.

1. +81
  −98
  −21
  + 2
  −55
	2. −46
	  −89
	  −32
	  −19
	  −20
		3. +51 + 65 − 8 + 43 − 22 = _____
		
		4. +$21.88 − $44.78 − $71.11 = _____

Solutions: **1.** −91; **2.** −206; **3.** +129; **4.** −$94.01

2.7 Terms Used in Multiplication

Multiplication is the mathematical procedure for finding the product of two numbers. The number to be multiplied is called the ***multiplicand***, and the number that indicates how many times to multiply is the ***multiplier***. The result or the answer is known as the *product*. Sometimes the multiplicand and multiplier are referred to as *factors*. Study the following example:

EXAMPLE _____

$$
\begin{array}{r}
473 \leftarrow \text{multiplicand} \\
\text{multiplication sign} \rightarrow \underline{\times\ 28} \leftarrow \text{multiplier} \\
13{,}244 \leftarrow \text{product}
\end{array} \Big\} \text{factors}
$$

2.8 Learning to Multiply Whole Numbers and Decimals

To find the product of two numbers, you would:

STEPS _____

1. Align the numbers as you would for addition and subtraction. Note the alignment of the preceding example.

2. Multiply the multiplicand by the first digit on the right (8 in the example) in the multiplier. Place the product below the line:

$$(473 \times 8 = 3784) \qquad \text{Written:} \quad \begin{array}{r} 473 \\ \underline{\times\ \ 8} \\ 3784 \end{array}$$

3. Multiply the multiplicand by the next digit to the left (2 in the example), placing the product on a separate line one place to the left under the first product:

$$(473 \times 2 = 946) \qquad \text{Written:} \quad \begin{array}{r} 473 \\ \underline{\times\ \ 28} \\ 3784 \\ \underline{946} \\ 13{,}244 \end{array}$$

4. Draw a horizontal line, then add.

   Note: Each time a number is multiplied, the product is written under the previous product one place to the left.

   Solution: 13,244

To check an answer in multiplication, reverse the multiplier and multiplicand and multiply. You will obtain the same product.

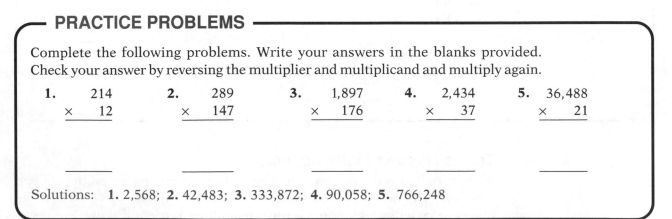

— **PRACTICE PROBLEMS** —

Complete the following problems. Write your answers in the blanks provided. Check your answer by reversing the multiplier and multiplicand and multiply again.

1. 214	**2.** 289	**3.** 1,897	**4.** 2,434	**5.** 36,488
× 12	× 147	× 176	× 37	× 21
_____	_____	_____	_____	_____

Solutions: **1.** 2,568; **2.** 42,483; **3.** 333,872; **4.** 90,058; **5.** 766,248

Practical Math Applications

Multiplication with Decimals

When multiplying numbers that contain decimals, multiply as previously explained. Then count the number of decimal places in the multiplicand and the multiplier and mark off that many places in the product, counting from right to left.

EXAMPLES

$$
\begin{array}{r}
1.9201 \text{ (4 decimal places)} \\
\times \quad 0.6 \text{ (1 decimal place)} \\
\hline
1.15206 \text{ (5 decimal places)}
\end{array}
$$

$$
\begin{array}{r}
\$17.95 \text{ (2 decimal places)} \\
\times \quad 15 \text{ (0 decimal places)} \\
\hline
8975 \\
1795 \\
\hline
\$269.25 \text{ (2 decimal places)}
\end{array}
$$

PRACTICE PROBLEMS

Complete the following problems. Write your answers in the blanks provided, showing maximum decimal places.

■ **TIP**   Remember to count the number of decimal places in each problem.

1.	2.	3.	4.
15.75	$75.38	436.29	0.19647
× 0.137	× 242	× 0.198	× 0.3333

Solutions: **1.** 2.15775; **2.** $18,241.96; **3.** 86.38542; **4.** 0.065483451

Difficulty with Zeros

Two instances where zeros occur in multiplication need explanation: (1) where zeros appear at the end of a number and (2) where they appear in the middle of a number. Follow these steps.

EXAMPLE

$$
\begin{array}{r}
2,150 \quad \text{Zeros at end of numbers} \\
\times \quad 1,300
\end{array}
$$

STEPS

1. When multiplying numbers ending in 0, simply multiply the number and ignore the zeros. Use the numbers from the previous example as follows:

$$
\text{Example:} \quad
\begin{array}{r}
215 \\
\times \quad 13 \\
\hline
2,795
\end{array}
\qquad
\text{Completed:} \quad
\begin{array}{r}
215 \\
\times \quad 13 \\
\hline
645 \\
215 \\
\hline
2,795
\end{array}
$$

2. Count the number of zeros ignored (1 in the multiplicand and 2 in the multiplier = 3) and add three zeros to the right of the product:

Example: 2,795,000

This procedure is called annexing or appending zeros.

Solution: 2,795,000

When the multiplier contains a zero within it, use the following steps to multiply:

EXAMPLE _____

$$\begin{array}{r} 136 \\ \times\ 206 \end{array}$$ Multiplier contains zero

STEPS

1. Apply the rules learned in multiplication until you get to the zero in the multiplier. Ignore it and multiply by the next number (2).

2. When you write the second product, move one extra place to the left for every zero in the multiplier. Study this example:

$$\begin{array}{r} 136 \\ \times\ 206 \\ \hline 816 \\ 272\ \ \ \\ \hline 28{,}016 \end{array}$$ $(136 \times 6 : \text{first product})$
(ignore 0, 136×2 : second product;
move to the position you're multiplying in)

Solution: 28,016

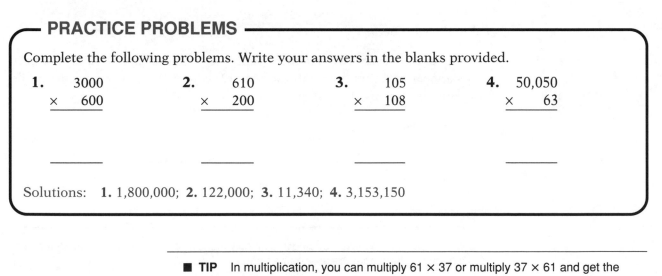

PRACTICE PROBLEMS

Complete the following problems. Write your answers in the blanks provided.

1.	2.	3.	4.
3000	610	105	50,050
× 600	× 200	× 108	× 63

_____   _____   _____   _____

Solutions: **1.** 1,800,000; **2.** 122,000; **3.** 11,340; **4.** 3,153,150

■ **TIP**   In multiplication, you can multiply 61 × 37 or multiply 37 × 61 and get the same product.

Example: 61 × 37 = 2,257
37 × 61 = 2,257

2.9 Accumulation of Products

In business, it is often necessary to accumulate the totals of several products. An example is when you make several purchases while shopping. Follow these steps.

EXAMPLE _____

10 dozen eggs @ $1.05 dozen   = _____

6 cans corn @ $0.53 each   = _____

18 apples @ $0.10 each   = _____

Total = _____

STEPS _____

1. Multiply 10 × $1.05.   (Product: $10.50)
2. Multiply 6 × $0.53.   (Product: $3.18)
3. Multiply 18 × $0.10.   (Product: $1.80)
4. Add the three products.   (Solution: $15.48)

— PRACTICE PROBLEMS —

Complete the following problems and write your answers in the blanks provided. Add the answers obtained in **1.** through **5.** to obtain a total for **6.** and add the answers obtained in **7.** through **9.** to obtain a total for **10.** Since you are working with dollar amounts, show 2 decimal places.

1. $3.47 × 121   = _____   **7.** $44.82 × 6   = _____

2. $1.56 × 17   = _____   **8.** $12.95 × 3   = _____

3. $0.95 × 118   = _____   **9.** $47.88 × 14   = _____

4. $4.11 × 175   = _____   **10.**   Total = _____

5. $1.06 × 140   = _____

6.   Total = _____

Solutions: **1.** $419.87; **2.** $26.52; **3.** $112.10; **4.** $719.25 **5.** $148.40; **6.** $1,426.14
7. $268.92; **8.** $38.85; **9.** $670.32; **10.** $978.09

2.10 Terms Used in Division

Division is the process of determining how many times one number is contained in another. The *dividend* is the number that is to be divided by another number. The *divisor* is the number by which to divide, and the *quotient* is the solution or answer to a division problem. The division symbol, $\overline{)\ \ }$, is drawn over the dividend or the symbol, ÷, is used to indicate you are to divide. If a dividend cannot be divided evenly, the number left over is known as the *remainder* and is often placed over the divisor, such as $\frac{5}{7}$. Study this example.

$$\text{quotient} \to \begin{array}{r} 13 \\ 6\overline{)79} \end{array} \leftarrow \text{dividend}$$

quotient → 13
divisor → 6)79 ← dividend
         6
partial dividend → 19
         18
remainder → 1

Solution: $13\frac{1}{6}$

2.11 Learning to Divide Whole Numbers and Decimals

Let's assume you made the following grades on 5 tests: 80, 92, 78, 91, and 84. To determine your overall grade, add the 5 grades (425) and divide the total by 5 (85). This procedure is called *averaging*. Using this example, let's learn to divide.

5)425

STEPS

1. Look at the dividend and determine what part is greater than the divisor (4 is not greater than 5, but 42 is). Determine the number of times this part can be divided by the divisor — 8 (42 ÷ 5 = 8; 8 × 5 = 40). Write this number (8) over the partial dividend, and write the product (40) under the partial dividend as shown in the following example. Subtract the product from the partial dividend.

      8
5)425 ← partial dividend
    40
     2

2. Bring down the next number in the dividend (5) and place it beside the remainder (2), creating a second partial dividend (25). Determine the number of times the second partial dividend can be divided by the divisor — 5 (25 ÷ 5 = 5; 5 × 5 = 25). Write this number (5) next to the 8, and write the product (25) under the partial dividend as shown:

      85
5)425
   40
    25 ← partial dividend
    25

Solution: 85 (Notice the divisor (5) divides evenly into the dividend (425) with no remainder.)

Division can be checked by multiplying the quotient by the divisor and adding any remainder to the product. The result will be the dividend. Study this example:

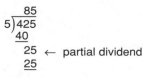

49)1,372   Check: 49 × 28 = 1,372

Estimating is another useful way of checking division. It allows you to determine if your quotient is close enough to the right answer without working through the entire division process. To do this, round the divisor and the dividend to such positions that you can easily divide mentally. Study the following examples.

984 ÷ 53 rounded to 1,000 ÷ 50
1.98 ÷ 30 rounded to 2.00 ÷ 30
1,525 ÷ 52 rounded to 1,500 ÷ 50

By using estimation, you can determine if a quotient is reasonable or not.

PRACTICE PROBLEMS

Complete the following problems. All answers will divide evenly. Use estimating to determine if your answers are reasonable. Check your answers by multiplying the quotient by the divisor.

1. $26\overline{)1,508}$ **2.** $175\overline{)19,250}$ **3.** $49\overline{)4,312}$ **4.** $18\overline{)1,152}$

Solutions: **1.** 58; **2.** 110; **3.** 88; **4.** 64

Division with Remainders

Notice when 425 is divided by 5, the divisor (5) divides the dividend (425) evenly; that is, without a remainder. In the following example, study the division process and note how a remainder is treated.

EXAMPLE _____

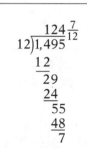

$$
\begin{array}{r}
124\frac{7}{12} \\
12\overline{)1,495} \\
\underline{12} \\
29 \\
\underline{24} \\
55 \\
\underline{48} \\
7
\end{array}
$$

The remainder (7) is shown placed over the divisor, such as $\frac{7}{12}$. This number is a fraction. You will learn more about fractions in Chapters 3-5.

Usually a remainder is shown as a decimal. To do this, you add one zero to the dividend for each decimal place desired. Let's assume in another example 3 decimal places are needed; you carry out the answer 4 places and round the answer to 3 places. See page 5 about rounding in Chapter 1.

EXAMPLE _____

$$
\begin{array}{r}
97.0666 \\
15\overline{)1,456.0000}
\end{array}
$$
 = 97.067 rounded to the thousandths position

← first partial dividend

$\underline{1\ 35}$

106   ← second partial dividend

$\underline{105}$

1 00   ← third partial dividend

$\underline{90}$

100   ← fourth partial dividend

$\underline{90}$

100   ← fifth partial dividend

$\underline{90}$

10

As shown in the third partial dividend, a zero was brought down, but the partial dividend (10) was still smaller than the divisor (15). Therefore, a zero was placed in the quotient and another zero was brought down from the dividend to the third partial dividend.

PRACTICE PROBLEMS

Complete the following problems. Show the remainder over the divisor as a fraction.

1. $21\overline{)3,872}$ 2. $7\overline{)596}$ 3. $41\overline{)1,386}$ 4. $10\overline{)957}$

Solutions: 1. $184\frac{8}{21}$; 2. $85\frac{1}{7}$; 3. $33\frac{33}{41}$; 4. $95\frac{7}{10}$

Division with Decimals

When the dividend contains decimal places, you simply place a decimal point in the quotient at the point above the decimal point in the dividend.

EXAMPLE

$$\begin{array}{r} 2.02 \\ 5\overline{)10.10} \\ \underline{10} \\ 10 \\ \underline{10} \end{array}$$

When the divisor has a decimal, however, it must be changed to a whole number. Therefore, in the example

$$1.95\overline{)975.56}$$

the divisor (1.95) must be changed to a whole number. To do this, you move the decimal point over two places to the right, changing the divisor to 195. When you move the decimal in the divisor, you must also move the decimal in the dividend the same number of places. You then divide as usual.

EXAMPLE

$$1{\scriptstyle\smile}95.\overline{)975{\scriptstyle\smile}56.}$$

In this example, because the decimal point in the divisor was moved two places to the right, the decimal point in the dividend was moved two places to the right.

PRACTICE PROBLEMS

Complete the following problems. Carry all problems to 3 decimal places and round to 2 places.

1. $1.77\overline{)79.65}$ 2. $5.21\overline{)1,976.98}$ 3. $1.98\overline{)369}$

Solutions: 1. 45.00; 2. 379.46; 3. 186.36

MATH ALERT

1. In the following sales report for Benson's Welding, add horizontally total sales for each week, than add vertically total sales for each month. The final total amount in the lower right-hand corner (**h.**) should be the same total both horizontally and vertically. The process of verifying vertical and horizontal addition is called *footing* (or *cross-footing*).

Benson's Welding: Sales for First Quarter

Week	January	February	March	Total Sales
1	$ 10,198	$ 17,533	$ 17,633	a.
2	11,467	16,922	15,171	b.
3	12,121	13,867	14,116	c.
4	15,863	19,730	11,169	d.
Totals	e.	f.	g.	h.

2. In the following problem, the goods in the warehouse at Fine Office Furniture were available for sale (inventory) on January 1 and are listed by dollar amount. The ending inventory as of January 31 has been determined and entered on the form. Determine the cost of goods sold during January by subtracting ending inventory from beginning inventory. Check your work by adding down each column and then subtracting the second column from the first.

Fine Office Furniture
Cost of Goods Sold: January, 19--

Item	Inventory January 1	Inventory January 31	Cost of Goods Sold
Desk, #2199	$ 22,120	$ 8,080	a.
Chair, #0201	975	195	b.
Lamp, #2931	3,258	629	c.
Table, #0875	577	0	d.
Totals	e.	f.	g.

3. Mr. Johnson gave a midterm exam in his Business Math class. The following list shows the scores of his 24 students. Total the scores and obtain an average grade for the class by dividing the total by the number of students taking the test. Round the average to a whole number.

99	72	52	77	98	88	82	76
55	71	77	90	100	60	89	65
88	86	97	100	96	65	82	77

Total Score: _____   Average: _____

4. Complete the following sales slip for Lucky Western Wear by *extending* (multiplying quantity by price and writing the product in the Amount column) and then totaling the Amount column. Write the total at the bottom of the sales slip to indicate a grand total. Since you are working with dollar amounts, round your answers to 2 decimal places.

Lucky Western Wear

No. 292001

2866 Hines Boulevard
Dallas, TX 75201-6328

Phone: (214) 555-2229

Customer order no.: *A1109* **Date:** *10-20-19--*

Sold to: *J. R. Barnes Corp.*

Address: *1406 56th Street*

Dallas, TX 75241-2201

Terms: *N/A* **Sales representative:** *R.L.*

Qty.	Stock No.	Description	Price	Amount
18	B2245	Boot, Leather	79.99	1.
6	H2933	Belt, Leather	28.95	2.
20	H2935	Belt Buckle, Silver	34.95	3.
15	I9986	Shirt, Long Sleeve	34.95	4.
6	K2916	Hat	75.00	5.
3	K1468	Hat	82.50	6.
24	L1456	Jeans	54.00	7.
15	M129	Hatband	67.50	8.
12	N1698	Bootjack	·12.95	9.
144	P1698	Hatpin, Assorted	3.95	10.
		Total		11.

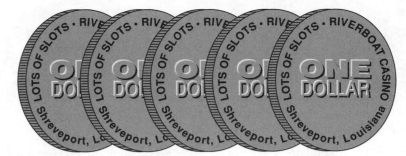

5. Beth and Mary decided to check out the casino boat gambling in Shreveport, Louisiana. Both women decided they would take $200 each to gamble. Beth decided to begin with buying 4 rolls of nickels at $2 a roll, 3 rolls of quarters at $10 a roll, and one roll of dollar coins at $20 a roll. Mary decided to buy 2 rolls of dollar coins.

a. How much money in coins did Beth buy? _____

b. How much money did Beth have left out of her $200 in currency? _____

c. How much money did Mary buy? _____

d. How much money did Mary have left out of her $200 in currency? _____

Mary played the dollar slot machines during her time at the casino. She put in $10 and hit triple red, white, and blue sevens and won $240. She continued playing and put $38 in coins back in the machine and won $80. She played another $45 in coins and hit for $600.

e. By the end of their time there, how much money did Mary end up with over and above her $200 investment? _____

No need to calculate Beth's winnings—she lost it all!

Study Guide

I. Terminology

page 16	addition	The process of combining two or more numbers and arriving at a larger number.
page 16	addend	Each number added.
page 16	sum, total, or amount	The solution to an addition problem.
page 17	subtraction	The process of determining the difference or remainder between two numbers.
page 17	difference or remainder	The result of subtracting the subtrahend from the minuend.
page 17	minuend	The number from which another number is being subtracted; written as the top number.
page 17	subtrahend	The number being subtracted from the minuend; written as the bottom number.
page 18	negative number	The difference or remainder that results from subtracting a larger number (subtrahend) from a smaller number (minuend). The difference is shown in brackets, in red, or by writing CR beside the answer.
page 18	regrouping or borrowing	Occurs in subtraction when any single number in the subtrahend is smaller than the number below it in the minuend. The process of subtracting 10 from the position to the left of the number in the subtrahend. The 10 is added to the number in the subtrahend so the number in the minuend can be subtracted from it.
page 21	multiplication	The mathematical procedure for finding the product of two numbers.
page 21	multiplicand	The number to be multiplied.
page 21	multiplier	The number that indicates how many times to multiply.
page 22	product	The result or answer to a multiplication problem.
page 22	factors	A reference to the multiplicand and multiplier.
page 24	annexing/appending	The process of ignoring zeros that appear to the right in a multiplication problem; multiplying; then adding the zeros to the right in the product.
page 25	division	The process of determining how many times one number is contained in another.
page 25	dividend	The number that is to be divided by another number.
page 25	divisor	The number by which to divide.
page 25	quotient	The solution or answer to a division problem.
page 25	division symbols	The symbol, $\overline{)}$, drawn over the dividend, and $\div$ for division.
page 25	remainder	The number left over when a dividend cannot be divided evenly by the divisor.
page 26	partial dividend	Any part of the dividend the divisor will divide into.

II. Addition

page 16   Align units above units, tens above tens, hundreds above hundreds, and so on. Place commas from right to left every three digits. If decimals are included, decimal points are aligned.

Whole numbers:	110,446	Decimal numbers:	310.45
	9,321		8.111
	25		0.2177
	129		89.01

page 16   *Adding:* Add down the units column a unit at a time arriving at subtotals after each unit. Write the unit part of the answer under the unit column and write the number to be carried over at the top of the tens column. Repeat this process for the remaining columns.

III. Subtraction

page 18   *Subtracting:* Align the minuend over the subtrahend. Subtract from right to left subtracting each position. When a number in the minuend is larger than the number above it in the subtrahend, you must borrow ten from the position to its left, reducing that position ten and adding ten to the number in the subtrahend and continue subtracting.

page 18   *Checking Subtraction:* Subtraction can be checked by adding the difference or remainder and the subtrahend.

page 20   *Subtracting Negative and Positive Numbers:* Add all positive numbers; add all negative numbers; subtract the larger number from the smaller. If the negative number is larger, it is a negative number. If the positive number is larger it is a positive number.

IV. Multiplication

page 22   *Multiplying:* Align the numbers or decimals; multiply the multiplicand by each digit working right to left in the multiplier; draw a horizontal line; add.

$$
\begin{array}{r}
473 \\
\times\ \ 28 \\
\hline
3784 \\
946\ \ \\
\hline
13,244 \\
\end{array}
$$

Example:

page 22   *Checking Multiplication:* Reverse the multiplier and multiplicand and multiply again.

page 23   *Multiplication with Decimals:* Multiply in the usual manner. Count the number of decimal places in the multiplicand and multiplier and mark that many places in the product, counting from right to left.

page 23   *Multiplying with Zeros:* When multiplying by numbers ending in 0, multiply ignoring the zeros. Count the number of zeros ignored and add them to the right side of the product.

page 25   *Accumulation of Products:* Multiply each multiplicand and multiplier and obtain the product. Add the products.

V. Division

page 26   *Dividing:* Place the dividend under the division symbol and the divisor to the left. Determine what part of the dividend the divisor will divide into (partial dividend) and how many times. Write this number over the partial dividend and write the product under the partial dividend. Subtract the product from the partial dividend. Bring down the next number in the dividend and place it by the remainder. Continue dividing, repeating the process until the desired number of decimals are determined.

$$\text{Example:} \quad 5\overline{)425} \atop \begin{array}{r} 85 \\ \hline 40 \\ \hline 25 \\ \underline{25} \end{array}$$

page 26   *Checking Division:* Multiply the divisor by the quotient.

page 26   *Estimating:* Round the divisor and the dividend to such positions that you can easily divide mentally, then compare the answers with the actual quotient to determine if the quotient is reasonable.

page 27   *Division with Remainders:* Place any remainder over the divisor as a fraction such as $\frac{2}{12}$.

page 28   *Division with Decimals:* When the dividend contains a decimal, place a decimal point in the quotient at the point above the decimal in the dividend. When the divisor has a decimal, change it to a whole number by moving the decimal to the right the desired number of places; move the decimal to the right in the dividend the same number of places.

Study Skills

Where to Study

Places to study vary from the library, home, in a lounge, in your car, at the student center, or in a lab at school, just to name a few.

Tips

1. Plan a place that is most suited to your habits—in your room, at the kitchen table, home office, den, or library. Identify it as your personal study place to other family members.
2. Have available at your study area reference materials, office supplies, calculator, clock, and various other supplies as needed. If you choose the kitchen table, prepare a box that can be taken to the table and later put away that has all your supplies in it.
3. Use the study area only for studying.
4. Make sure the room is a comfortable temperature and is well lighted.
5. Make sure you have plenty of surface area on which to work.
6. Choose a comfortable chair—one with armrests—and make certain you sit upright while studying.

Assignment 1 Name_____ Date_____

Complete the following problems. Write your answers beneath the problems. Add dollar signs and commas to mark off thousands where necessary. Show maximum decimal places.

A. Add.

| 1. | 3,226
231
50
+ 1,257 | 2. | 45,112
3,785
221
+ 9,353 | 3. | 70,001
3,233
75
+ 15,280 | 4. | 398
4,567
197
+ 8,199 | 5. | 33,723
37
1,445
+ 25,256 |

| 6. | 98,000
7,455
290
+ 14,230 | 7. | 0.455
0.23
1.096
+ 0.905 | 8. | 1.223
0.6
0.9855
+ 2.0505 | 9. | 77.44
2.9
36.97
+ 85.31 | 10. | 9.99
5.32
0.355
+ 6.999 |

| 11. | 7.222
0.579
0.166
+ 1.02 | 12. | 4.9123
0.288
0.76
+ 7.987 | 13. | $27.50
15.95
1.99
+ 83.11 | 14. | $1.98
3.01
7.55
+ 5.05 | 15. | $98.88
7.27
90.10
+ 27.01 |

B. Subtract vertically.

| 16. | 600
− 234 | 17. | 7,090
− 3,156 | 18. | $900.00
− 436.48 | 19. | 5,000,000
− 90,768 |

| 20. | 11,908
− 9,090 | 21. | 8,007
− 5,970 | 22. | 380.25
− 198.07 | 23. | 30,080,193
− 29,090,009 |

| 24. | $200.05
− 10.09 | 25. | $90.90
− 70.60 | 26. | $3,000,100
− 999,999 | 27. | $769,487
− 10,807 |

C. Subtract horizontally.

28. 345 − 290 = _____

29. $59.50 − $5.07 = _____

30. 4,200 − 387 = _____

31. 4,489 − 379 = _____

32. $24.69 − $3.02 = _____

33. 1,467,902 − 9,999 = _____

Assignment 2

Name_____ Date _____

Complete the following problems. Write your answers beneath the problems or in the blanks provided. Be sure to place commas and dollar signs as needed. Show maximum decimals.

A. Add vertically.

1.	2.	3.	4.	5.
4,590	15,020	$190.00	349.12	$899.35
328	987	214.99	21.33	654.76
5,326	10,440	43.55	2.19	333.77
+10,430	+22,436	+653.80	+321.03	+496.13

6.	7.	8.	9.	10.
3,711	15,215	$180.00	321.00	9,000,214
6,156	2,647	507.20	32.76	8,976,021
830	218	129.90	235.22	6,865,038
+8,964	+14,657	+482.13	+598.14	+1,904,320

B. Add horizontally.

11. 3,975 + 125 + 1,890 + 25 = _____

12. $27.95 + $15.44 + $18.01 + $10 = _____

13. 125 + 600 + 150 + 45 + 11 + 2 = _____

14. 1,224 + 7,993 + 1,000 + 511 = _____

15. 1.238 + 13.05 + 0.9090 + 4.555 = _____

16. 3.88 + 0.1956 + 2.1 + 6 = _____

C. Subtract vertically.

17.	18.	19	20.	21.
1,557	18.99	47.960	$88.75	10.45
– 999	– 25.40	– 33.111	– 96.16	– 9.19

22.	23.	24.	25.	26.
$396.50	43.120	43.9091	273.10	$12,889.22
– 451.77	– 21.05	– 51.9214	– 111.96	– 6,021.67

D. Subtract horizontally.

27. $231.04 – $18.95 = _____

28. 43.19 – 11.76 = _____

29. 1,234 – 987 = _____

30. 1.235 – 8.303 = _____

31. 76.111 – 32.93 = _____

32. 18.18 – 12.12 = _____

E. Combine addition and subtraction.

33. $27.55 – $18 + $23.10 + $56.33 – $22.02 – $4.04 = _____

34. 21.99 + 31.25 + 289.77 – 28.99 – 43.12 – 45.33 = _____

35. –17.11 + 27.90 – 32.87 – 39.90 – 41.35 + 3.56 = _____

36. 1.909 + 0.991 + 1.966 + 9.123 – 8.438 + 15.290 = _____

Assignment 3

Name_____ Date _____

**Complete the following problems. Write your answers in the blanks provided.
Be sure to place commas and dollar signs in all problems where needed.
Round amounts to 2 places. (See page 5 in Chapter 1 about rounding.)**

1. $\begin{array}{r} 0.3952 \\ \times\ 0.4834 \\ \hline \end{array}$

2. $\begin{array}{r} 76.9 \\ \times\ 48.2 \\ \hline \end{array}$

3. $\begin{array}{r} 1.111 \\ \times\ 1.3 \\ \hline \end{array}$

4. $\begin{array}{r} \$86.92 \\ \times\ 0.05 \\ \hline \end{array}$

_____ _____ _____ _____

5. $\begin{array}{r} \$12.37 \\ \times\ 0.0008 \\ \hline \end{array}$

6. $\begin{array}{r} 1.38 \\ \times\ 0.19 \\ \hline \end{array}$

7. $\begin{array}{r} 3.92 \\ \times\ 0.1619 \\ \hline \end{array}$

8. $\begin{array}{r} 5.15 \\ \times\ 0.482 \\ \hline \end{array}$

_____ _____ _____ _____

9. $\begin{array}{r} 69.1 \\ \times\ 0.3 \\ \hline \end{array}$

10. $\begin{array}{r} 41.0909 \\ \times\ 0.008 \\ \hline \end{array}$

11. $\begin{array}{r} 4.6 \\ \times\ 1.8 \\ \hline \end{array}$

12. $\begin{array}{r} 6.15 \\ \times\ 0.59 \\ \hline \end{array}$

_____ _____ _____ _____

13. $\begin{array}{r} 1.566 \\ \times\ 0.20 \\ \hline \end{array}$

14. $\begin{array}{r} \$8.11 \\ \times\ 0.25 \\ \hline \end{array}$

15. $\begin{array}{r} \$104.05 \\ \times\ 0.101 \\ \hline \end{array}$

16. $\begin{array}{r} 6.13 \\ \times\ 2.5 \\ \hline \end{array}$

_____ _____ _____ _____

17. $\begin{array}{r} 7.01 \\ \times\ 22 \\ \hline \end{array}$

18. $\begin{array}{r} 376 \\ \times\ 0.24 \\ \hline \end{array}$

19. $\begin{array}{r} 5.59 \\ \times\ 1.6 \\ \hline \end{array}$

20. $\begin{array}{r} \$295.01 \\ \times\ 0.49 \\ \hline \end{array}$

_____ _____ _____ _____

Assignment 4

Name_____ Date _____

**Complete the following problems. Write your answers in the blanks provided.
Be sure to place commas and dollar signs in all problems where needed.
Round all answers to two decimal places.**

A. Division using whole numbers

1. $2,118 \div 736$ = _____

2. $\$4,536 \div 4$ = _____

3. $10,754 \div 378$ = _____

4. $\$30,478 \div 10$ = _____

5. $3,372 \div 14$ = _____

6. $6,257 \div 27$ = _____

7. $2,756 \div 260$ = _____

8. $33,632 \div 39$ = _____

9. $7,104 \div 3$ = _____

10. $\$4,032 \div 191$ = _____

11. $34,038 \div 92$ = _____

12. $3,922 \div 523$ = _____

13. $6,868 \div 10$ = _____

14. $\$55,726 \div 24$ = _____

15. $2,184 \div 8$ = _____

16. $10,282 \div 50$ = _____

17. $3,656 \div 396$ = _____

18. $62,892 \div 601$ = _____

19. $1,536 \div 72$ = _____

20. $1,870 \div 291$ = _____

21. $469 \div 83$ = _____

22. $3,089 \div 206$ = _____

Write answers as shown.

23. $97\overline{)9,409}$ (answer 97 shown above)

24. $125\overline{)15,625}$

25. $61\overline{)5,063}$

26. $254\overline{)2,794}$

B. Division containing decimals–Round all answers to 2 places.

27. $11.1 \div 0.38$ = _____

28. $601 \div 2.221$ = _____

29. $0.25 \div 0.413$ = _____

30. $433.87 \div 21$ = _____

31. $\$75.40 \div 1.40$ = _____

32. $1.85 \div 0.131$ = _____

33. $\$1.99 \div 0.501$ = _____

34. $2,997 \div 0.77$ = _____

35. $67.5 \div 9.87$ = _____

36. $623 \div 88.11$ = _____

37. $2.22 \div 0.111$ = _____

38. $\$200 \div 5.11$ = _____

39. $32.87 \div 199$ = _____

40. $\$21.00 \div 98$ = _____

41. $21.998 \div 0.77$ = _____

42. $7.865 \div 0.65$ = _____

43. $\$2.98 \div 69$ = _____

44. $8.090 \div 0.006$ = _____

Assignment 5

Name_____ Date _____

Complete the following problems. Write your answers in the blanks provided. Be sure to place commas and dollar signs where needed. Round amounts to two places. (See page 5 in Chapter 1 about rounding.)

1. Madison Furniture Company ordered 4 sofa-sleepers at $599 each, 3 love seats at $199 each, and 3 ottomans at $79 each. What was the total purchase price of all the furniture ordered? _____

2. Meredith worked part time sacking groceries and checking at the cash register for a local chain food store. When she sacked groceries, she was paid $4 per hour; when she checked, she was paid $5.50 per hour. Monday she checked and worked 8 hours, Tuesday she sacked and worked 4 hours, Wednesday she did not work, Thursday she checked and worked 8 hours, Friday and Saturday she sacked and worked 8 hours each day. What were Meredith's total wages earned before taxes for the week? _____

3. Jesse Morton sold 5 cars he had restored to like-new condition. He received $8,000 each for the cars. What was the total price he received for the 5 cars? _____

4. An accident occurred, causing damages to 2 cars. Each owner took his car to Carlos' Repair Shop for estimates. Carlos estimated each car needed a new fender at $420 per fender, a new bumper at $450 each, and 1 pair of headlights each at $50 a pair. What was the total estimate for 1 car? _____ For 2 cars? _____

5. Erin went to Best Burgers and bought 5 hamburgers at $1.79 each, 5 small soft drinks at $.89 each, and 5 large orders of French fries at $.79 each. What did the purchase cost Erin before tax? _____

6. Fuji Moore paid $467 per month for 48 months on her automobile. What did the car cost her? _____

7. Dale has been working for Very-Clean Carpet Cleaners. A bonus is paid for each additional service he sells during the week. This week he sold 5 heavy-duty treatments and received a $25 bonus for each, 6 extra rooms and received a $10 bonus per room, 10 poly-care protective coatings and received an $8 bonus for each. How much money did he make in bonuses this week? _____

8. Kim received a raise of $.70 per hour. Her old hourly wage was $4.80. How much per 40-hour week was Kim making before her raise? _____ After her raise? _____

9. Rosita ordered several board feet of lumber. The order included 12 1" x 4s" at $2.50 each, 9 1" x 2s" at $1.30 each, 12 sheets of plywood at $10 each. What was Rosita's total bill?

    _____

10. Rick, Dave, Bryan, and John decided to purchase a used bass boat for $4,550 and share the cost equally. How much would each person pay? _____

11. Ruth and Merri ate lunch together at Berries Restaurant. Since Berries serves such large amounts, they decided to buy one order and split the cost. Their lunch came to $12.74, and they left a $2 tip. How much did lunch (with tip) cost each person? _____

12. The Navajo National Bank loaned John Morgan $150,000 including interest for a new home he and his wife were purchasing. The loan is to be paid over a period of 25 years. How much would John's monthly payments be? _____

Assignment 6

Name _____ Date _____

Complete the following problems. Write your answers in the blanks provided. Be sure to place commas and dollar signs where needed. Round amounts to two places. (See page 5, in Chapter 1 about rounding.)

1. Randy sells bread to the local grocery stores. He sells daily 144 loaves of wheat at $.75 each and 35 loaves of french bread at $.55 each. What are Randy's total daily sales?

   _____

2. Ann Trujillo can type 80 words per minute on the computer. How many words can she type in an 8 hour day if she never stopped?

   _____

3. Boris works in a dry cleaning store. He worked 40 hours at $5.50 and 23 hours at $8.25 for overtime. What was his total pay check? _____

4. Julio played all three courses at the miniature golf course. His scores were 42, 36 and 48. What was the average of all three scores?

   _____

5. Winston has paid $268.34 for 60 months on his automobile. What did the car cost Winston? _____

6. Bill bowled on a league every Monday night. He bowled 4 games with the following scores: 237, 222, 190, and 256. What is his average score for the night? _____

7. Redmond Brick Company delivered 4,368 bricks for a new home to be built. There are 14 workers ready to begin. How many bricks will each worker use? _____

8. Gloria went shopping and bought 4 dresses at $49.99 each, 5 pairs of jeans at $35.00 each and 3 pairs of shoes at $29.99 each. How much did Gloria spend? _____

9. Hector has 55 pounds of cans to recycle at $0.55 per pound. He also has 28 pounds of newspaper to recycle at $0.34 per pound. How much will the recycling plant pay Hector?

   _____

10. Sally jogged 5 miles on Monday, 2.5 miles on Tuesday, 3.4 miles on Wednesday, 4.1 miles on Thursday, and 3.1 miles on Friday. What is her average distance for the week?

    _____

11. Leticia decided to pay for a shampoo and set and a manicure for her bridesmaids the day of her wedding. A shampoo and set at Hair 'n' Nail Masters is $18, and a manicure is $25. Leticia has asked six friends to be her bridesmaids. How much will she pay Hair 'n' Nail Masters for her bridesmaids' trip to the beauty salon? _____

12. Bob travels for his company. He must estimate the expenses he believes he will have for a trip to San Francisco: Hotel for 5 nights at $105 per night; food for 5 days at $36 per day; taxi and tips for 5 days at $25 per day; car rental for 5 days at $46 per day; 2 tanks of gasoline at $20 per tank; round-trip airfare at $458. What will Bob's estimated expenses be? _____

13. Over a span of 8 weeks last winter, David recorded the following snowfall amounts per week in Alamosa, Colorado: 6 inches, 14 inches, 22 inches, 4 inches, 12 inches, 29 inches, 3 inches, and 8 inches. What was Alamosa's average snowfall over the 8-week period? _____

Proficiency Quiz
R E V I E W

Name_____ Date_____

$$\frac{\text{Student's Score}}{\text{Maximum Score}} = \frac{}{70} = \text{Grade}_____$$

Complete the following problems. Write your answers in the blanks.

A. Numbers

1. What is the position the underlined number represents: 6<u>7</u>9 _____

2. Write the following in numbers: ninety-three and eleven hundredths _____

3. Round the following number to the nearest ten: $87.56 _____

4. Round the following number to the nearest tenth: 0.65901 _____

5. Round the following number to the nearest ten thousandth: 0.98356 _____

6. What position does the underlined number represent: 104.99<u>5</u> _____

7. Write the following in numbers: nine and thirty-seven thousandths _____

8. Round the following number to the nearest ten thousand: 48,694 _____

9. Round the following number to the nearest tenth: 3,894.763 _____

10. Round the following number to the nearest unit: 465.87 _____

Complete the following problems. Write your answers in the blanks. Place dollar signs in answers dealing with money and mark off thousands by commas.

B. Addition

11.	12.	13.	14.
4,611	389.141	$869.42	0.9
386	23.865	17.50	8.8
5,865	111.862	0.66	57.008
5	21.844	1.00	0.026

15. $13,491 + 24,090 + 211 + 93,488 =$ _____

16. $4.32 + 8.95 + 4.10 + 8.01 =$ _____

17. $14.95 + 67.82 + 29.04 + 4.11 =$ _____

18. $923.66 + 873.75 + 763.50 =$ _____

19. $27 + 86 + 41 + 33 + 69 =$ _____

20. $1.165 + 2.8 + 4.49 + 0.006 =$ _____

21. Diana purchased several office supplies. She purchased a box of labels for $1.49; a box of staples for $0.99; a box of paper clips for $0.87; a roll of masking tape for $1.29; and a roll of postage stamps for $32. What was the total of Diana's purchases? _____

22. Brian worked overtime several days this week. On Monday he worked 6 hours overtime; Tuesday, 7 hours overtime; Thursday, 3 hours overtime; and Friday, 2 hours overtime. How many overtime hours did Brian work? _____

C. Subtraction

23.
$$\begin{array}{r} 56{,}189 \\ -\,38{,}106 \\ \hline \end{array}$$

24.
$$\begin{array}{r} 1{,}009 \\ -\,467 \\ \hline \end{array}$$

25.
$$\begin{array}{r} \$809.01 \\ -\,146.92 \\ \hline \end{array}$$

26.
$$\begin{array}{r} 0.139 \\ -\,0.87 \\ \hline \end{array}$$

27. $465 - 382 =$ _____

28. $\$189.90 - \$0.06 =$ _____

29. $\$892.11 - \$5.95 =$ _____

30. $\$239.60 - \$223.59 =$ _____

31. Hans bought a full page scanner, a mouse, and three new software programs totaling $1,395. Hans decided to return the mouse costing $129. How much was Hans' purchase after he returned the mouse? _____

32.
$$\begin{array}{r} +97 \\ -49 \\ -14 \\ -87 \\ +99 \\ +21 \\ \hline \end{array}$$

33.
$$\begin{array}{r} +\ 97 \\ -\ 149 \\ +388 \\ -\ 211 \\ +423 \\ +\ 53 \\ \hline \end{array}$$

34.
$$\begin{array}{r} -269 \\ +382 \\ -101 \\ -266 \\ +856 \\ +217 \\ \hline \end{array}$$

35.
$$\begin{array}{r} +\ 1{,}114 \\ -\ \ 801 \\ -\ \ \ 91 \\ +3{,}890 \\ -\ 1{,}500 \\ +\ \ 305 \\ \hline \end{array}$$

36.
$$\begin{array}{r} -472 \\ -311 \\ -899 \\ -\ 66 \\ -901 \\ -321 \\ \hline \end{array}$$

D. Multiplication (Show maximum decimal places.)

37. 136
$\times$ 12
_____

38. 324
$\times$ 322
_____

39. 916
$\times$ 4
_____

40. 512
$\times$ 100
_____

41. 722
$\times$ 204
_____

42. 1.33
$\times$ 0.21
_____

43. 8.102
$\times$ 1.1
_____

44. 129.36
$\times$ 0.2122
_____

45. $3.99
$\times$ 55
_____

46. $27.75
$\times$ 0.03
_____

47. 12 dozen pencils @ $0.99 a dozen = _____

5 packages envelopes @ $1.29 per package = _____

12 writing pads @ $5.95 per dozen = _____

Total = _____

48. 22 $\times$ 45 = _____

49. 110 $\times$ 45 = _____

50. 102 $\times$ 61 = _____

51. 2,004 $\times$ 5 = _____

52. 23 $\times$ 46 $\times$ 88 = _____

53. 103 $\times$ 21 $\times$ 902 _____

54. Jackie Moore paid $300 per month for 48 months on her automobile. What did the car cost Jackie? _____

55. Johnson & Johnson, attorneys, purchased new office furniture for their client waiting room. They purchased 2 tables for $495 each, 4 chairs for $799 each, and 3 flower arrangements for $60 each. What was the total of all the items purchased? _____

E. Division (Round your answers to 4 decimal places.)

56. $38\overline{)466}$ **57.** $9\overline{)9,018}$ **58.** $12\overline{)24,336}$ **59.** $102\overline{)10,550}$

60. $1.5\overline{)456.96}$ **61.** $0.22\overline{)808}$ **62.** $0.18\overline{)0.9446}$ **63.** $0.5\overline{)1,000}$

64. $32,146 \div 23 =$ _____   **65.** $\$9,403 \div 369 =$ _____

66. $2,479 \div 8 =$ _____   **67.** $35,642 \div 30 =$ _____

68. $23,211 \div 45 =$ _____   **69.** $3,891 \div 175 =$ _____

70. A 382-page report has to be proofread and facts in the report verified. John, Morgan, Sue, and David must complete the job by noon. How many pages must each proofread and check? _____

CHAPTER

Introduction to Fractions

3

Introduction to Fractions

OBJECTIVES

After completing this chapter, you will be able to:

1. Convert improper fractions to mixed or whole numbers.
2. Convert mixed numbers to improper fractions.
3. Reduce fractions to lowest terms.
4. Raise fractions to higher terms.
5. Convert fractions to decimals.
6. Convert decimals to fractions.

You probably are familiar with reading or hearing about common fractions and decimal fractions because they are used everyday. Here is how you might use common fractions: $\frac{3}{4}$ hour overtime worked last night, $\frac{1}{2}$ stick of butter for the recipe, or $\frac{7}{8}$ yard of blue silk sewn into a scarf. Decimal fractions are used in this manner: eggs cost $1.05 per dozen or the microcomputer's monthly rental fee is $129.25. To be successful in business transactions, you must have a clear understanding of fractions. This chapter provides terminology, explanations of types of fractions, and methods of working with fractions.

3.1 Terms Used in Fractions

When a whole number has been divided, the parts can be expressed as *fractions*. A fraction contains one number written above another number, separated by a bar, as shown here:

$$\frac{1}{3}, \frac{2}{3}, \frac{5}{4}$$

The bottom number is the ***denominator*** and expresses the number of equal parts by which a whole number is divided. The top number is the ***numerator*** and expresses the number of equal parts of the whole number represented. The line or bar that separates the numerator from the denominator means "divided by." Study the following example.

EXAMPLE ──────────────────────────────────

$$\text{fraction bar} \rightarrow \frac{1}{2} \begin{array}{l} \leftarrow \text{numerator} \\ \leftarrow \text{denominator} \end{array}$$

3.2 Fractions and Mixed Numbers

There are three types of fractions: proper, improper, and mixed. A *proper fraction* is one in which the numerator is less than the denominator. Therefore, a proper fraction expresses less than one whole number, such as $\frac{1}{2}$, $\frac{3}{4}$, or $\frac{1}{8}$.

$$\frac{1}{2} \quad \text{numerator is smaller than denominator}$$

An *improper fraction* is one in which the numerator is equal to or greater than the denominator and which expresses one or more whole numbers, such as $\frac{8}{5}$, $\frac{5}{5}$, or $\frac{12}{6}$.

$$\frac{8}{5} \quad \text{numerator is greater than denominator}$$

A *mixed number* consists of a whole number and a fraction, such as $3\frac{2}{5}$, $2\frac{1}{3}$, or $6\frac{7}{8}$.

$$\text{whole number part} \rightarrow 3\frac{2}{5} \leftarrow \text{fraction part}$$

PRACTICE PROBLEMS

Label the following as either proper fractions, improper fractions, or mixed numbers. Write your answers in the blanks provided. Use **P** for proper, **I** for improper, or **M** for mixed number.

1. $\frac{2}{3}$ ___ 2. $\frac{10}{8}$ ___ 3. $4\frac{1}{2}$ ___ 4. $\frac{1}{5}$ ___ 5. $2\frac{5}{8}$ ___ 6. $\frac{7}{5}$ ___

Solutions: **1.** P; **2.** I; **3.** M; **4.** P; **5.** M; **6.** I

3.3 Converting an Improper Fraction to a Whole Number

You can convert certain improper fractions to whole numbers. For example, the improper fraction $\frac{16}{4}$ can be converted to the whole number 4 by dividing the numerator by the denominator as shown in the following example.

EXAMPLE _____

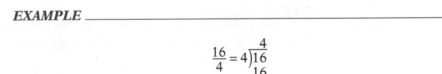

$$\frac{16}{4} = 4\overline{)16} \quad \frac{4}{16}$$

PRACTICE PROBLEMS

Convert the following improper fractions to whole numbers. Write your answers in the blanks provided.

1. $\frac{180}{30}$ ___   2. $\frac{880}{40}$ ___   3. $\frac{120}{15}$ ___   4. $\frac{72}{9}$ ___

Solutions: **1.** 6; **2.** 22; **3.** 8; **4.** 8

3.4 Converting an Improper Fraction to a Mixed Number

Most improper fractions can be converted to mixed numbers by dividing the numerator by the denominator. Let's convert $\frac{35}{6}$ to a mixed number.

STEPS

1. Divide the numerator by the denominator.

$$\begin{array}{r} 5 \\ 6\overline{)35} \\ 30 \\ \hline 5 \end{array}$$

2. Place the remainder over the divisor.

$\frac{5}{6}$ 5 remainder from division
   6 original denominator (divisor)

3. Write the mixed number as $5\frac{5}{6}$.

PRACTICE PROBLEMS

Convert the following improper fractions to mixed numbers. Write your answers in the blanks provided.

1. $\frac{19}{6}$ _____ **2.** $\frac{38}{3}$ _____ **3.** $\frac{132}{7}$ _____ **4.** $\frac{109}{2}$ _____

Solutions: **1.** $3\frac{1}{6}$; **2.** $12\frac{2}{3}$; **3.** $18\frac{6}{7}$; **4.** $54\frac{1}{2}$

3.5 Converting a Mixed Number to an Improper Fraction

A mixed number can be converted to an improper fraction by multiplying the denominator of the fraction by the whole number, adding the numerator to that product, and placing the sum over the original denominator. Convert $9\frac{1}{4}$ to an improper fraction.

STEPS

1. Multiply the denominator of the fraction by the whole number.

$$4 \times 9 = 36$$

2. Add the numerator of the fraction and the product.

$$36 + 1 = 37$$

3. Place the sum over the denominator.

$$\frac{37}{4}$$

Convert the following mixed numbers to improper fractions. Write your answers in the blanks provided.

1. $24\frac{7}{8}$ _____   **2.** $4\frac{8}{12}$ _____   **3.** $6\frac{9}{10}$ _____   **4.** $11\frac{3}{4}$ _____

Solutions: **1.** $\frac{199}{8}$; **2.** $\frac{56}{12}$; **3.** $\frac{69}{10}$; **4.** $\frac{47}{4}$

3.6 Reducing Fractions to Lowest Terms

Sometimes proper fractions, such as $\frac{88}{100}$ and $\frac{75}{125}$, are too large to work with in solving a math problem. Therefore, it is helpful to reduce them to their lowest terms. Reducing a fraction to lowest terms means that you must find the smallest numerator and denominator possible without changing the original value of the fraction being reduced. To do this, you must find a number or factor that will divide evenly both the numerator and denominator.

Trial-and-Error Method

In many cases, fractions can be reduced by using a trial-and-error method. Always look for a common number that will divide evenly both the numerator and denominator. Let's reduce $\frac{75}{125}$ to lowest terms. The common number 25 is found for this example.

EXAMPLE _____

$$\frac{75}{125} \begin{array}{c} \div 25 \\ \div 25 \end{array} = \frac{3}{5}$$

■ **TIP** 1. If the numerator will divide the denominator evenly, the numerator is the greatest common factor as shown in the following example:

$$\frac{9}{18} \begin{array}{c} \div 9 \\ \div 9 \end{array} = \frac{1}{2}$$

2. If the numerator and denominator are both even numbers, they can be divided by 2 to reduce the fraction. This may take several steps. Study this example:

$$\frac{18}{32} \begin{array}{c} \div 2 \\ \div 2 \end{array} = \frac{9}{16}$$

3. If both the numerator and denominator end in 0, they can be divided by 10 to arrive at the lowest terms as shown here:

$$\frac{40}{50} \begin{array}{c} \div 10 \\ \div 10 \end{array} = \frac{4}{5}$$

4. If both the numerator and denominator end in 5, the lowest terms can be determined by dividing each by 5. Study this example:

$$\frac{25}{35} \begin{array}{c} \div 5 \\ \div 5 \end{array} = \frac{5}{7}$$

5. If the numerator and denominator can be divided by 3, you can find the common factor, in multiples of 3—such as 3, 6, 9, 12, 15, and so on. Study the example:

$$\frac{18}{33} \begin{array}{c} \div 3 \\ \div 3 \end{array} = \frac{6}{11}$$

Greatest Common Factor

If none of the trial-and-error rules applies, then a greatest common factor can be determined by these steps.

STEPS

1. Divide the denominator by the numerator.

2. If there is a remainder, divide the original divisor by the remainder.

3. Continue to divide the last divisor by the remainder until there is no remainder, if possible.

4. If there is no remainder, the last divisor used in this process is the greatest common factor.

5. Divide both the numerator and denominator by the greatest common factor. The answer will be the lowest terms to which the fraction can be reduced.

6. If there is ever a remainder of 1, the fraction is already at its lowest terms.

Using the greatest-common-factor method, reduce the fraction $\frac{268}{460}$ to its lowest terms by following these steps.

STEPS

1. Divide the denominator by the numerator.

$$\begin{array}{r} 1 \\ 268\overline{)460} \\ \underline{268} \\ 192 \end{array}$$

2. Divide the original divisor by the remainder.

$$\begin{array}{r} 1 \\ 192\overline{)268} \\ \underline{192} \\ 76 \end{array}$$

3. Continue the division process until there is no remainder, as shown in (d).

(a) $\begin{array}{r} 2 \\ 76\overline{)192} \\ \underline{152} \\ 40 \end{array}$ (b) $\begin{array}{r} 1 \\ 40\overline{)76} \\ \underline{40} \\ 36 \end{array}$ (c) $\begin{array}{r} 1 \\ 36\overline{)40} \\ \underline{36} \\ 4 \end{array}$ (d) $\begin{array}{r} 9 \\ 4\overline{)36} \\ \underline{36} \\ 0 \end{array}$

4. The last divisor, 4, is the greatest common factor and is used to divide 268 and 460.

$$\begin{array}{l} 268 \div 4 = 67 \\ \overline{460 \div 4 = 115} \end{array}$$

In the next example the fraction $\frac{278}{455}$ is already reduced to its lowest terms because the last divisor, 1, in the last division step (f), is the greatest common factor.

$$\text{(a)} \quad 278\overline{)455} \atop \underline{278} \atop 177 \qquad\qquad \begin{array}{r} 1 \\ \end{array}$$

(a) 278)455
        278
        177

(d) 76)101
        76
        25

(b) 177)278
        177
        101

(e) 25)76
        75
        1

(c) 101)177
        101
        76

(f) 1)25
        2
        5
        5

PRACTICE PROBLEMS

Reduce the following common fractions to their lowest terms using the greatest-common-factor method. Write your answers in the blanks provided.

1. $\frac{145}{240} =$ _____ **2.** $\frac{134}{245} =$ _____ **3.** $\frac{168}{198} =$ _____

Solutions: **1.** Greatest common factor is 5 after dividing 5 times, $\frac{29}{48}$.

2. After dividing 6 times there is a remainder of 1; therefore, this fraction is at its lowest terms.

3. Greatest common number that divides evenly into both parts of the fraction is 6. The fraction can be reduced to $\frac{28}{33}$.

3.7 Raising Fractions to Higher Terms

Just as fractions can be reduced to lowest terms, they can also be raised to higher terms. Fractions are raised to higher terms to obtain a desired denominator, usually for adding or subtracting fractions. Let's raise $\frac{2}{3}$ to a fraction with a denominator of 12 by following these steps.

EXAMPLE ——————————————————————

$$\frac{2}{3} = \frac{?}{12}$$

STEPS

1. Divide the desired denominator by the original denominator of the fraction to be raised, such as

$$12 \div 3 = 4$$

2. Multiply the numerator by the quotient found in Step 1.

$$2 \times 4 = 8$$

3. Write the new numerator over the desired denominator.

$$\frac{8}{12}$$

The result, $\frac{8}{12}$, is sometimes referred to as an *equivalent fraction* because it is equal to the original fraction, $\frac{2}{3}$, or $\frac{2}{3} = \frac{8}{12}$.

PRACTICE PROBLEMS

Raise the following fractions to higher terms using the indicated denominators. Write your answers in the blanks provided.

1. $\frac{7}{9} = \frac{?}{45}$ _____   **2.** $\frac{11}{12} = \frac{?}{72}$ _____   **3.** $\frac{8}{10} = \frac{?}{60}$ _____   **4.** $\frac{3}{5} = \frac{?}{105}$ _____

Solutions: **1.** $\frac{35}{45}$; **2.** $\frac{66}{72}$; **3.** $\frac{48}{60}$; **4.** $\frac{63}{105}$

3.8 Converting a Common Fraction to a Decimal Fraction

Because fractions and decimals both represent parts of a whole number, a fraction can be converted to a decimal, and a decimal can be converted to a fraction. The conversion may simplify some calculations when you are solving a particular math problem. For example, Mrs. Brewer purchased 25 shares of stock at $19\frac{5}{8}$ per share for a total of $490.75. It is easier and quicker to find the total by expressing $19\frac{5}{8}$ as its decimal equivalent, $19.625 or $19.63.

■ **TIP**   Most calculators cannot solve problems using fractions, but can solve a problem using decimal equivalents. Therefore, it is important to convert fractions to decimals when using fractions in machine calculations.

To convert a common fraction to a decimal, divide the numerator by the denominator. This example shows how $\frac{7}{8}$ is converted to a decimal, carrying the answer to three decimal places (nearest thousandth).

$$
\begin{array}{r}
0.875 \\
8\overline{)7.000} \\
\underline{6\,4} \\
60 \\
\underline{56} \\
40 \\
\underline{40}
\end{array}
$$

Sometimes a denominator may not divide evenly a numerator, as shown here. Convert $\frac{5}{12}$ to a decimal.

$$
\begin{array}{r}
0.4166 \\
12\overline{)5.0000} \\
\underline{4\ 8} \\
20 \\
\underline{12} \\
80 \\
\underline{72} \\
8
\end{array}
$$

Decimal places have been added to continue the division process; however, in this particular problem, there will always be a remainder because the denominator, 12, will not divide evenly the numerator, 5. You may round your answer to any desired number of places. In this example, the decimal equivalent 0.4166 is shown rounded to 2 and 3 decimal places. You could also place a line above the last digit to indicate this digit will always repeat.

$$\frac{5}{12} = 0.42 \text{ (nearest hundredth)}$$

$$\frac{5}{12} = 0.417 \text{ (nearest thousandth)}$$

PRACTICE PROBLEMS

Convert the following common fractions to decimal fractions and round to the nearest thousandth. Write your answers in the blanks provided.

1. $\frac{1}{8} =$ _____   **2.** $\frac{1}{3} =$ _____   **3.** $\frac{1}{6} =$ _____   **4.** $\frac{5}{8} =$ _____

Solutions: **1.** 0.125; **2.** 0.333; **3.** 0.167; **4.** 0.625

3.9 Decimal Equivalents Chart

Rather than taking time to convert a fraction to its decimal equivalent, you should become familiar with common decimal equivalents. A list follows. Memorize these decimal equivalents if you use them often.

Fraction	Decimal Equivalent		Fraction	Decimal Equivalent
$\frac{1}{5}$	=	0.2	$\frac{5}{8}$	=	0.625
$\frac{1}{4}$	=	0.25	$\frac{2}{3}$	=	0.6667
$\frac{1}{3}$	=	$0.333\overline{3}$	$\frac{3}{4}$	=	0.75
$\frac{1}{2}$	=	0.5	$\frac{7}{8}$	=	0.875

3.10 Converting a Mixed Number to a Decimal Fraction

A mixed number is converted to a decimal by the same method used to convert a common fraction to a decimal. Let's convert $5\frac{5}{8}$ to a decimal following these steps.

STEPS

1. Convert the fractional part by dividing the numerator by the denominator.

$$\frac{5}{8} = 8\overline{)\begin{array}{r} 0.625 \\ 5.000 \\ \underline{48} \\ 20 \\ \underline{16} \\ 40 \\ \underline{40} \end{array}}$$

2. Write the whole number 5 with the decimal equivalent 0.625.

$$5.625$$

Of course you can round to the nearest hundredth (5.63).

PRACTICE PROBLEMS

Convert the following mixed numbers to decimals and round your answers to the nearest hundredth. Write your answers in the blanks provided.

1. $8\frac{5}{12} =$ _____   **2.** $36\frac{4}{5} =$ _____   **3.** $2\frac{3}{7} =$ _____   **4.** $4\frac{2}{3} =$ _____

Solutions: **1.** 8.42; **2.** 36.80; **3.** 2.43; **4.** 4.67

3.11 Converting a Decimal to a Fraction

There are several methods of converting a decimal to a fraction. One easy method is to think of the decimal in its word form as you learned in Chapter 1. In the following example, 1.35 is read as one and thirty-five hundredths. Convert 1.35 to a common fraction using these steps.

STEPS

1. Write the decimal in its fractional form. $\frac{35}{100}$

2. Reduce to its lowest terms, if necessary. $\frac{35}{100} = \frac{7}{20}$

Therefore, the fractional equivalent of 1.35 is $1\frac{7}{20}$.

■ **TIP**   When converting decimals to fractions, write the denominator as 1 and add as many zeros as there are places in the numerator.
For example: $0.333 = \frac{333}{1000}$.

Convert the following decimals to fractions, and reduce them to their lowest terms, if necessary. Write your answers in the blanks provided.

1. 0.45 _____   **2.** 0.65 _____   **3.** 4.444 _____   **4.** 0.105 _____

5. 0.32 _____   **6.** 2.583 _____   **7.** 0.09 _____   **8.** 0.584 _____

Solutions: **1.** $\frac{9}{20}$; **2.** $\frac{13}{20}$; **3.** $4\frac{111}{250}$; **4.** $\frac{21}{200}$; **5.** $\frac{8}{25}$; **6.** $2\frac{583}{1000}$; **7.** $\frac{9}{100}$; **8.** $\frac{73}{125}$

MATH ALERT

The following baseball scores were posted. The **Pct.** column represents percent of the games won, shown in decimals. The **GB** column shows fractions for games behind.

AL WEST	W	L	Pct.	GB
Minnesota	70	48	.593	—
Chicago	66	51	.564	$3\frac{1}{2}$
Oakland	64	54	.542	6
Seattle	63	54	.538	$6\frac{1}{2}$
Kansas City	61	54	.530	$7\frac{1}{2}$
Texas	59	56	.513	$9\frac{1}{2}$
California	57	59	.491	12

AL EAST	W	L	Pct.	GB
Toronto	64	54	.542	—
Detroit	62	56	.525	2
Boston	58	59	.496	$5\frac{1}{2}$
Milwaukee	54	63	.462	$9\frac{1}{2}$
New York	53	62	.461	$9\frac{1}{2}$
Baltimore	48	59	.410	$15\frac{1}{2}$
Cleveland	38	78	.328	25

1. Change the following decimals to fractions.

	Decimal	**Fraction**
a. Minnesota	0.593	_____
b. California	0.491	_____
c. Boston	0.496	_____
d. Cleveland	0.328	_____
e. New York	0.461	_____

2. Change the Chicago GB fraction to a decimal: $3\frac{1}{2}$ = _____

Study Guide

I. Terminology

page 46	fraction	Part of a whole number.
page 46	denominator	The bottom number below the dividing bar of a fraction.
page 46	numerator	The top number above the dividing bar of a fraction.
page 47	proper fraction	A fraction that contains a numerator smaller than the denominator.
page 47	improper fraction	A fraction that contains a numerator greater than or equal to the denominator.
page 47	mixed number	A number consisting of a whole number part and a fraction part.
page 50	greatest common factor	A number that will divide evenly the numerator and denominator of a fraction.
page 52	decimal fraction	A fraction expressed as a decimal number.
page 53	decimal equivalent	The decimal number equal to a fraction.

II. Converting Fractions

page 47 *Converting an Improper Fraction to a Whole Number:* Divide the numerator by the denominator—$\frac{16}{4}$ would be 16 divided by 4 = the whole number 4.

page 48 *Converting an Improper Fraction to a Mixed Number:* Divide the numerator by the denominator. If there is a remainder, write the whole number and then write the fraction. $\frac{35}{6}$ would be 35 divided by 6 = 5 with a remainder of 5. Place the remainder of 5 over the divisor—$\frac{5}{6}$; then place the whole number 5 beside the fraction—$5\frac{5}{6}$.

page 48 *Converting a Mixed Number to an Improper Fraction:* Multiply the denominator of the fraction by the whole number, add the numerator, place the sum over the original denominator.
Example: $9\frac{1}{4} = 9 \times 4 = 36 + 1 = \frac{37}{4}$

III. Reducing Fractions to Lowest Terms

page 49 *Trial-and-Error Method:*

    a. Look for a common number that will divide evenly both the numerator and denominator. If the numerator will divide the denominator evenly, the numerator is the greatest common factor—$\frac{9}{18} = \frac{1}{2}$.

    b. If the numerator and denominator are both even numbers, they can be divided by 2 to reduce the fraction to its lowest terms—$\frac{18}{32} = \frac{9}{16}$.

    c. If both the numerator and denominator end in 0, they can be divided by 10 to arrive at the lowest terms—$\frac{40}{50} = \frac{4}{5}$.

    d. If both the numerator and denominator end in 5, the lowest terms can be determined by dividing each by 5—$\frac{25}{35} = \frac{5}{7}$.

    e. If the numerator and denominator can be divided by 3, you can find the common factor, in multiples of 3—such as 3, 6, 9, 12, 15, and so on—$\frac{18}{33} = \frac{6}{11}$.

Greatest Common Factor: Divide the denominator by the numerator. If there is a remainder, divide the original divisor by the remainder. Continue dividing the last divisor by the last remainder until there is no remainder, if possible. If there is no remainder, the last divisor used in this process is the greatest common factor. Divide both the numerator and denominator by the greatest common factor. The answer will be the lowest terms to which the fraction can be reduced. If there is ever a remainder of 1, the fraction is already at its lowest terms.

IV. Raising Fractions to Higher Terms

Fractions are raised to higher terms to obtain a desired denominator, usually for adding or subtracting fractions.

Example: Raise $\frac{2}{3}$ to a fraction with a denominator of 12.

Divide the desired denominator by the original denominator of the fraction to be raised— $12 \div 3 = 4$. Multiply the numerator by the quotient found— $2 \times 4 = 8$. Write the new numerator over the desired denominator— $\frac{8}{12}$.

V. Decimal Fractions

Converting a Common Fraction to a Decimal Fraction: Divide the numerator by the denominator—7 divided by 8 and carried three places = 0.875

Converting a Mixed Number to a Decimal Fraction: Convert the fractional part by dividing the numerator by the denominator; then write the whole number to the left of the decimal— $5\frac{5}{8}$ would be 5 divided by 8 = 0.625; then 5.625.

Converting a Decimal to a Fraction: Think of the decimal in its word form. For example, 1.35 would be read as one and thirty-five hundredths. Since you identified hundredths as the divisor, the fraction would be written as $1\frac{35}{100}$ (or $1\frac{7}{20}$ in lowest terms).

Study Skills

When to Study

Some of you are morning people, and some of you are night people; and some may feel sluggish in the afternoon. Choose the time that you feel the most alert to study.

Tips

1. Do your most demanding tasks at times when you are the most alert.
2. Save easy, no-pressure, routine tasks when you feel less productive.
3. Listen to your body. If you feel sleepy and tired, quit and go to sleep. If you must get a task done, take frequent breaks and move around to keep from becoming fatigued.

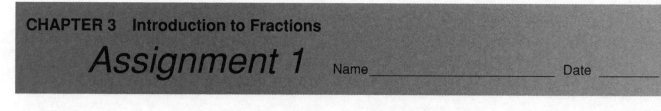

CHAPTER 3   Introduction to Fractions

Assignment 1

Name_____ Date _____

Complete the following problems. Write your answers in the blanks provided.

A. Identify each of the following numbers as a proper fraction, an improper fraction, or a mixed number. Indicate your answer with P for a proper fraction, I for an improper fraction, or M for a mixed number.

1. $\frac{33}{6}$ _____

2. $\frac{5}{13}$ _____

3. $2\frac{1}{6}$ _____

4. $\frac{3}{5}$ _____

5. $\frac{8}{3}$ _____

6. $1\frac{2}{3}$ _____

7. $\frac{5}{8}$ _____

8. $\frac{7}{8}$ _____

B. Convert the following improper fractions to mixed numbers or whole numbers. Reduce to its lowest terms where necessary.

9. $\frac{75}{8}$ _____

10. $\frac{8}{3}$ _____

11. $\frac{7}{5}$ _____

12. $\frac{110}{89}$ _____

13. $\frac{99}{8}$ _____

14. $\frac{59}{12}$ _____

15. $\frac{128}{16}$ _____

16. $\frac{145}{120}$ _____

C. Change the following mixed numbers to improper fractions.

17. $6\frac{1}{4}$ _____

18. $36\frac{3}{4}$ _____

19. $5\frac{1}{3}$ _____

20. $4\frac{3}{8}$ _____

21. $75\frac{3}{8}$ _____

22. $37\frac{1}{2}$ _____

23. $4\frac{3}{4}$ _____

24. $12\frac{2}{3}$ _____

D. Convert the following improper fractions to whole numbers.

25. $\frac{12}{3}$ _____

26. $\frac{120}{40}$ _____

27. $\frac{60}{12}$ _____

28. $\frac{88}{8}$ _____

29. $\frac{36}{6}$ _____

30. $\frac{132}{4}$ _____

31. $\frac{168}{8}$ _____

32. $\frac{210}{5}$ _____

Assignment 2

Name_____ Date _____

Complete the following problems. Write your answers in the blanks provided.

A. Reduce these fractions to their lowest terms.

1. $\frac{9}{18}$ _____

2. $\frac{36}{124}$ _____

3. $\frac{8}{48}$ _____

4. $\frac{228}{314}$ _____

5. $\frac{12}{28}$ _____

6. $\frac{6}{9}$ _____

7. $\frac{183}{366}$ _____

8. $\frac{150}{365}$ _____

B. Raise the following fractions to higher terms using the indicated denominators.

9. $\frac{10}{25} = \frac{?}{125}$ _____

10. $\frac{11}{18} = \frac{?}{36}$ _____

11. $\frac{5}{24} = \frac{?}{96}$ _____

12. $\frac{5}{8} = \frac{?}{72}$ _____

13. $\frac{4}{19} = \frac{?}{57}$ _____

14. $\frac{7}{17} = \frac{?}{51}$ _____

15. $\frac{7}{10} = \frac{?}{50}$ _____

16. $\frac{7}{8} = \frac{?}{24}$ _____

C. Find the decimal equivalents of the following fractions. Round your answers to the nearest hundredth.

17. $\frac{8}{50}$ _____

18. $5\frac{2}{7}$ _____

19. $9\frac{5}{12}$ _____

20. $8\frac{12}{100}$ _____

21. $\frac{4}{9}$ _____

22. $\frac{9}{20}$ _____

23. $7\frac{1}{7}$ _____

24. $\frac{7}{12}$ _____

D. Convert the following decimals to fractions or mixed numbers and reduce to their lowest terms.

25. 0.75 _____

26. 0.8 _____

27. 0.03 _____

28. 0.2482 _____

Assignment 3   Name_____ Date _____

Complete the following problems. Write your answers in the blanks provided.

A. Convert the following improper fractions to mixed or whole numbers. Reduce to lowest terms where necessary.

1. $\frac{59}{12}$ _____

2. $\frac{5}{2}$ _____

3. $\frac{132}{12}$ _____

4. $\frac{62}{20}$ _____

5. $\frac{25}{4}$ _____

6. $\frac{141}{6}$ _____

7. $\frac{39}{9}$ _____

8. $\frac{45}{8}$ _____

B. Change the following mixed numbers to improper fractions. Do not reduce to lowest terms.

9. $4\frac{15}{30}$ _____

10. $7\frac{3}{5}$ _____

11. $4\frac{9}{16}$ _____

12. $11\frac{5}{16}$ _____

13. $3\frac{5}{16}$ _____

14. $8\frac{5}{10}$ _____

15. $2\frac{20}{7}$ _____

16. $6\frac{2}{3}$ _____

C. Find the fraction equivalents of the following decimals. Reduce to lowest terms where necessary.

17. 3.6 _____

18. 2.08 _____

19. 8.039 _____

20. 1.206 _____

21. 9.469 _____

22. 7.5 _____

23. 3.66 _____

24. 0.54 _____

D. Write the following as decimals. Round answers to 2 decimal places.

25. $\frac{4}{5}$ _____

26. $\frac{3}{18}$ _____

27. $2\frac{2}{7}$ _____

28. $5\frac{5}{8}$ _____

29. $4\frac{3}{8}$ _____

30. $\frac{7}{8}$ _____

31. $3\frac{3}{7}$ _____

32. $\frac{4}{20}$ _____

Assignment 4

Name_____ Date _____

Complete the following problems. Write your answers in the blanks provided. Be sure to place commas and dollar signs where needed. Express fractional parts in lowest terms. Round amounts to the nearest hundredth.

1. John, a basketball center for the Lewis Lions, made 12 of 20 shots.

   a. Write the fraction, $\frac{12}{20}$, as a decimal

   _____

   b. Write the fraction, $\frac{12}{20}$, as an improper fraction and convert the improper fraction to a mixed number.

   _____

2. Roscoe, a basketball guard for Lake Houston's Gators, made 18 of 23 throws.

   a. Write the fraction, $\frac{18}{23}$, as a decimal.

   _____

   b. Write the fraction, $\frac{18}{23}$, as improper fraction and convert the improper fraction to a mixed number.

   _____

3. George was asked to convert the following ounces to mixed numbers.

   a. 28.4 ounces   _____

   b. 36.8 ounces   _____

   c. 64.5 ounces   _____

   d. 32.6 ounces   _____

   e. 8.12 ounces   _____

4. Debbie was asked to prepare a report for the Personnel Department, using the following information: $\frac{1}{5}$ of all employees were rated as superior performers; $\frac{2}{5}$ performed in a competent manner; $\frac{1}{5}$ performed at a satisfactory level, and $\frac{1}{5}$ of all employees performed less than satisfactory and are subject to termination. Convert each of these fractions to decimals.

   a. $\frac{1}{5}$ _____   **b.** $\frac{2}{5}$ _____

5. In a personnel report, the following items were reported: $\frac{2}{3}$ of all employees hired in the last month were Asian and $\frac{1}{3}$ were Hispanic. Convert each of the fractions to decimals.

   a. $\frac{2}{3}$ _____   **b.** $\frac{1}{3}$ _____

6. You have been asked to keep a record of the following stocks reported in today's newspaper. Convert the stock prices to dollars and cents. Round your answers to the nearest hundredth.

   a. $26 $\frac{3}{4}$ _____   **b.** $20 $\frac{1}{4}$ _____

   c. $39 $\frac{1}{8}$ _____   **d.** $42 $\frac{1}{2}$ _____

   e. $29 $\frac{5}{8}$ _____

7. On your job, you are asked to convert these decimals to fractions.

   a. 0.54 _____   **b.** 0.491 _____

   c. 1.34 _____   **d.** 0.328 _____

   e. 32.163 _____

Assignment 5

Name_____ Date _____

Solve the following problems. Write your answers in the blanks provided. Be sure to place commas and dollar signs where needed. Express fractional parts in lowest terms. Round amounts to the nearest hundredth.

1. Janice was asked to prepare a weekly stock report on the Employee Stock Share Program. Here is a list of prices for the week:

    Mon: 34\frac{1}{2}$ _____   Tue: 36\frac{1}{8}$ _____

    Wed: 38\frac{3}{8}$ _____   Thur: 37\frac{2}{5}$ _____

    Fri: 35\frac{3}{4}$ _____

    Convert the stock prices to dollars and cents. Round your answers to the nearest hundredth.

2. Bob Townsend, a quarterback for the Washington Bears, completed 23 out of 25 passes for the last game.

    a. What was the fraction for completed passes?

        _____

    b. Rewrite the fraction as a decimal:

        _____

    c. Rewrite the fraction as an improper fraction and convert the improper fraction to a mixed number.

        _____

3. Rita was asked to convert the following ounces to fractions for her weekly chemistry report:

    a. 7.15 oz _____   b. 23.4 oz _____

    c. 2.75 oz _____   d. 48.8 oz _____

    e. 69.6 oz _____

4. On your job, you are asked to convert these fractions to decimals.

    a. $\frac{5}{100}$ _____   b. $\frac{8}{64}$ _____

    c. $\frac{34}{90}$ _____   d. $\frac{17}{119}$ _____

    e. $\frac{25}{400}$ _____

5. Margo placed an order for pizza for the office Christmas party. One-sixth of the office wanted pepperoni pizza, $\frac{1}{3}$ wanted hamburger pizza, and $\frac{1}{2}$ wanted sausage pizza. Convert these fractions to decimals.

    a. $\frac{1}{6}$ _____   b. $\frac{1}{3}$ _____   c. $\frac{1}{2}$ _____

6. The office softball team batting figures are as follows: Convert these decimals to fractions.

Decimals	Fractions
Smith–0.338	_____
Perez–0.225	_____
Clark–0.125	_____
Hunter–0.386	_____
Martin–0.263	_____

7. A company expects to lay off 59 of its 246 employees. What fractional part of workers are expected to be laid off?

    _____

Assignment 6

Name_____ Date _____

Solve the following problems. Write your answers in the blanks provided. Be sure to place dollar signs as needed. Express fractional parts in lowest terms. Round amounts to the nearest hundredth.

1. On your job, you are asked to convert these decimals to fractions.

Decimals	Fractions
a. 79.124	_____
b. 0.8276	_____
c. 0.55	_____
d. 3.85	_____
e. 0.25	_____

2. Amy has been asked to keep a record of the following stocks reported in the daily newspaper. Convert the stock prices to dollars and cents. Round your answers to the nearest hundredth.

   a. $49\frac{2}{3}$ _____ b. $17\frac{1}{7}$ _____

   c. $96\frac{8}{9}$ _____ d. $66\frac{1}{8}$ _____

   e. $84\frac{4}{5}$ _____

3. In a marketing department, a survey was done that revealed $\frac{1}{4}$ of the employees liked red cars, $\frac{1}{2}$ liked blue cars, $\frac{1}{8}$ liked green cars and $\frac{2}{16}$ liked brown cars. Convert each of these fractions to decimals.

   a. $\frac{1}{4}$ _____ b. $\frac{1}{2}$ _____

   c. $\frac{1}{8}$ _____ d. $\frac{2}{16}$ _____

4. Convert each of these fractions to decimals for a personnel report:

   a. $\frac{2}{3}$ _____ b. $\frac{4}{5}$ _____

   c. $\frac{1}{9}$ _____ d. $\frac{2}{7}$ _____

   e. $\frac{3}{8}$ _____

5. Rhoda's hobby is skeet shooting. Yesterday, she hit 35 out of 80 clay disks. Write the fraction that shows her score. _____ or _____ Rewrite the fraction as a decimal: _____

6. Randy was asked to convert the following liters to fractions.

   a. 32.8 liters _____

   b. 1.46 liters _____

   c. 19.5 liters _____

   d. 97.2 liters _____

   e. 83.1 liters _____

7. Convert the following improper fractions to mixed or whole numbers; then the mixed or whole numbers to decimals.

Improper Fraction	Mixed Numbers	Decimals
a. $\frac{144}{12}$	_____	_____
b. $\frac{167}{13}$	_____	_____
c. $\frac{295}{6}$	_____	_____
d. $\frac{499}{3}$	_____	_____
e. $\frac{7}{2}$	_____	_____

CHAPTER
Adding and Subtracting Fractions
4

CHAPTER

4

Adding and Subtracting Fractions

Not only should you be able to convert fractions to decimals and decimals to fractions, but you should also be able to add and subtract fractions. Figuring the amount of yardage needed to make curtains, such as $12\frac{7}{8}$ yards, or the amount of paneling needed to remodel your office, such as 8 ft $4\frac{1}{2}$ inch panels, are just two examples of the need to understand fractions.

4.1 Terms Used in Adding and Subtracting Fractions

Fractions such as $\frac{1}{4}$ and $\frac{3}{4}$ are called *like fractions* because they have a common denominator. Fractions such as $\frac{1}{4}$ and $\frac{1}{8}$ are called *unlike fractions* because they have different denominators. The *lowest common denominator* is the smallest number that can be divided evenly by all the unlike denominators. The *inspection method* is used to determine if the largest denominator can be divided by the unlike denominators. A *prime number* is a whole number larger than 1 that can be divided evenly only by itself and by 1.

4.2 Adding Like Fractions

To add like fractions—those with common denominators—add all numerators and then place the sum of the numerators over the common denominator to form a new fraction. When needed, reduce the answer to its lowest terms.

EXAMPLE _____

$$\frac{5}{9}+\frac{2}{9}=\frac{7}{9}$$

When like fractions are added, the sum may be an improper fraction, as shown below.

EXAMPLE _____

$$\frac{3}{5} + \frac{4}{5} = \frac{7}{5}$$

The sum of the fractions, $\frac{7}{5}$, is then converted to a mixed number.

$$\frac{7}{5} = 1\frac{2}{5}$$

PRACTICE PROBLEMS

Add these like fractions. Write your answers in the blanks provided. Remember to convert any improper fraction to a mixed number. Reduce answers to lowest terms as needed.

1. $\frac{5}{9} + \frac{7}{9} =$ _____

2. $\frac{21}{50} + \frac{12}{50} =$ _____

3. $\frac{12}{30} + \frac{24}{30} =$ _____

4. $\frac{15}{45} + \frac{16}{45} =$ _____

Solutions: **1.** $1\frac{1}{3}$; **2.** $\frac{33}{50}$; **3.** $1\frac{1}{5}$; **4.** $\frac{31}{45}$

4.3 Adding Unlike Fractions

As you have just learned, fractions such as $\frac{1}{4}$ and $\frac{3}{4}$ are like fractions and can be added as they appear because they have a common denominator. Fractions such as $\frac{1}{4}$ and $\frac{1}{2}$ are called unlike fractions because they have different denominators. Because unlike fractions have different denominators, they cannot be added as they appear. The denominators must be changed so that they are the same. To add unlike fractions, you must first find a number that can be divided evenly by all the denominators. This number is called a *common denominator*

4.4 Finding the Lowest Common Denominator

The lowest common denominator, sometimes referred to as the *LCD*, is the smallest number that can be divided evenly by all the unlike denominators. For example, the lowest common denominator for the fractions $\frac{1}{2}$, $\frac{3}{4}$, and $\frac{2}{3}$ is 12 because 12 is the smallest common number that all the unlike denominators, 2, 4, and 3, will divide evenly.

$$12 \div 2 = 6$$
$$12 \div 4 = 3$$
$$12 \div 3 = 4$$

There are at least two methods of finding the lowest common denominator. The two methods shown in this chapter are the inspection and prime number methods.

4.5 Using the Inspection Method

The inspection method is quick and easy when the denominators involved are small. Using the inspection method, let's find the lowest common denominator in the following example.

EXAMPLE _____

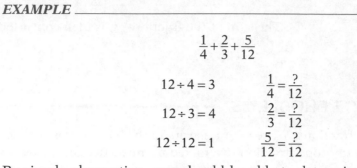

$$\frac{1}{4}+\frac{2}{3}+\frac{5}{12}$$

$$12 \div 4 = 3 \qquad \frac{1}{4}=\frac{?}{12}$$

$$12 \div 3 = 4 \qquad \frac{2}{3}=\frac{?}{12}$$

$$12 \div 12 = 1 \qquad \frac{5}{12}=\frac{?}{12}$$

By simple observation, you should be able to determine that the lowest common denominator is 12 because 12 is the smallest number the denominators 4, 3, and 12 will divide evenly.

PRACTICE PROBLEMS

Determine the lowest common denominators for the following problems. Write your answers in the blanks provided.

1. $\frac{1}{2},\frac{1}{3},\frac{5}{6}$ _____ **2.** $\frac{1}{3},\frac{5}{8},\frac{7}{12}$ _____ **3.** $\frac{1}{2},\frac{2}{3},\frac{5}{6}$ _____ **4.** $\frac{1}{2},\frac{3}{8},\frac{4}{16}$ _____

Solutions: **1.** 6; **2.** 24; **3.** 6; **4.** 16

4.6 Using Prime Numbers

In some cases it is difficult to determine the lowest common denominator using the inspection method. This method may be difficult to use when fractions have large denominators or when there are many fractions to be added. The second method of determining the lowest common denominator is to use prime factorization. A ***prime number*** is a whole number larger than 1 that can be divided evenly only by itself and by 1. The six smallest prime numbers are 2, 3, 5, 7, 11, and 13. The *prime factorization* of a number is the number written as a product of primes.

Using the prime factorization method, let's determine the lowest common denominator for the following fractions and then add.

EXAMPLE _____

$$\frac{3}{8}+\frac{7}{12}+\frac{11}{20}$$

STEPS

1. Write the denominators horizontally across the page, such as

    8   12   20

2. Divide each of these denominators by a prime number common to two or more of the numbers. Write the quotient above each denominator.

    $$\begin{array}{r} 4 \quad 6 \quad 10 \\ 2)\overline{\,8 \quad 12 \quad 20\,} \end{array}$$

3. Each of the new quotients—4, 6, and 10—can still be divided by 2; therefore, continue the division process, using the same prime number 2.

    $$\begin{array}{r} 2 \quad 3 \quad 5 \\ 2)\overline{\,4 \quad 6 \quad 10\,} \\ 2)\overline{\,8 \quad 12 \quad 20\,} \end{array}$$

4. Continue to divide by the prime number 2. The new quotient 2 can still be divided by the prime number 2, but the quotients 3 and 5 cannot be divided evenly by 2. Because 3 and 5 cannot be divided evenly by 2, just write these amounts in the row of answers. Continue to divide by the prime number 2.

    $$\begin{array}{r} 1 \quad 3 \quad 5 \\ 2)\overline{\,2 \quad 3 \quad 5\,} \\ 2)\overline{\,4 \quad 6 \quad 10\,} \\ 2)\overline{\,8 \quad 12 \quad 20\,} \end{array}$$

5. Because the prime number 2 will not evenly divide these last quotients in the row, try the next prime number, which is 3. Because 1 and 5 are not divisible by 3, simply write these amounts in the row of answers.

    $$\begin{array}{r} 1 \quad 1 \quad 5 \\ 3)\overline{\,1 \quad 3 \quad 5\,} \\ 2)\overline{\,2 \quad 3 \quad 5\,} \\ 2)\overline{\,4 \quad 6 \quad 10\,} \\ 2)\overline{\,8 \quad 12 \quad 20\,} \end{array}$$

6. Because the prime number 3 will not divide evenly these last quotients, try the next prime number, which is 5. The last quotient can be divided once by 5. Divide until all quotients are 1.

    $$\begin{array}{r} 1 \quad 1 \quad 1 \\ 5)\overline{\,1 \quad 1 \quad 5\,} \\ 3)\overline{\,1 \quad 3 \quad 5\,} \\ 2)\overline{\,2 \quad 3 \quad 5\,} \\ 2)\overline{\,4 \quad 6 \quad 10\,} \\ 2)\overline{\,8 \quad 12 \quad 20\,} \end{array}$$

7. Because no prime number can divide evenly these last quotients, you can now determine the lowest common denominator by multiplying the five divisors (2, 2, 2, 3, and 5—prime numbers) used in this example.

    $$2 \times 2 \times 2 \times 3 \times 5 = 120$$

    120 is the lowest common denominator of $\frac{3}{8}$, $\frac{7}{12}$, and $\frac{11}{20}$.

Now that you have determined the lowest common denominator, you can complete the division process by raising each fraction to 120ths and then adding. For example:

$$\frac{3}{8} = \frac{45}{120} \quad (120 \div 8 = 15; 15 \times 3 = 45)$$

$$\frac{7}{12} = \frac{70}{120} \quad (120 \div 12 = 10; 10 \times 7 = 70)$$

$$+\frac{11}{20} = +\frac{66}{120} \quad (120 \div 20 = 6; 6 \times 11 = 66)$$

$$\frac{181}{120} = 1\frac{61}{120}$$

The sum, $\frac{181}{120}$, is converted to a mixed number, $1\frac{61}{120}$.

PRACTICE PROBLEMS

Find the lowest common denominator for each of these fractions, using the prime number method. Write your answers in the blanks provided.

1. $\frac{7}{12} + \frac{7}{28} + \frac{15}{42} =$ _____ 2. $\frac{3}{15} + \frac{7}{12} =$ _____ 3. $\frac{5}{12} + \frac{5}{8} + \frac{9}{20} =$ _____ 4. $\frac{15}{32} + \frac{5}{8} + \frac{7}{18} =$ _____

Solutions: **1.** 84; **2.** 60; **3.** 120; **4.** 288

4.7 Adding Mixed Numbers

To add several mixed numbers with **like** fractions, follow these steps:

STEPS

1. Add the fractional parts.

2. Add the whole numbers.

3. Add the sum of the fractional parts and the sum of the whole numbers together.

4. Reduce answer to lowest terms as needed.

Suppose you are a sales clerk who cuts two pieces of trimming for a customer. One piece measured $12\frac{4}{7}$ inches and the other measured $15\frac{2}{7}$ inches. Add the two pieces together to get the total measurement for the order.

$$12\frac{4}{7} + 15\frac{2}{7}$$

STEPS

1. Add the fractional parts.

$$\frac{\frac{4}{7}}{+\frac{2}{7}} \over \frac{6}{7}$$

2. Add the whole numbers.

$$\begin{array}{r} 12 \\ +15 \\ \hline 27 \end{array}$$

3. Add the two sums.

$$\begin{array}{r} \frac{6}{7} \\ +27 \\ \hline 27\frac{6}{7} \end{array}$$

In some cases, the sum of the fractional parts of the mixed number is greater than 1, making it an improper fraction. Add these mixed numbers following the steps shown.

EXAMPLE

$$11\frac{9}{12} + 3\frac{7}{12}$$

STEPS

1. Add the fractional parts and convert the improper fraction to a mixed number.

$$\frac{\frac{9}{12}}{+\frac{7}{12}} \over \frac{16}{12} = 1\frac{4}{12} = 1\frac{1}{3}$$

2. Add the whole numbers.

$$\begin{array}{r} 11 \\ +\ 3 \\ \hline 14 \end{array}$$

3. Add the mixed number and the sum of the whole numbers.

$$\begin{array}{r} 1\frac{1}{3} \\ +14 \\ \hline 15\frac{1}{3} \end{array}$$

To add mixed numbers with *unlike* fractions, follow the same steps for adding unlike fractions. Remember to determine the lowest common denominator.

$$13\frac{3}{5}+2\frac{3}{10}$$

$$13\frac{3}{5} = 13\frac{6}{10}$$
$$+\ 2\frac{3}{10} =\ 2\frac{3}{10}$$
$$15\frac{9}{10}$$

PRACTICE PROBLEMS

Add these mixed numbers. Write your answers in the blanks provided.

1. $11\frac{2}{4}+6\frac{3}{12}=$ _____ **2.** $24\frac{1}{2}+17\frac{3}{8}=$ _____ **3.** $11\frac{5}{7}+6\frac{4}{21}=$ _____ **4.** $12\frac{2}{3}+17\frac{3}{8}=$ _____

Solutions: **1.** $17\frac{3}{4}$; **2.** $41\frac{7}{8}$; **3.** $17\frac{19}{21}$; **4.** $30\frac{1}{24}$

4.8 Subtracting Fractions

As with addition of fractions, subtraction of fractions can only occur when there is a common denominator. The basic steps for subtracting fractions are as follows:

STEPS

1. Find the lowest common denominator as necessary.

2. Subtract the numerators, and place the difference over the common denominator.

3. Reduce to lowest terms as necessary.

Study these examples illustrating the steps involved in subtracting fractions.

1. Subtract $\frac{3}{8}$ from $\frac{5}{8}$.

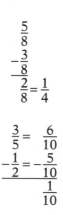

$$\frac{5}{8}$$
$$-\frac{3}{8}$$
$$\frac{2}{8}=\frac{1}{4}$$

2. Subtract $\frac{1}{2}$ from $\frac{3}{5}$.

$$\frac{3}{5} = \frac{6}{10}$$
$$-\frac{1}{2} = -\frac{5}{10}$$
$$\frac{1}{10}$$

Practical Math Applications

3. To subtract $\frac{1}{6}$ from the whole number 9, borrow 1 from the 9 and express it in sixths as in $\frac{6}{6}$. The 9 reduced by 1 is an 8 because you borrowed 1 from it. Place the 8 beside the $\frac{6}{6}$ then subtract the fractions.

$$\begin{array}{r} 9 = 8\frac{6}{6} \\ -\ \frac{1}{6} = -\ \frac{1}{6} \\ \hline 8\frac{5}{6} \end{array}$$

4. Subtract $4\frac{1}{5}$ from $12\frac{3}{10}$.

$$\begin{array}{r} 12\frac{3}{10} = 12\frac{3}{10} \\ -4\frac{1}{5} = -4\frac{2}{10} \\ \hline 8\frac{1}{10} \end{array}$$

5. Subtract $8\frac{7}{8}$ from $13\frac{3}{4}$.

a. The common denominator is 8 which makes the problem read $8\frac{7}{8}$ subtracted from $13\frac{6}{8}$.

$$\begin{array}{r} 13\frac{3}{4} = 13\frac{6}{8} \\ -8\frac{7}{8} = -8\frac{7}{8} \end{array}$$

b. Before $\frac{7}{8}$ can be subtracted from $\frac{6}{8}$, borrow 1 from 13 reducing 13 to 12. When this happens, $\frac{6}{8}$ becomes $\frac{14}{8}$ because $\frac{8}{8} + \frac{6}{8} = \frac{14}{8}$. Now the problem is read as $8\frac{7}{8}$ subtracted from $12\frac{14}{8}$.

$$\begin{array}{r} 13\frac{3}{4} = 13\frac{6}{8} = 12\frac{14}{8} \\ -8\frac{7}{8} = -8\frac{7}{8} = -8\frac{7}{8} \\ \hline 4\frac{7}{8} \end{array}$$

6. Subtract $1\frac{2}{3}$ from 3. Borrow 1 from 3, leaving $2\frac{3}{3}$.

$$\begin{array}{r} 3 = 2\frac{3}{3} \\ -1\frac{2}{3} = -1\frac{2}{3} \\ \hline 1\frac{1}{3} \end{array}$$

PRACTICE PROBLEMS

Subtract these fractions. Write your answers in the blanks provided.

1. Subtract $15\frac{1}{3}$ from 510   = _____   **2.** Subtract $9\frac{3}{8}$ from $17\frac{2}{7}$ = _____

3. Subtract $7\frac{3}{8}$ from $9\frac{11}{12}$ = _____   **4.** Subtract $3\frac{5}{8}$ from $6\frac{1}{2}$ = _____

5. Subtract $3\frac{5}{7}$ from $9\frac{3}{7}$ = _____   **6.** Subtract $6\frac{2}{3}$ from $9\frac{1}{2}$ = _____

Solutions: **1.** $494\frac{2}{3}$; **2.** $7\frac{51}{56}$; **3.** $2\frac{13}{24}$; **4.** $2\frac{7}{8}$; **5.** $5\frac{5}{7}$; **6.** $2\frac{5}{6}$

MATH ALERT

1. Figure the total hours worked in the drafting department of Santiago, Inc. for last week. Write your answers in the blanks provided. Show the total as a fraction; then convert the fraction to a decimal fraction.

   Carmen   45

   Jose $40\frac{1}{2}$

   Akemi $42\frac{3}{4}$

   Joan $38\frac{1}{2}$

   a. Total hours worked (as a fraction) = _____

   b. Total hours worked (as a decimal) = _____

2. Figure the total weight of fresh fruits and vegetables purchased at the market on Monday: $3\frac{3}{4}$ lb Idaho potatoes; $2\frac{1}{4}$ lb watermelon; 3 lb carrots; $2\frac{1}{2}$ lb cantaloupe; $1\frac{1}{3}$ lb tomatoes; $1\frac{3}{4}$ lb squash. Write the total weight as a fraction; then convert the fraction to a decimal fraction carried to three places.

   a. Total weight (as a fraction) = _____

   b. Total weight (as a decimal) = _____

3. The owner of Channes Designs has asked you to total an order for fabric for one of the store's clients. Show the total order as a fraction; then convert the fraction to a decimal fraction.

   Fabric for drapes $36\frac{7}{8}$ yd

   Fabric for couch $21\frac{5}{8}$ yd

   Fabric for pillows $3\frac{3}{8}$ yd

   a. Total order (as a fraction) = _____

   b. Total order (as a decimal) = _____

   The client returned $3\frac{3}{8}$ yards of fabric because it was defective. Figure the new total fabric order at this point, showing the order as a fraction.

   c. Revised order (as a fraction) = _____

Study Guide

I. Terminology

like fractions Fractions that have like denominators.

unlike fractions Fractions with different denominators.

lowest common denominator The smallest number that can be divided by all the unlike denominators.

inspection method Used to determine if the largest denominator can be divided by the unlike denominators.

prime number A number larger than 1 that can be divided evenly only by itself and 1. The first six prime numbers are 2, 3, 5, 7, 11, and 13.

II. Adding Fractions

Adding Like Fractions: Add all the numerators and then place the sum of the numerators over the common denominator to form a new fraction.

Example: $\frac{5}{9} + \frac{2}{9} = \frac{7}{9}$

Adding Unlike Fractions: Once you have determined the lowest common denominator, you can complete the addition process by raising each fraction to the denominator and adding.

Example:

$$\frac{3}{8} = \frac{?}{120} \quad (120 \text{ divided by } 8 = 15;\ 15 \times 3 = 45) = \frac{45}{120}$$

$$\frac{7}{12} = \frac{?}{120} \quad (120 \text{ divided by } 12 = 10;\ 10 \times 7 = 70) = \frac{70}{120}$$

$$+\frac{11}{20} = \frac{?}{120} \quad (120 \text{ divided by } 20 = 6;\ 6 \times 11 = 66) = \frac{66}{120}$$

$$\text{Answer} = 1\frac{61}{120}$$

Finding the lowest common denominator: Referred to also as the LCD, the lowest common denominator can be determined by using two methods—the inspection method and using prime factorization.

a. *Inspection Method:* By observation, determine which number is the smallest number the denominators will divide evenly. Using the multiplication tables will help—begin with 2s, then 3s, then 4s, and so on until you determine the smallest number that can be divided evenly by all denominators.

b. *Using Prime Numbers:* Write the denominators horizontally across the page; divide each by any prime number common to two or more of the numbers; write the quotients above each denominator. Continue dividing each of the quotients until all quotients are equal to 1. The lowest common denominator is determined by multiplying all of the divisors used.

page 70

Adding Mixed Numbers: Add the fractional parts; add the whole numbers; add the sum of the fractional parts and the sum of the whole numbers; reduce answer to lowest terms as needed.

Example: $12\frac{4}{7} + 15\frac{2}{7} =$

$\frac{4}{7} + \frac{2}{7} = \frac{6}{7}$

$12 + 15 = 27$

$27 + \frac{6}{7} = 27\frac{6}{7}$

When the sum of the fractional parts of the mixed numbers is greater than 1, thus making it an improper fraction, add the fractional parts and convert the improper fraction to a mixed number and then add the sum of the whole numbers to it.

Example: $11\frac{9}{12} + 13\frac{7}{12} =$

$\frac{9}{12} + \frac{7}{12} = \frac{16}{12} = 1\frac{4}{12} = 1\frac{1}{3}$

$11 + 13 = 24$

$24 + 1\frac{1}{3} = 25\frac{1}{3}$

III. Subtracting Fractions

page 72

Find the lowest common denominator as necessary; subtract the numerators, and place the difference over the common denominator; reduce to lowest terms as necessary.

When the numerator in the fraction of the minuend is smaller than the numerator of the fraction in the subtrahend as in $2\frac{2}{10} - \frac{5}{10}$, borrow $\frac{10}{10}$ from the whole number 2 and reduce it to 1; add the $\frac{10}{10}$ to the $\frac{2}{10}$ making it $\frac{12}{10}$ and subtract $\frac{5}{10}$ from it as usual ($1\frac{12}{10} - \frac{5}{10} = 1\frac{7}{10}$).

Study Skills

How to Study Math

Math may be a more difficult area to study because it is abstract—something you can't tangibly touch or feel; therefore, it involves analytical thinking.

Tips

1. Be patient. Math is a building process. You build a firm foundation; then continue to build with each new learning—always build on what you last learned. For instance, you learned to add, subtract, and multiply before you learned to divide. Division requires knowledge of the first three skills to be successful.
2. Repetition is the key. The more practice you get the quicker you will learn and remember. Work through an assignment. If you have extra unassigned problems in the text, do those as well; if not, make up some of your own.
3. Don't stop when you understand the principle. Doing one pushup won't make you able to do a hundred. You must continue the practice to reinforce the learning.
4. Stay organized. After class, outline your notes. Make certain you understand everything you wrote down. Look up unfamiliar terms and procedures. Use 3" × 5" cards with examples and illustrations to use as a quick reference.
5. Get help if you need it. When you are having trouble, don't wait! Get help immediately from your instructor, math lab assistant, classmate, or tutor.

Assignment 1   Name_____ Date _____

Complete the following problems. Write your answers in the blanks provided. Express answers in lowest terms.

A. Find the lowest common denominator of the following fractions.

1. $\frac{2}{15}, \frac{3}{5}, \frac{6}{25}$ = _____

2. $\frac{5}{12}, \frac{6}{10}$ = _____

3. $\frac{3}{5}, \frac{1}{2}, \frac{3}{4}$ = _____

4. $\frac{2}{3}, \frac{6}{7}, \frac{10}{21}$ = _____

5. $\frac{3}{4}, \frac{5}{6}, \frac{2}{3}$ = _____

6. $\frac{5}{18}, \frac{4}{9}, \frac{6}{36}$ = _____

7. $\frac{2}{5}, \frac{3}{15}, \frac{6}{25}$ = _____

8. $\frac{1}{3}, \frac{1}{2}, \frac{2}{6}$ = _____

B. Add the following fractions, using like denominators.

9. $\frac{6}{8} + \frac{5}{8} + \frac{3}{8}$ = _____

10. $\frac{4}{5} + \frac{3}{5} + \frac{2}{5}$ = _____

11. $\frac{11}{25} + \frac{8}{25}$ = _____

12. $\frac{15}{30} + \frac{12}{30}$ = _____

13. $\frac{1}{5} + \frac{2}{5} + \frac{4}{5}$ = _____

14. $\frac{3}{7} + \frac{6}{7}$ = _____

15. $\frac{1}{3} + \frac{2}{3} + \frac{1}{3}$ = _____

16. $\frac{15}{32} + \frac{27}{32}$ = _____

C. Add the following mixed numbers, using *like* denominators.

17. $71\frac{2}{8}$
  $3\frac{5}{8}$
  $+ \ 4\frac{3}{8}$

18. $13\frac{4}{5}$
  $+18\frac{3}{5}$

19. $12\frac{3}{8}$
  $+ \ 9\frac{1}{8}$

20. $15\ \frac{4}{5}$
  $+17\frac{11}{25}$

21. $5\frac{6}{35}$
  $+7\frac{8}{35}$

22. $17\frac{5}{12}$
  $13\frac{7}{12}$
  $+ \ 4\frac{5}{12}$

23. $2\frac{12}{13}$
  $5\frac{9}{13}$
  $+ \ 3\frac{6}{13}$

24. $11\frac{9}{16}$
  $7\frac{5}{16}$
  $+ \ 2\frac{7}{16}$

D. Add these fractions, using *unlike* denominators.

25. $\frac{1}{4} + \frac{5}{8}$ = _____

26. $\frac{3}{4} + \frac{6}{7} + \frac{4}{28}$ = _____

27. $\frac{7}{12} + \frac{3}{7}$ = _____

28. $\frac{5}{18} + \frac{19}{24}$ = _____

29. $\frac{5}{6} + \frac{1}{3}$ = _____

30. $\frac{4}{9} + \frac{3}{8}$ = _____

31. $\frac{9}{8} + \frac{15}{16}$ = _____

32. $\frac{13}{25} + \frac{1}{30} + \frac{19}{30}$ = _____

E. Add the following mixed numbers, using *like* denominators.

33. $44\frac{5}{8}$
$+ \ 3\frac{5}{12}$

34. $16\frac{5}{6}$
$+10\frac{6}{7}$

35. $13\frac{4}{9}$
$+11\frac{3}{8}$

36. $12\frac{5}{8}$
$+16\frac{1}{7}$

37. $9\frac{5}{8}$
$+13\frac{4}{7}$

38. $71\frac{2}{3}$
$54\frac{5}{8}$
$+ \ 3\frac{5}{12}$

39. $27\frac{1}{2}$
$84\frac{3}{5}$
$+ \ 10\frac{2}{10}$

40. $17\frac{3}{8}$
$1\frac{1}{4}$
$+ \ 5\frac{5}{16}$

F. Add the following fractions.

41. $\frac{1}{4} + \frac{11}{12} + \frac{7}{16}$ = _____

42. $\frac{7}{8} + \frac{3}{4}$ = _____

43. $\frac{1}{2} + \frac{7}{8}$ = _____

44. $\frac{2}{5} + \frac{5}{6}$ = _____

45. $\frac{4}{5} + \frac{3}{4}$ = _____

46. $7\frac{1}{2} + 3\frac{3}{8}$ = _____

Assignment 2   Name_____ Date _____

Complete the following problems. Write your answers in the blanks provided. Reduce to lowest terms when possible.

A. Subtract these fractions, using *like* denominators.

1. $\dfrac{11}{12} - \dfrac{5}{12}$ = _____

2. $\dfrac{5}{9} - \dfrac{2}{9}$ = _____

3. $\dfrac{11}{15} - \dfrac{8}{15}$ = _____

4. $\dfrac{19}{30} - \dfrac{7}{30}$ = _____

5. $\dfrac{7}{8} - \dfrac{3}{8}$ = _____

6. $\dfrac{17}{25} - \dfrac{12}{25}$ = _____

7. $\dfrac{11}{16} - \dfrac{7}{16}$ = _____

8. $\dfrac{13}{24} - \dfrac{7}{24}$ = _____

B. Subtract these mixed numbers, using *like* denominators.

9. $11\dfrac{11}{16}$
    $-\ 5\dfrac{9}{16}$

10. $6\dfrac{3}{4}$
     $-\ 4\dfrac{1}{4}$

11. $10\dfrac{16}{21}$
     $-\ 6\dfrac{5}{21}$

12. $14\dfrac{4}{5}$
     $-\ 6\dfrac{1}{5}$

13. $18\dfrac{7}{8}$
     $-\ 9\dfrac{3}{8}$

14. $12\dfrac{15}{32}$
     $-10\dfrac{7}{32}$

15. $112\dfrac{5}{8}$
     $-\ 10\dfrac{3}{8}$

16. $35\dfrac{11}{13}$
     $-27\dfrac{3}{13}$

C. Subtract the following fractions, using *unlike* denominators.

17. $\dfrac{3}{4} - \dfrac{1}{3}$ = _____

18. $\dfrac{5}{15} - \dfrac{7}{25}$ = _____

19. $\dfrac{3}{4} - \dfrac{3}{8}$ = _____

20. $\dfrac{6}{7} - \dfrac{11}{21}$ = _____

21. $\dfrac{1}{2} - \dfrac{2}{5}$ = _____

22. $\dfrac{5}{8} - \dfrac{5}{12}$ = _____

23. $\dfrac{4}{5} - \dfrac{2}{3}$ = _____

24. $\dfrac{5}{6} - \dfrac{7}{9}$ = _____

D. Subtract the following mixed numbers using *unlike* denominators.

25. $23\dfrac{3}{5}$
     $-\ 11\dfrac{2}{7}$

26. $9\dfrac{5}{8}$
     $-7\dfrac{7}{12}$

27. $15\dfrac{5}{7}$
     $-10\dfrac{1}{3}$

28. $37\dfrac{4}{7}$
     $-\ 21\dfrac{5}{9}$

29. $2\frac{3}{4}$
    $-\,1\frac{2}{3}$

30. $12\frac{5}{8}$
    $-\,\,9\frac{1}{2}$

31. $25\frac{5}{6}$
    $-\,19\frac{5}{9}$

32. $7\frac{5}{6}$
    $-\,4\frac{2}{3}$

D. Subtract the following fractions.

33. $\frac{13}{16}-\frac{2}{3}$ = _____

34. $\frac{1}{2}-\frac{1}{3}$ = _____

35. $\frac{5}{6}-\frac{2}{3}$ = _____

36. $\frac{5}{8}-\frac{1}{2}$ = _____

37. $\frac{4}{7}-\frac{5}{9}$ = _____

38. $\frac{5}{7}-\frac{1}{3}$ = _____

F. Subtract the following fractions.

39. $4\frac{5}{6}$
    $-\,3\frac{1}{3}$

40. $4\frac{1}{2}$
    $-\,3\frac{5}{7}$

41. $4\frac{5}{12}$
    $-\,1\frac{1}{3}$

42. $8\frac{5}{8}$
    $-\,3\frac{3}{4}$

43. $13\frac{4}{5}$
    $-\,4\frac{1}{3}$

44. $16\frac{5}{8}$
    $-\,12\frac{2}{3}$

45. $3\frac{1}{2}$
    $-\,1\frac{1}{4}$

46. $4\frac{1}{2}$
    $-\,3\frac{6}{7}$

Assignment 3   Name_____ Date _____

Complete the following problems. Write your answers in the blanks provided. Reduce to lowest terms when possible

A.   Subtract these whole and mixed numbers.

1. $5 - \frac{1}{4}$ = _____

2. $23 - 11\frac{2}{5}$ = _____

3. $26 - 11\frac{1}{6}$ = _____

4. $35 - 22\frac{1}{3}$ = _____

5. $112 - 15\frac{1}{3}$ = _____

6. $8 - 4\frac{3}{5}$ = _____

7. $3 - 1\frac{1}{2}$ = _____

8. $27 - 13\frac{2}{7}$ = _____

B.   Subtract these mixed numbers, using *unlike* denominators. Remember to borrow if the numerator is smaller than the denominator.

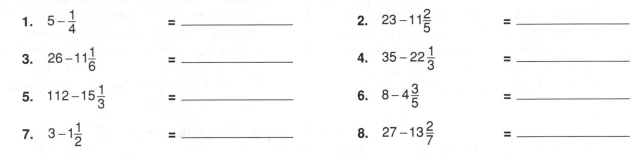

9. $18\frac{1}{2}$
     $- 5\frac{4}{9}$

10. $8\frac{1}{2}$
     $- 3\frac{5}{8}$

11. $25\frac{2}{7}$
     $- 13\frac{3}{4}$

12. $16\frac{19}{30}$
     $- 9\frac{3}{4}$

13. $8\frac{3}{7}$
     $- 3\frac{9}{10}$

14. $15\frac{3}{4}$
     $- 7\frac{5}{6}$

15. $13\frac{1}{5}$
     $- 6\frac{3}{4}$

16. $21\frac{1}{4}$
     $- 4\frac{15}{16}$

C.   Subtract the following fractions.

17. $6 - 3\frac{3}{4}$ = _____

18. $7\frac{1}{2} - 5\frac{4}{9}$ = _____

19. $13\frac{1}{3} - 6\frac{5}{9}$ = _____

20. $9 - 4\frac{9}{10}$ = _____

21. $27\frac{3}{4} - 18\frac{5}{6}$ = _____

22. $6\frac{3}{4} - 2\frac{7}{8}$ = _____

23. $10 - 7\frac{5}{11}$ = _____

24. $32\frac{1}{12} - 9\frac{3}{10}$ = _____

Assignment 4

Name_____ Date _____

Complete the following word problems. Write your answers in the blanks provided. Be sure to place dollar signs as needed. Express fractional parts in lowest terms.

1. Hector Perez purchased a total of 22 yards of fabric to upholster his office couch but only needs $19\frac{5}{8}$ yards. In addition to upholstering his couch, he wants to cover 3 pillows. How many total yards of fabric will be left for making pillows after he upholsters the couch?

   _____

2. Last week John Barton worked the following number of hours: 8, $6\frac{3}{4}$, $9\frac{1}{4}$, $5\frac{1}{2}$, and 6. How many hours did John work last week?

   _____

3. During the month of May, the Liang Catering Service filled the following tropical punch orders: $5\frac{3}{4}$ gallons, $7\frac{1}{2}$ gallons, $9\frac{1}{4}$ gallons, and $15\frac{2}{3}$ gallons. How many gallons were filled during May.

   _____

4. Part-time student workers restocked the college bookstore and worked a total of 50 hours last week. They worked $10\frac{1}{2}$ hours on Monday, $8\frac{1}{4}$ hours on Tuesday, $9\frac{3}{4}$ hours on Wednesday, and $11\frac{1}{4}$ hours on Thursday. Calculate the hours the students worked on Friday.

   _____

5. Before Chikara left on his trip, he read the mileage that showed on his car's odometer and wrote down $33,150\frac{2}{10}$. When he returned home, he again read his car's odometer and wrote down the mileage as $35,970\frac{8}{10}$. How many miles did Chikara drive on his trip? Because mileage is shown in tenths, do not reduce your answer.

   _____

6. Sutton's Produce delivered the following produce on Thursday morning to your catering shop: $6\frac{1}{2}$ pounds of green leafy lettuce, 4 pounds of tomatoes, $2\frac{1}{4}$ pounds of cucumbers, $3\frac{3}{4}$ pounds of white onions, 3 pounds of fresh mushrooms, and $1\frac{3}{4}$ pounds of parsley. Total the produce order.

   a. Total produce order   _____

   After your supervisor came into the shop, you discovered that $1\frac{3}{4}$ pounds of parsley was delivered in error; therefore, Sutton's Produce will pick up the parsley. Recalculate the produce order:

   b. Revised produce order   _____

   Now your supervisor tells you that an additional order was filled for the following items: $3\frac{1}{2}$ pounds of carrots, $2\frac{1}{4}$ pounds of celery, and 2 pounds of bell peppers. Calculate the additional order.

   c. Total additional order   _____

7. In your report of stock prices, you notice the changes below. Calculate the difference between the changes from yesterday to today. Indicate if the change was a gain (+) or a loss (−) beside each answer.

	Yesterday	Today	Difference
a.	$56\frac{7}{8}$	$52\frac{1}{4}$	_____
b.	$38\frac{1}{4}$	$36\frac{3}{8}$	_____
c.	$42\frac{1}{3}$	$41\frac{2}{3}$	_____
d.	$101\frac{5}{8}$	$99\frac{1}{4}$	_____
e.	$24\frac{1}{2}$	$22\frac{5}{8}$	_____

Assignment 5

Name_____ Date _____

**Complete the following word problems. Write your answers in the blanks provided.
Be sure to place dollar signs as needed. Express fractional parts in lowest terms.**

1. Julie Cravens bought $2\frac{4}{5}$ pounds of meat, $1\frac{1}{2}$ pounds of cheese, 2 pounds of bread and $2\frac{3}{4}$ pounds of baked beans for the company picnic. What is the total weight of her purchases?

   _____

2. This week, Matt Horsley, an employee of the State Sign Company, worked the following hours: Monday, $7\frac{3}{4}$ hours; Tuesday $5\frac{1}{2}$ hours; Wednesday, $8\frac{1}{5}$ hours; and Thursday, 7 hours. How many hours must Matt work on Friday to complete his 40 hour week?

   _____

3. The Wilson Cement Company purchased the following yards of sand this week: Monday, $5\frac{1}{8}$; Tuesday, $7\frac{2}{3}$; Wednesday, $8\frac{1}{2}$; Thursday, 3; and Friday, $8\frac{1}{6}$. What was the total number of yards of sand purchased by the Wilson Cement Company this week?

   _____

4. Claudia purchased $16\frac{5}{8}$ yards of lining for drapes. She made an error in her calculations and needed only $12\frac{1}{4}$ yards. How many yards did she overpurchase?

   _____

5. The office manager of Truett, Inc. needs to know how much paper is being used per day by her office staff. On Monday, $\frac{1}{3}$ ream was used; Tuesday, $\frac{1}{2}$ ream was used; Wednesday, $\frac{1}{4}$ ream was used; Thursday, $\frac{2}{3}$ ream was used; and on Friday, $\frac{1}{6}$ ream was used. How many reams of paper were used this week?

   _____

6. The Ashton Bakery used the following quantities of sugar during a recent week: Monday, $65\frac{3}{4}$ lbs.; Tuesday, $97\frac{2}{3}$ lbs.; Wednesday, $88\frac{3}{8}$ lbs.; Thursday, $57\frac{3}{8}$ lbs.; Friday, $55\frac{2}{3}$ lbs.; and on Saturday, $64\frac{1}{4}$ lbs. How many pounds of sugar did the bakery use during the week?

   _____

7. During a recent week, Johnny Weekly worked the following hours: $4\frac{1}{2}$, $5\frac{3}{4}$, $6\frac{1}{2}$, and $3\frac{1}{2}$. How many hours must Johnny work to complete his regular 25 hour week?

   _____

8. Three pieces of lumber measure $12\frac{3}{8}$ feet, $7\frac{1}{2}$ feet, and $5\frac{3}{8}$ feet. What is the total length of the lumber?

   _____

9. Permeate, Inc.'s stock reached an all-time high of $\$47\frac{1}{8}$ per share on Tuesday. At the end of the day, the stock fell to $\$32\frac{5}{8}$. How much did the stock fall from its high on Tuesday?

   _____

10. Eric Hess, a local designer, purchased the following increments of fabric for a special design: $\frac{1}{2}$, $\frac{1}{3}$, $2\frac{1}{4}$, and $\frac{5}{8}$ yards. He had $2\frac{1}{3}$ yards left over after he completed his design. How much fabric did Eric use?

   _____

Assignment 6

Name_____ Date _____

**Complete the following word problems. Write your answers in the blanks provided.
Be sure to place dollar signs as needed. Express fractional parts in lowest terms.**

1. Calculate Fred Quimby's overtime for the week: $\frac{1}{2}$ hour on Monday, $2\frac{3}{4}$ hours on Tuesday, $1\frac{1}{2}$ hours on Wednesday, 3 hours on Thursday, and 0 hours on Friday.

    Total overtime for the week: _____

2. Three types of fabric are needed for curtains. The valance requires $6\frac{2}{3}$ yards, the curtain requires $10\frac{1}{2}$ yards, and the lining requires $16\frac{3}{8}$ yards. How many total yards of fabric are needed?

    _____

3. A board $4\frac{6}{7}$ feet long must be sawed from an 8-foot board. How long is the remaining piece?

    _____

4. Shelly Moore worked the following hours during a week: $8\frac{1}{4}$, $6\frac{3}{4}$, $7\frac{1}{2}$, 8, and $8\frac{3}{4}$. Paula Stoutt worked $40\frac{1}{4}$ hours.

    Who worked the most hours?   _____

    How many more?   _____

5. Funtime Party Goods sold $\frac{1}{8}$ gross of invitations on July, 1, $\frac{3}{7}$ gross on July 8, and $\frac{2}{3}$ gross on July 15. How many gross were sold altogether on these three days?

    _____

6. The proposed land development project has four sides that measure $\frac{7}{8}$ mile, $\frac{1}{6}$ mile, $\frac{2}{5}$ mile, and $\frac{11}{12}$ mile. What is the total distance around the proposed project?

    _____

7. Catlin Cross bought 3 shares of stock. The prices for 2 of the shares were $82\frac{1}{3}$ and $78\frac{1}{8}$. Find the price of the third share if she paid a total of $225\frac{1}{4}$.

    _____

8. Three sides of a field are $99\frac{2}{3}$ feet, $87\frac{6}{7}$ feet, and $79\frac{3}{4}$ feet. If the distance around the field is $408\frac{1}{3}$ feet, find the length of the fouth side.

    _____

9. Four pieces of lumber measure $3\frac{1}{2}$ feet, $4\frac{3}{4}$ feet, $6\frac{5}{8}$ feet, and $1\frac{3}{8}$ feet. What is the total length of lumber?

    _____

10. Yolanda joined a weight-loss clinic to lose weight. In week 1, she lost $4\frac{1}{4}$ pounds; week 2, $3\frac{1}{2}$ pounds; week 3, $\frac{1}{4}$ pounds; week 4, $2\frac{1}{2}$ pounds; and week 5, $3\frac{1}{2}$ pounds. On week 6, she gained $2\frac{1}{4}$ pounds. What is Yolanda's net loss?

    _____

Practical Math Applications

CHAPTER 5

Multiplying and Dividing Fractions

5

Multiplying and Dividing Fractions

OBJECTIVES

After completing this chapter, you will be able to:

1. Multiply fractions, whole numbers, and mixed numbers.
2. Divide fractions, whole numbers, and mixed numbers.

You probably already have encountered multiplying and dividing fractions in some of your day-to-day activities. For example, suppose you have been asked to pick up 30 gallons of popcorn for the annual company picnic. Your committee has decided to have several $2\frac{1}{2}$-gallon cans placed at different points on the picnic grounds. This type of situation requires that you know how many $2\frac{1}{2}$-gallon cans to bring to fill with popcorn from the 30-gallon supply. You will learn the skills necessary to solve this problem in this chapter.

5.1 Multiplying Common Fractions

Mario has $\frac{2}{3}$ yard of fabric. He used $\frac{7}{8}$ of it to cover an office seat cushion. How much fabric did he use? To answer this question, let's multiply the two fractions using these basic steps.

> **EXAMPLE** _____
>
> $$\frac{2}{3} \times \frac{7}{8}$$

STEPS

1. Multiply the numerators.

    $$2 \times 7 = 14$$

2. Multiply the denominators.

    $$3 \times 8 = 24$$

3. Write the new numerator and new denominator as a fraction.

    $$\frac{14}{24}$$

4. Reduce to lowest terms as necessary.

    $$\frac{14}{24} = \frac{7}{12}$$

Notice the following example has three common fractions. Multiply the fractions using these steps.

EXAMPLE

$$\frac{5}{8} \times \frac{3}{5} \times \frac{1}{3} =$$

STEPS

1. Multiply the numerators of the first two fractions, then multiply that product by the numerator of the third fraction.

    $$5 \times 3 \times 1 = 15$$

2. Multiply the denominators of the first two fractions, then multiply that product by the denominator of the third fraction.

    $$8 \times 5 \times 3 = 120$$

3. Rewrite the fraction.

    $$\frac{15}{120}$$

4. Reduce to lowest terms as needed.

    $$\frac{15}{120} = \frac{1}{8}$$

PRACTICE PROBLEMS

Multiply these common fractions. Write your answers in the blanks provided.

1. $\frac{5}{12} \times \frac{2}{3} =$ _____

2. $\frac{3}{8} \times \frac{1}{2} =$ _____

3. $\frac{5}{8} \times \frac{3}{5} \times \frac{2}{3} =$ _____

4. $\frac{5}{8} \times \frac{1}{8} \times \frac{7}{8} =$ _____

Solutions: **1.** $\frac{5}{18}$; **2.** $\frac{3}{16}$; **3.** $\frac{1}{4}$; **4.** $\frac{35}{512}$

5.2 Multiplying Fractions by Using Cancellation

Multiplying and dividing fractions can be simplified by using cancellation. *Cancellation* is the process of determining a common number that will divide evenly any one of the numerators and any one of the denominators in the fractions being multiplied or divided.

Follow these steps to multiply two fractions in this example.

EXAMPLE

$$\frac{2}{7} \times \frac{5}{8}$$

STEPS

1. Determine the common factor that will divide evenly any one of the numerators and any one of the denominators. In the preceding example, the numbers 2 and 8 are divisible by 2. The original numerator and denominator are marked with a diagonal line to cancel them as shown.

$$\frac{\cancel{2}}{7} \times \frac{5}{\cancel{8}}$$

2. Divide the numerator (2) and the denominator (8) by the common factor (2). The quotients become the new numerator and denominator.

$$\frac{\overset{1}{\cancel{2}}}{7} \times \frac{5}{\underset{4}{\cancel{8}}}$$

3. Multiply the numerators, then the denominators.

$$\frac{1}{7} \times \frac{5}{4} = \frac{5}{28}$$

Let's multiply three fractions in this next example using the same steps.

EXAMPLE _____

$$\frac{1}{2} \times \frac{6}{15} \times \frac{5}{12}$$

STEPS

1. $\dfrac{1}{\underset{1}{\cancel{2}}} \times \dfrac{\overset{3}{\cancel{6}}}{\underset{3}{\cancel{15}}} \times \dfrac{\overset{1}{\cancel{5}}}{12}$

2. $1 \times 3 \times 1 = 3$

3. $1 \times 3 \times 12 = 36$

4. $\dfrac{3}{36} = \dfrac{1}{12}$

At this point, make sure you understand the difference between canceling and reducing fractions. Canceling is a method of "lowering" any numerator and/or denominator before multiplying or dividing. Canceling is a shortcut.

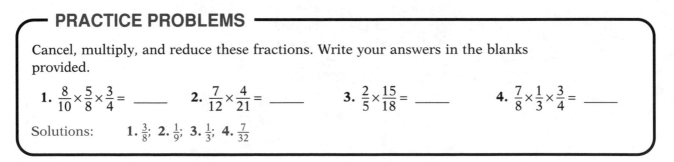

PRACTICE PROBLEMS

Cancel, multiply, and reduce these fractions. Write your answers in the blanks provided.

1. $\dfrac{8}{10} \times \dfrac{5}{8} \times \dfrac{3}{4} =$ _____ **2.** $\dfrac{7}{12} \times \dfrac{4}{21} =$ _____ **3.** $\dfrac{2}{5} \times \dfrac{15}{18} =$ _____ **4.** $\dfrac{7}{8} \times \dfrac{1}{3} \times \dfrac{3}{4} =$ _____

Solutions: **1.** $\frac{3}{8}$; **2.** $\frac{1}{9}$; **3.** $\frac{1}{3}$; **4.** $\frac{7}{32}$

5.3 Multiplying Whole Numbers and Fractions

Follow these steps to multiply whole numbers and common fractions, using this example.

EXAMPLE _____

$$5 \times \frac{2}{3} \times \frac{7}{8}$$

STEPS

1. Convert the whole number to an improper fraction and cancel any numerators and denominators if possible.

$$\frac{5}{1} \times \frac{\overset{1}{\cancel{2}}}{3} \times \frac{7}{\underset{4}{\cancel{8}}}$$

2. Multiply the numerators, then the denominators.

$$\frac{5}{1} \times \frac{1}{3} \times \frac{7}{4} = \frac{35}{12}$$

3. Convert the improper fraction to a mixed number. Make sure that the fractional part of the mixed number is in lowest terms.

$$2\frac{11}{12}$$

EXAMPLE _____

Let's assume that Phillip Lawrence earns $2,345 a month. He spends one-fifth of his monthly salary for rent. What does he pay each month for rent? When a math problem uses the word "of" as in "He spends one-fifth *of* his monthly salary," the word *of* means to multiply. To answer this question, multiply, following these steps.

$$\frac{1}{5} \times \$2,345$$

STEPS

1. $\frac{1}{5} \times \$2,345 =$

2. $\frac{1}{\underset{1}{\cancel{5}}} \times \frac{\overset{469}{\cancel{2,345}}}{1} =$

3. $\frac{1}{1} \times \frac{469}{1} = \$469$

PRACTICE PROBLEMS

Multiply these whole numbers and fractions. Write your answers in the blanks provided. Reduce to lowest terms as needed.

1. $\frac{2}{5} \times \frac{3}{8} \times 15 =$ _____   **2.** $13 \times \frac{3}{4} \times \frac{2}{8} =$ _____   **3.** $85 \times \frac{3}{4} \times \frac{1}{8} =$ _____   **4.** $\frac{1}{6} \times 2 \times \frac{5}{12} =$ _____

Solutions: **1.** $2\frac{1}{4}$;   **2.** $2\frac{7}{16}$;   **3.** $7\frac{31}{32}$;   **4.** $\frac{5}{36}$

5.4   Multiplying Mixed Numbers

The steps for multiplying mixed numbers are basically the same steps used in the multiplication of whole numbers and fractions. Let's study the following example.

EXAMPLE _____

$$6\frac{2}{7} \times 3\frac{1}{8}$$

STEPS

1. Convert the mixed numbers to improper fractions.

$$6\frac{2}{7} = \frac{44}{7} \quad \times \quad 3\frac{1}{8} = \frac{25}{8}$$

2. Rewrite the problem and cancel any numerators and denominators if possible.

$$\frac{\overset{11}{\cancel{44}}}{7} \times \frac{25}{\underset{2}{\cancel{8}}}$$

3. Multiply the numerators, then the denominators.

$$\frac{11}{7} \times \frac{25}{2} = \frac{275}{14}$$

4. Convert the improper fraction to a mixed number.

$$\frac{275}{14} = 19\frac{9}{14}$$

PRACTICE PROBLEMS

Multiply the following mixed numbers. Write your answers in the blanks provided.

1. $30\frac{1}{8} \times 7\frac{2}{3} =$ _____   **2.** $8\frac{1}{4} \times 16\frac{1}{3} =$ _____   **3.** $13\frac{5}{6} \times 6\frac{2}{3} =$ _____   **4.** $14\frac{2}{3} \times 1\frac{2}{3} =$ _____

Solutions: **1.** $230\frac{23}{24}$;   **2.** $134\frac{3}{4}$;   **3.** $92\frac{2}{9}$;   **4.** $24\frac{4}{9}$

5.5 Multiplying Mixed Numbers and Whole Numbers

One yard of chain costs \$12. Find the cost of $2\frac{7}{8}$ yards of chain. You are already familiar with the basic steps used for multiplying mixed numbers. The same basic steps will be used here.

EXAMPLE _____

$$12 \times 2\frac{7}{8}$$

STEPS

1. Change the whole number to a fraction and convert the mixed number to an improper fraction.

$$\frac{12}{1} \times \frac{23}{8}$$

2. Cancel any numerators and denominators and multiply.

$$\frac{\overset{3}{\cancel{12}}}{1} \times \frac{23}{\underset{2}{\cancel{8}}} = \frac{69}{2}$$

3. Convert the improper fraction to a mixed number.

$$\frac{69}{2} = 34\frac{1}{2} \text{ or } \$34.50$$

PRACTICE PROBLEMS

Multiply these mixed numbers and whole numbers. Write your answers in the blanks provided.

1. $7 \times 3\frac{3}{8} =$ _____   **2.** $2 \times 3\frac{1}{6} =$ _____   **3.** $1\frac{3}{5} \times 7 =$ _____   **4.** $7\frac{7}{8} \times 3 =$ _____

Solutions: **1.** $23\frac{5}{8}$; **2.** $6\frac{1}{3}$; **3.** $11\frac{1}{5}$; **4.** $23\frac{5}{8}$

5.6 Dividing Fractions

A common method of dividing fractions is to invert the divisor, change the division symbol to a multiplication symbol, and then multiply. To invert is simply to reverse the order of the fraction. For example, a divisor such as $\frac{2}{3}$ is inverted to $\frac{3}{2}$. Let's divide some fractions following these steps.

EXAMPLE _____

$$\frac{5}{6} \div \frac{2}{3}$$

STEPS

1. Invert the divisor, then multiply. (The dividend, $\frac{5}{6}$, remains the same.)

$$\frac{5}{\cancel{6}_2} \times \frac{\cancel{3}^1}{2}$$

2. Convert any improper fraction to a mixed number or reduce if necessary.

$$\frac{5}{4} = 1\frac{1}{4}$$

PRACTICE PROBLEMS

Divide these fractions. Write your answers in the blanks provided.

1. $\frac{3}{4} \div \frac{1}{2} =$ _____   2. $\frac{7}{8} \div \frac{7}{8} =$ _____   3. $\frac{8}{9} \div \frac{2}{3} =$ _____   4. $\frac{5}{7} \div \frac{1}{3} =$ _____

Solutions: **1.** $1\frac{1}{2}$; **2.** 1; **3.** $1\frac{1}{3}$; **4.** $2\frac{1}{7}$

5.7   Dividing Whole Numbers and Fractions

Follow these basic steps to divide whole numbers and fractions.

EXAMPLE

$$6 \div \frac{3}{4}$$

STEPS

1. Express the whole number as a fraction by placing it over the denominator 1, and inverting the divisor.

$$\frac{6}{1} \times \frac{4}{3}$$

2. Cancel, if necessary, and multiply.

$$\frac{\cancel{6}^2}{1} \times \frac{4}{\cancel{3}_1} = \frac{8}{1} = 8$$

■ **TIP**   When a whole number is divided by a proper fraction, the answer will be larger than the whole number, because this is the same as multiplying by an improper fraction.

PRACTICE PROBLEMS

Divide these whole numbers and fractions. Write your answers in the blanks provided.

1. $15 \div \frac{7}{8} =$ _____   2. $8 \div \frac{2}{3} =$ _____   3. $12 \div \frac{4}{5} =$ _____   4. $6 \div \frac{3}{5} =$ _____

Solutions: **1.** $17\frac{1}{7}$; **2.** 12; **3.** 15; **4.** 10

5.8 Dividing Mixed Numbers

Let's divide the mixed numbers in the next example by using these steps.

EXAMPLE

$$14\frac{4}{5} \div 9\frac{1}{3}$$

STEPS

1. Convert each mixed number to an improper fraction.

$$14\frac{4}{5} = \frac{74}{5} \qquad 9\frac{1}{3} = \frac{28}{3}$$

2. Invert the divisor, cancel if necessary, and multiply.

$$\frac{\overset{37}{\cancel{74}}}{5} \times \frac{3}{\underset{14}{\cancel{28}}} = \frac{111}{70} = 1\frac{41}{70}$$

PRACTICE PROBLEMS

Divide these mixed numbers. Write your answers in the blanks provided.

1. $1\frac{3}{5} \div 2\frac{3}{4} =$ _____ **2.** $2\frac{2}{3} \div 4\frac{1}{3} =$ _____ **3.** $1\frac{6}{10} \div 2\frac{2}{3} =$ _____ **4.** $2\frac{1}{8} \div 4\frac{3}{8} =$ _____

Solutions: **1.** $\frac{32}{55}$; **2.** $\frac{8}{13}$; **3.** $\frac{3}{5}$; **4.** $\frac{17}{35}$

5.9 Dividing Whole Numbers and Mixed Numbers

Let's divide some whole numbers and mixed numbers by following these steps. (Notice that in this first problem the whole number is the dividend.)

EXAMPLE

$$9 \div 5\frac{3}{4}$$

STEPS

1. Express the whole number as a fraction by placing it over the denominator 1, and convert the mixed number to an improper fraction.

$$\frac{9}{1} \div \frac{23}{4}$$

2. Invert the divisor, cancel if necessary, and multiply.

$$\frac{9}{1} \times \frac{4}{23} = \frac{36}{23}$$

3. Convert any improper fraction to a mixed number or reduce if necessary.

$$\frac{36}{23} = 1\frac{13}{23}$$

In the next problem, notice the whole number is the divisor.

$$2\frac{1}{8} \div 5$$

Divide by using the same basic steps.

STEPS

1. $\frac{17}{8} \div \frac{5}{1}$

2. $\frac{17}{8} \times \frac{1}{5} = \frac{17}{40}$

PRACTICE PROBLEMS

Divide the following whole numbers and mixed numbers. Write your answers in the blanks provided.

1. $7\frac{2}{3} \div 8 =$ _____   **2.** $6 \div 2\frac{1}{3} =$ _____   **3.** $1\frac{2}{3} \div 8 =$ _____   **4.** $5 \div 3\frac{5}{9} =$ _____

Solutions: **1.** $\frac{23}{24}$; **2.** $2\frac{4}{7}$; **3.** $\frac{5}{24}$; **4.** $1\frac{13}{32}$

MATH ALERT

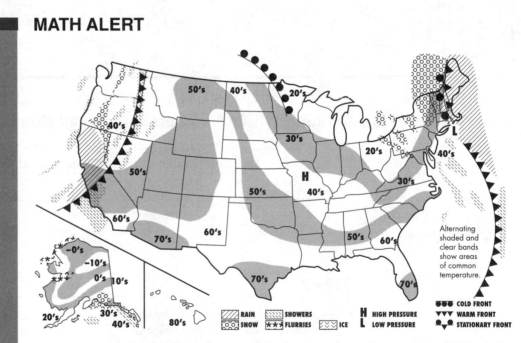

Have you ever listened to the weather forecast on television and heard the forecaster say something like this: "We have had a total of $69\frac{1}{2}$ inches of rainfall in our area. Our average rainfall per month is $13\frac{9}{10}$ inches." Now that you have worked with fractions, you should be able to make simple calculations like this one.

Determine the average monthly rainfall if the total rainfall was $50\frac{1}{4}$ inches and today's date was July 1.

Practical Math Applications

Study Guide

I. Terminology

page 87   *Cancellation:* The process of determining a common number that will divide evenly any one of the numerators and any one of the denominators in the fractions being multiplied or divided.

Example: $\dfrac{\overset{1}{\cancel{2}}}{7} \times \dfrac{5}{\underset{4}{\cancel{8}}}$

II. Multiplying Fractions

page 86   *Multiplying Common Fractions:* Multiply the numerators; multiply the denominators; write the new numerator and new denominator as a fraction; reduce to the lowest terms.

Example: $\dfrac{2}{3} \times \dfrac{7}{8} =$
$2 \times 7 = 14$
$3 \times 8 = 24$
$\dfrac{14}{24}$ reduced to $\dfrac{7}{12}$

page 87   *Multiplying Fractions by Cancellation:* Determine the common factor that will divide evenly any one of the numerators and any of the denominators; draw a line through each; divide the numerator and the denominator by the common factor and write the quotient above each; the quotients become the new numerator and denominator; multiply the numerators, then the denominators.

Example: $\dfrac{2}{7} \times \dfrac{5}{8} = \dfrac{\overset{1}{\cancel{2}}}{7} \times \dfrac{5}{\underset{4}{\cancel{8}}} = \dfrac{1}{7} \times \dfrac{5}{4} = \dfrac{5}{28}$

page 89   *Multiplying Whole Numbers and Fractions:* Convert the whole number to an improper fraction and cancel any numerators and denominators if possible; multiply the numerators, then the denominators; convert the improper fraction to a mixed number.

Example: $5 \times \dfrac{2}{3} \times \dfrac{7}{8} = \dfrac{5}{1} \times \dfrac{\overset{1}{\cancel{2}}}{3} \times \dfrac{7}{\underset{4}{\cancel{8}}} = \dfrac{35}{12} = 2\dfrac{11}{12}$

page 90   *Multiplying Mixed Numbers:* Convert the mixed numbers to improper fractions; rewrite the problem and cancel any numerators and denominators if possible; multiply the numerators, then the denominators; convert the improper fraction to a mixed number.

Example: $6\dfrac{2}{7} \times 3\dfrac{1}{8} = 6\dfrac{2}{7} = \dfrac{44}{7} \times 3\dfrac{1}{8} = \dfrac{25}{8}$

$\dfrac{\overset{11}{\cancel{44}}}{7} \times \dfrac{25}{\underset{2}{\cancel{8}}} = \dfrac{275}{14} = 19\dfrac{9}{14}$

page 91   *Multiplying Mixed Numbers and Whole Numbers:* Change the whole number to a fraction and convert the mixed number to an improper fraction; cancel any numerators and denominators and multiply; convert an improper fraction to a mixed number.

Example: $12 \times 2\dfrac{7}{8} = \dfrac{\overset{3}{\cancel{12}}}{1} \times \dfrac{23}{\underset{2}{\cancel{8}}} = \dfrac{69}{2} = 34\dfrac{1}{2}$

III. Dividing Fractions

page 91   *Dividing Fractions:* Invert the divisor; then multiply; convert any improper fraction to a mixed number or reduce if necessary.

Example: $\dfrac{5}{6} \div \dfrac{2}{3} = \dfrac{5}{\overset{}{\underset{2}{6}}} \times \dfrac{\overset{1}{3}}{2} = \dfrac{5}{4} = 1\dfrac{1}{4}$

page 92   *Dividing Whole Numbers and Fractions:* Express the whole number as a fraction by placing it over the denominator 1, and inverting the divisor; cancel, if necessary, and multiply.

Example: $6 \div \dfrac{3}{4} = \dfrac{\overset{2}{6}}{1} \times \dfrac{4}{\underset{1}{3}} = \dfrac{8}{1} = 8$

page 93   *Dividing Mixed Numbers:* Convert each mixed number to an improper fraction; invert the divisor, cancel, if necessary, and multiply.

Example: $14\dfrac{4}{5} \div 9\dfrac{1}{3}$ $(14 \times 5) + 4 = 74$ $(9 \times 3) + 1 = 28$

$\dfrac{\overset{37}{74}}{5} \times \dfrac{3}{\underset{14}{28}} = \dfrac{111}{70} = 1\dfrac{41}{70}$

page 93   *Dividing Whole Numbers and Mixed Numbers:* Express the whole number as a fraction by placing it over the denominator 1, and convert the mixed number to an improper fraction; invert the divisor, cancel, if necessary, and multiply; convert any improper fraction to a mixed number or reduce if necessary.

Example: $9 \div 5\dfrac{3}{4} = \dfrac{9}{1} \div \dfrac{23}{4} = \dfrac{9}{1} \times \dfrac{4}{23} = \dfrac{36}{23} = 1\dfrac{13}{23}$

Study Skills

Studying For a Test

Perhaps this is best called "Managing Test Anxiety" because we all get nervous and worried before a test. That is normal. Some say they hyperventilate, have clammy hands, perspire, or have a knot in their stomach. To avoid this anxiety, you should have practiced good study skills prior to the test.

Tips

1. Replace negative thoughts with positive ones: "I feel good about this test. I'm going to do my best. I know I'll do well."
2. Arrive for the test early. Get a good seat—alone—not next to a friend. Spend a few minutes mentally reviewing the material.
3. Do not talk to classmates or discuss various possible test information. If a classmate mentions what they think is a major point, and one for which you have not prepared, you may panic.
4. Take seven to ten deep breaths, holding each one about five seconds; then release all the air possible after each breath and count to five. Tell yourself to relax and be calm.

Assignment 1   Name_____ Date _____

**Complete the following problems. Write your answers in the blanks provided.
Reduce your answers to lowest terms as needed.**

Multiply these fractions, whole numbers, and mixed numbers.

1. $\frac{2}{3} \times \frac{5}{6} \times \frac{1}{4} =$ _____

2. $\frac{4}{5} \times \frac{3}{8} =$ _____

3. $\frac{2}{4} \times \frac{2}{3} =$ _____

4. $\frac{5}{6} \times \frac{7}{8} =$ _____

5. $\frac{6}{9} \times \frac{5}{8} =$ _____

6. $6\frac{4}{9} \times \frac{5}{7} =$ _____

7. $2 \times \frac{5}{6} =$ _____

8. $\frac{2}{3} \times 6 =$ _____

9. $\frac{5}{9} \times 1\frac{2}{3} =$ _____

10. $4 \times \frac{7}{8} =$ _____

11. $\frac{5}{9} \times 5 =$ _____

12. $7\frac{5}{6} \times 2\frac{5}{8} =$ _____

13. $2\frac{7}{8} \times 4\frac{1}{3} =$ _____

14. $4\frac{2}{3} \times 4\frac{1}{3} =$ _____

15. $2\frac{7}{8} \times 2\frac{1}{2} =$ _____

16. $\frac{6}{7} \times \frac{3}{8} =$ _____

17. $5\frac{5}{6} \times 5 =$ _____

18. $\frac{1}{4} \times \frac{3}{8} =$ _____

19. $\frac{7}{9} \times \frac{6}{7} =$ _____

20. $\frac{4}{6} \times \frac{9}{10} =$ _____

21. $\frac{16}{15} \times \frac{5}{32} =$ _____

22. $\frac{3}{12} \times \frac{15}{30} =$ _____

23. $1\frac{1}{2} \times 4\frac{7}{8} =$ _____

24. $2\frac{6}{9} \times 1\frac{1}{8} =$ _____

25. $6\frac{1}{8} \times 3\frac{1}{4} =$ _____

26. $8\frac{11}{10} \times 6\frac{3}{5} =$ _____

27. $\frac{2}{7} \times \frac{3}{4} =$ _____

28. $\frac{1}{9} \times \frac{4}{8} =$ _____

29. $\frac{5}{6} \times \frac{3}{8} =$ _____

30. $\frac{2}{3} \times \frac{18}{22} =$ _____

31. $\frac{1}{4} \times \frac{4}{12} =$ _____

32. $3\frac{1}{9} \times 4\frac{3}{12} =$ _____

33. $5\frac{1}{5} \times 6\frac{3}{5} =$ _____

34. $8\frac{1}{10} \times 7\frac{2}{5} =$ _____

35. $\frac{2}{10} \times \frac{3}{7} =$ _____

36. $\frac{4}{8} \times \frac{6}{9} =$ _____

37. $\frac{1}{6} \times \frac{7}{8} =$ _____

38. $\frac{6}{9} \times \frac{3}{12} =$ _____

39. $\frac{8}{10} \times \frac{7}{9} =$ _____

40. $6\frac{8}{9} \times 2\frac{3}{7} =$ _____

41. $3\frac{1}{3} \times 4\frac{1}{4} =$ _____

42. $\frac{4}{5} \times \frac{1}{2} =$ _____

43. $\frac{1}{5} \times \frac{3}{5} =$ _____

44. $\frac{2}{3} \times \frac{1}{4} =$ _____

45. $\frac{5}{6} \times \frac{7}{8} =$ _____

46. $2\frac{1}{2} \times 8\frac{1}{4} =$ _____

47. $5\frac{9}{10} \times 4\frac{1}{2} =$ _____

48. $1\frac{1}{2} \times 3 =$ _____

49. $\frac{3}{4} \times 7 =$ _____

50. $5 \times \frac{2}{3} =$ _____

51. $3\frac{2}{3} \times 1\frac{1}{8} =$ _____

52. $4\frac{2}{3} \times 3 =$ _____

Assignment 2   Name_____ Date _____

Complete the following problems. Write your answers in the blanks provided. Reduce to lowest terms as needed.

Divide these fractions, whole numbers, and mixed numbers.

1. $2 \div \frac{3}{4} =$ _____

2. $\frac{1}{9} \div \frac{1}{3} =$ _____

3. $\frac{2}{8} \div \frac{2}{7} =$ _____

4. $\frac{3}{8} \div \frac{1}{4} =$ _____

5. $\frac{4}{9} \div \frac{1}{4} =$ _____

6. $\frac{1}{2} \div \frac{3}{8} =$ _____

7. $5 \div 3\frac{1}{6} =$ _____

8. $6\frac{1}{3} \div 3 =$ _____

9. $4 \div 1\frac{2}{3} =$ _____

10. $8\frac{3}{8} \div 4 =$ _____

11. $6\frac{2}{3} \div 2\frac{3}{5} =$ _____

12. $6\frac{1}{4} \div 7\frac{2}{5} =$ _____

13. $2\frac{1}{3} \div 3\frac{1}{8} =$ _____

14. $3\frac{1}{3} \div 1\frac{5}{8} =$ _____

15. $\frac{1}{9} \div \frac{1}{7} =$ _____

16. $\frac{3}{11} \div \frac{4}{7} =$ _____

17. $\frac{2}{3} \div \frac{4}{8} =$ _____

18. $\frac{3}{8} \div \frac{1}{4} =$ _____

19. $\frac{6}{7} \div \frac{3}{9} =$ _____

20. $\frac{1}{6} \div \frac{3}{10} =$ _____

21. $\frac{15}{45} \div \frac{1}{5} =$ _____

22. $\frac{1}{12} \div \frac{5}{6} =$ _____

23. $1\frac{1}{2} \div \frac{3}{4} =$ _____

24. $3\frac{1}{4} \div \frac{6}{8} =$ _____

25. $4\frac{1}{2} \div \frac{1}{2} =$ _____

26. $\frac{2}{11} \div \frac{1}{9} =$ _____

27. $6\frac{1}{2} \div 1\frac{1}{10} =$ _____

28. $2\frac{1}{8} \div 5\frac{1}{6} =$ _____

29. $3\frac{1}{9} \div 8\frac{1}{7} =$ _____

30. $4\frac{1}{5} \div 4\frac{1}{12} =$ _____

31. $8\frac{1}{4} \div \frac{1}{5} =$ _____

32. $2\frac{7}{8} \div \frac{5}{6} =$ _____

33. $1\frac{7}{8} \div 1\frac{9}{10} =$ _____

34. $\frac{1}{9} \div \frac{2}{3} =$ _____

35. $\frac{1}{3} \div \frac{1}{4} =$ _____

36. $\frac{2}{7} \div \frac{1}{8} =$ _____

37. $\frac{3}{8} \div \frac{1}{3} =$ _____

38. $\frac{4}{5} \div \frac{1}{7} =$ _____

39. $\frac{1}{4} \div \frac{3}{4} =$ _____

40. $\frac{3}{9} \div \frac{1}{3} =$ _____

41. $\frac{1}{5} \div \frac{1}{3} =$ _____

42. $\frac{6}{8} \div \frac{6}{7} =$ _____

43. $\frac{3}{9} \div \frac{1}{2} =$ _____

44. $\frac{1}{2} \div \frac{1}{10} =$ _____

45. $\frac{1}{6} \div \frac{1}{12} =$ _____

46. $\frac{3}{11} \div \frac{6}{7} =$ _____

47. $\frac{4}{6} \div \frac{10}{12} =$ _____

48. $\frac{7}{15} \div \frac{10}{20} =$ _____

49. $\frac{1}{14} \div \frac{8}{9} =$ _____

50. $\frac{3}{9} \div \frac{4}{6} =$ _____

51. $\frac{2}{3} \div \frac{1}{4} =$ _____

52. $3\frac{1}{5} \div 14 =$ _____

Assignment 3   Name_____ Date _____

**Complete the following problems. Write your answers in the blanks provided.
Reduce to lowest terms as needed.**

A.   Multiply or divide these fractions, whole numbers, and mixed numbers.

1. $\frac{2}{8} \div \frac{4}{7} =$ _____

2. $\frac{1}{9} \div \frac{8}{10} =$ _____

3. $\frac{7}{8} \times \frac{4}{9} =$ _____

4. $\frac{6}{7} \times \frac{4}{8} =$ _____

5. $\frac{8}{9} \times \frac{11}{15} =$ _____

6. $1\frac{1}{2} \div 1\frac{2}{10} =$ _____

7. $5 \div \frac{1}{3} =$ _____

8. $8 \times \frac{1}{6} =$ _____

9. $\frac{5}{8} \div 1\frac{1}{2} =$ _____

10. $4\frac{2}{9} \div 3\frac{2}{5} =$ _____

11. $\frac{8}{12} \times \frac{4}{64} =$ _____

12. $\frac{50}{150} \div \frac{25}{75} =$ _____

13. $\frac{3}{21} \div \frac{7}{9} =$ _____

14. $\frac{8}{20} \div \frac{6}{10} =$ _____

15. $\frac{4}{5} \times \frac{11}{12} =$ _____

16. $\frac{3}{4} \times \frac{1}{2} \times \frac{3}{8} =$ _____

17. $\frac{10}{12} \times \frac{1}{5} \times \frac{3}{4} =$ _____

18. $\frac{8}{9} \times \frac{4}{5} \times \frac{2}{3} =$ _____

19. $1\frac{1}{5} \times 2\frac{1}{2} \times 1\frac{1}{8} =$ _____

20. $2\frac{1}{3} \times 6 \times 4\frac{1}{2} =$ _____

21. $8\frac{1}{5} \div 6\frac{7}{8} =$ _____

22. $4\frac{1}{9} \div 6\frac{3}{12} =$ _____

23. $\frac{4}{7} \div \frac{3}{14} =$ _____

24. $\frac{5}{6} \times \frac{1}{3} =$ _____

25. $\frac{2}{3} \times 3\frac{1}{3} =$ _____

26. $5\frac{1}{2} \times 5\frac{4}{6} =$ _____

27. $\frac{3}{7} \times \frac{2}{7} =$ _____

28. $2 \times \frac{1}{10} =$ _____

29. $\frac{1}{2} \times \frac{2}{5} =$ _____

30. $3\frac{3}{4} \div 1\frac{1}{4} =$ _____

Assignment 4

Name_____ Date _____

Complete the calculations for the following problems by multiplying or dividing these fractions, whole numbers, and mixed numbers. Write your answers in the blanks. Reduce to lowest terms as needed.

1. Morgan, Smith, and Williams Construction Company own a $514\frac{1}{2}$-acre tract of land. They want to divide the land into 5-acre lots. How many lots can they form from this tract of land?

   _____

2. The Fabulous Fabric Shop made the following sales of material: $3\frac{3}{4}$ yd, $5\frac{1}{8}$ yd, $2\frac{1}{2}$ yd, $6\frac{1}{4}$ yd, $1\frac{1}{16}$ yd, and $\frac{5}{8}$ yd. The fabric sold for $125. What was the average selling price per yard of material?

   _____

3. Raoul owns and operates Martinez Landscape Company. He has contracted to put in a sprinkler system. He cut a piece of pipe into 5 lengths of $1\frac{1}{2}$ feet each. How long was the original piece of pipe?

   _____

4. Carswell Concrete Company doubled its orders for concrete this month over last month's figure of $75\frac{3}{4}$ loads. How many loads did Carswell deliver this month?

   _____

5. Reynolds Insurance Agency has a word processing pool to do typing for the agents. Rita typed a report for an agent that was $19\frac{1}{4}$ pages long in $2\frac{2}{3}$ hours. How many pages per hour did she type?

   _____

6. Seminole inherited $\frac{2}{3}$ of her grandmother's estate. The estate amounted to $78,000. How much money did Seminole receive?

   _____

7. Grady sold $\frac{1}{2}$ of his interest of $50,000 in the Randolph Manufacturing Company to one of his 6 partners. How much did the one partner receive?

   _____

8. If gold is selling for $402 an ounce, what is the cost of $24\frac{1}{2}$ ounces of gold?

   _____

9. How far can Diedre travel in her car on $14\frac{5}{8}$ gallons of gasoline if the car averages $22\frac{1}{2}$ miles per gallon?

   _____

10. If office space is renting for $1.25 per square foot, how much would it cost you to rent an office with $900\frac{2}{3}$ square feet?

    _____

11. One cord of wood sells for $120. What is the cost of $3\frac{3}{4}$ cords of wood?

    _____

12. If you drove your car 65 miles an hour for $3\frac{3}{4}$ hours, how many miles would you drive?

    _____

13. John and David have a business painting house numbers on curbs. When they were painting curbs on a long street that contained 24 houses, John painted $\frac{3}{4}$ of the curbs and David painted $\frac{1}{4}$. How many curbs did each paint?

    _____

Assignment 5

Name_____ Date _____

Complete the calculations for the following problems by multiplying or dividing these fractions, whole numbers, and mixed numbers. Write your answers in the blanks. Reduce to lowest terms as needed.

1. Gail bought $\frac{4}{5}$ of a yard of fabric to make a pillow. She used $\frac{2}{3}$ of it. How much fabric did she use?

    _____

2. Sandpiper Sightseeing Tours purchased a new van for the business. On the first trip, Jesse drove $4\frac{1}{5}$ hours at a speed of 40 mph. How far did Jesse travel?

    _____

3. Wayne served 30 pizzas at the office party, all of which were eaten. Each person ate $\frac{1}{4}$ of a pizza. How many people did Wayne serve at the party?

    _____

4. Polly makes a peach cobbler for every family gathering. The recipe calls for $1\frac{1}{2}$ cups flour, $1\frac{1}{3}$ cup sugar, $\frac{1}{2}$ cup margarine, $3\frac{1}{3}$ cups peaches, 3 teaspoons of baking powder, and 2 cups of milk. For this particular gathering, they are expecting more guests, so Polly needs to triple the recipe. How much of each ingredient should Polly use?

    a. flour   _____

    b. sugar   _____

    c. margarine   _____

    d. peaches   _____

    e. baking powder _____

    f. milk   _____

5. A tract of land containing $288\frac{3}{4}$ acres was divided into lots of $4\frac{1}{8}$ acres in size. How many building lots were formed from this tract of land?

    _____

6. A one-way trip to Colorado Springs, Colorado from Dallas, Texas is 700 miles. Assume you got $26\frac{1}{2}$ miles to the gallon of gasoline. How much gas did you use?

    _____

7. The price of a new Cadillac has increased by $\frac{4}{5}$ times its earlier price. Assume the original price of the Cadillac was $17,500. What is the new price?

    _____

8. Howard prepared a 12-foot submarine sandwich for a party. Howard decided to cut the sandwich into pieces of $\frac{3}{4}$ feet. How many pieces of the sandwich can Howard cut?

    _____

9. Southside Mechanic's new machine produces $15\frac{1}{4}$ tools each hour. If the machine runs 16 hours, how many tools will the machine produce?

    _____

10. On a recent business trip, Sherry traveled $246\frac{1}{4}$ miles. If her car averaged $18\frac{1}{2}$ miles per gallon, what was the total cost of gasoline for the trip at $1.19 per gallon?

    _____

11. Jacob, Miller, and Griffith are equal partners in a law firm. Jacob sells $\frac{1}{2}$ his interest to Miller and $\frac{1}{3}$ his interest to Griffith. If the business was worth $150,000, what is the value of the share left to Jacob?

    _____

Assignment 6

Name_____ Date _____

Complete the calculations for the following problems by multiplying or dividing these fractions, whole numbers, and mixed numbers. Complete your calculations beneath each problem in the space provided. Write your answers in the blanks. Reduce to lowest terms as needed.

1. Debbie West wants to make 20 hairbows to sell at the church bazaar. Each bow requires $3\frac{1}{8}$ yards of ribbon. Find the total number of yards of ribbon she will need.

    _____

2. Adam Tompkins bought 23 bottles of champagne for the New Year's Party, all of which was consumed. Each guest drank $\frac{1}{3}$ of a bottle. How many guests did Adam serve?

    _____

3. How many $\frac{1}{4}$ ounce vials can be filled with 12 ounces of medicine?

    _____

4. A company will divide $\frac{6}{7}$ of its profits evenly among 6 employees. What fraction of the total profits will each employee receive?

    _____

5. Leslie Bonn bought 25 shares of stock at $\$23\frac{1}{5}$ per share and 50 shares of stock at $\$45\frac{1}{4}$ per share. How much did she pay altogether?

    _____

6. The price of a new motorcycle has increased $\frac{3}{5}$. If the original price of the motorcycle was $5,400, what is the new price today?

    _____

7. Karen Deats is paid $90 per day. The company has a policy of no work-no pay. Karen became ill on Tuesday and had to leave after $\frac{4}{7}$ of the day. What did Karen earn on Tuesday?

    _____

8. A board 256 inches long is cut into pieces that are each $8\frac{5}{6}$ inches. How many pieces can be cut?

    _____

    What is the size of the board left over?

    _____

9. A trip to Los Angeles from San Diego will take you $6\frac{1}{3}$ hours. If you have traveled $\frac{1}{5}$ of the way, how much longer will the trip take?

    _____

10. A tract of land containing 143 acres was divided into building lots of $2\frac{3}{5}$ acres in size. How many building lots were formed from this tract of land?

    _____

11. The alterations shop sold $5\frac{1}{4}$ yd, $7\frac{1}{5}$ yd, $8\frac{1}{2}$ yd, and $6\frac{6}{7}$ yd at a total price of $89.89. What was the average selling price per yard?

    _____

12. On a recent business trip, Ropert Smart traveled $428\frac{1}{4}$ miles. If his car averaged $19\frac{1}{3}$ miles per gallon, what was the total cost of gasoline for the trip at $1.25 per gallon?

    _____

Name_____ Date_____

$$\frac{\text{Student's Score}}{\text{Maximum Score}} = \frac{\_\_\_\_}{93} = \text{Grade}\_\_\_\_\_$$

A. Convert improper fractions to mixed numbers or whole numbers. Reduce to lowest terms when possible.

1. $\frac{31}{5}$ = _____

2. $\frac{69}{2}$ = _____

3. $\frac{181}{9}$ = _____

4. $\frac{43}{8}$ = _____

5. $\frac{74}{4}$ = _____

6. $\frac{21}{4}$ = _____

7. $\frac{82}{6}$ = _____

8. $\frac{39}{3}$ = _____

9. $\frac{16}{7}$ = _____

10. $\frac{144}{12}$ = _____

B. Change mixed numbers to improper fractions.

11. $21\frac{3}{4}$ = _____

12. $6\frac{4}{9}$ = _____

13. $2\frac{5}{8}$ = _____

14. $13\frac{4}{7}$ = _____

15. $3\frac{9}{12}$ = _____

16. $1\frac{4}{8}$ = _____

17. $4\frac{3}{4}$ = _____

18. $8\frac{1}{2}$ = _____

19. $6\frac{2}{3}$ = _____

20. $5\frac{4}{5}$ = _____

C. Reduce fractions to lowest terms.

21. $\frac{50}{200}$ = _____

22. $\frac{144}{288}$ = _____

23. $\frac{36}{39}$ = _____

24. $\frac{16}{98}$ = _____

25. $\frac{75}{135}$ = _____

26. $\frac{42}{84}$ = _____

D. Raise the following fractions to higher terms using the indicated denominators.

27. $\frac{20}{60} = \frac{?}{180}$ = _____

28. $\frac{6}{13} = \frac{?}{52}$ = _____

29. $\frac{8}{9} = \frac{?}{72}$ = _____

30. $\frac{2}{6} = \frac{?}{96}$ = _____

E. Convert decimals to fractions or mixed numbers and reduce to lowest terms.

31. 0.61 = _____

32. 0.045 = _____

33. 0.30 = _____

34. 0.2986 = _____

35. 0.463 = _____

36. 0.003 = _____

37. 0.0193 = _____

38. 0.2 = _____

39. 0.4902 = _____

40. 0.32 = _____

F. Find the decimal equivalents of the following fractions. Carry your answers to three decimal places.

41. $\frac{6}{20}$ = _____

42. $3\frac{9}{36}$ = _____

43. $6\frac{1}{2}$ = _____

44. $\frac{3}{50}$ = _____

45. $\frac{3}{30}$ = _____

46. $\frac{6}{500}$ = _____

47. $\frac{28}{40}$ = _____

48. $\frac{16}{25}$ = _____

G. Find the lowest common denominator.

49. $\frac{3}{20}, \frac{9}{40}, \frac{30}{60}$ = _____

50. $\frac{8}{4}, \frac{12}{16}, \frac{3}{32}$ = _____

51. $\frac{6}{9}, \frac{1}{4}, \frac{8}{12}$ = _____

52. $\frac{2}{5}, \frac{4}{6}, \frac{7}{8}$ = _____

H. Add the following fractions and reduce to lowest terms when possible.

53. $\frac{2}{3} + \frac{4}{6} + \frac{5}{12}$ = _____

54. $\frac{8}{7} + \frac{9}{7}$ = _____

55. $\frac{2}{4} + \frac{6}{8} + \frac{1}{3}$ = _____

56. $6\frac{4}{5} + 7\frac{2}{9}$ = _____

57. $2\frac{3}{4} + 3\frac{7}{8}$ = _____

58. $\frac{8}{9} + \frac{4}{7}$ = _____

I. Subtract the following fractions and reduce to lowest terms when possible.

59. $\frac{9}{12} - \frac{3}{12}$ = _____

60. $\frac{32}{40} - \frac{2}{4}$ = _____

61. $\frac{2}{5} - \frac{1}{3}$ = _____

62. $5 - \frac{2}{3}$ = _____

63. $6\frac{1}{4} - 3\frac{2}{3}$ = _____

64. $2\frac{4}{9} - 1\frac{1}{3}$ = _____

65. $4\frac{1}{8} - 2\frac{6}{8}$ = _____

66. $\frac{8}{7} - \frac{8}{9}$ = _____

Practical Math Applications

J. Multiply the following fractions; cancel and reduce to lowest terms when possible.

67. $\frac{3}{4} \times \frac{1}{3} \times \frac{2}{9}$ = _____

68. $\frac{6}{7} \times \frac{8}{9}$ = _____

69. $2\frac{2}{3} \times 4$ = _____

70. $9\frac{1}{2} \times \frac{3}{4}$ = _____

71. $\frac{1}{3} \times \frac{3}{5} \times \frac{4}{10}$ = _____

72. $\frac{2}{9} \times \frac{6}{8} \times \frac{1}{2}$ = _____

73. $\frac{4}{9} \times \frac{3}{8}$ = _____

74. $\frac{2}{3} \times 8$ = _____

75. $3\frac{1}{9} \times 4\frac{3}{12}$ = _____

76. $\frac{3}{7} \times \frac{6}{8}$ = _____

K. Divide the following fractions; cancel and reduce to lowest terms when possible.

77. $\frac{5}{6} \div \frac{5}{6}$ = _____

78. $12 \div \frac{3}{4}$ = _____

79. $2\frac{2}{3} \div 1\frac{1}{3}$ = _____

80. $2\frac{6}{7} \div \frac{4}{5}$ = _____

81. $\frac{1}{2} \div \frac{1}{10}$ = _____

82. $3\frac{1}{9} \div \frac{1}{3}$ = _____

83. $\frac{1}{5} \div \frac{1}{4}$ = _____

84. $3\frac{1}{3} \div 3\frac{1}{4}$ = _____

85. $4 \div \frac{2}{3}$ = _____

86. $\frac{1}{6} \div \frac{5}{12}$ = _____

L. Solve the following word problems. Write your answers in the blanks provided. Be sure to place dollar signs as needed. Express fractional parts in lowest terms.

87. Last week, Tom Smith worked the following number of hours: 9, $2\frac{2}{3}$, $6\frac{1}{4}$, 8, and $3\frac{1}{3}$. How many hours did Tom work last week?   _____

88. Showoff Fabrics is having a sale on decorator fabric. Yesterday, they sold 10 yards of blue, $3\frac{1}{3}$ yards red, $8\frac{1}{2}$ yards green, and $6\frac{3}{8}$ yards of yellow. Calculate total yardage sold.   _____

89. If tin is selling for $1.50 per pound, what is the cost of $15\frac{1}{3}$ pounds of tin? Carry your answers to three places and round to the nearest hundredth.

90. Bob Jones has $6\frac{1}{2}$ cakes for his friend Mark's birthday party. There are 12 people attending the party. What portion of cake will each person receive if the cakes are divided equally?

91. The city had 120 garden plots to resod. If each plot was $4\frac{1}{4}$ feet wide and $8\frac{1}{2}$ feet long, how many square feet of sod will it take to resod each plot individually?

92. If sod costs $1.63 per square foot at Robert's Garden Supplies, what is the cost to resod all the plots?

93. What is $\frac{1}{120}$ of the total cost?

CHAPTER

Introduction to Percents

6

Introduction to Percents

OBJECTIVES

After completing this chapter, you will be able to:

1. Convert percents to decimals and fractions.
2. Convert decimals to percents.
3. Convert common fractions to percents.

Percents are widely used in business and in our everyday living. We are faced with an abundant supply of information relating to percents. For example, percents are used in reporting the cost of living, unemployment, discounts, tax rates, and interest rates. You should fully understand the concepts associated with percents in order to deal with their many applications.

6.1 Terms Used in Percents

A *percent*, like a fraction or a decimal, represents some part of a whole. *Percent* means hundredths or parts in 100. In other words, 25 percent expressed as a fraction means 25 parts of 100 parts or $\frac{25}{100}$. Expressed as a decimal, 25 percent is 0.25. As another example, thirty-three hundredths may be written 33% as a percent using the percent symbol (%), $\frac{33}{100}$ as a fraction, or 0.33 as a decimal. Percents must be converted to decimals or fractions before calculations can be completed.

6.2 Converting Percents to Decimals

The easiest way to convert a percent to a decimal is to drop the percent symbol and divide the number by 100 because the percent means hundreds. For example, you can think of 35% as 35 out of 100 which means 35 divided by 100.

$$100 \overline{)35.00}^{\,0.35} \qquad 35\% = 0.35$$

A shortcut for dividing the number by 100 is to drop the percent symbol and move the decimal point 2 places to the left. Let's convert 35% to its decimal equivalent using this method.

Percent				Decimal Equivalent
35 %	=	0.35	=	0.35

In the preceding example, it is understood that there is a decimal point to the right of the whole number 35, even though it may not appear.

EXAMPLES

Percent		Decimal Equivalent
75%	= 0.75ₓ =	0.75
125%	= 1.25ₓ =	1.25
8.6%	= 0.08ₓ6 =	0.086

In the last example, a zero has been added so the decimal may be moved two places to the left. Many businesses use mixed numbers, such as an interest rate of $8\frac{1}{4}$%. To convert a mixed number percent to its decimal equivalent, use these steps.

STEPS

1. Change the fractional part of the mixed number to a decimal.

$$8\tfrac{1}{4}\% = 8.25\%$$

2. Convert the percent (8.25 %) to its decimal equivalent.

$$0.08\,25\,\% = 0.0825$$

You can see that one zero has been added to express the decimal equivalent, 0.0825.

PRACTICE PROBLEMS

Convert the following percents to decimals. Write your answers in the blanks provided.

1. 112 % = _____   **2.** 92 % = _____

3. 210.3 % = _____   **4.** 0.0025 % = _____

5. 52 % = _____   **6.** $2\frac{3}{4}$% = _____

Solutions: **1.** 1.12; **2.** 0.92; **3.** 2.103; **4.** 0.000025; **5.** 0.52; **6.** 0.0275

6.3 Converting Decimals to Percents

Converting a decimal to a percent does not change the value of either one. To convert a decimal to a percent, multiply the decimal number by 100 and add the percent sign. For example, to convert 0.157 to a percent:

$$0.157 \times 100 = 15.7 = 15.7\,\%$$

A shortcut for multiplying the decimal number by 100 is to move the decimal point 2 places to the right and add a percent sign. Let's convert 0.05 to a percent using this method.

Decimal		Percent
0.05	= 0ₓ05. =	5 %

Study these examples illustrating the conversion of decimals to percents.

EXAMPLES _____

Decimal				Percent
0.6	=	0.60.	=	60 %
0.08	=	0.08.	=	8 %
2.10	=	2.10.	=	210 %
0.825	=	0.82.5	=	82.5 %
0.0034	=	0.00.34	=	0.34%

PRACTICE PROBLEMS

Convert the following decimals to percents. Write your answers in the blanks provided.

1. 0.0075 = _____   **2.** 0.4 = _____

3. 0.325 = _____   **4.** 0.123 = _____

5. 0.85 = _____   **6.** 0.03 = _____

Solutions: **1.** 0.75 %;   **2.** 40 %;   **3.** 32.5 %;   **4.** 12.3 %;   **5.** 85 %;   **6.** 3 %

6.4   Converting Percents to Fractions

To convert percents to fractions, drop the percent sign and write a fraction, placing the percent as the numerator and 100 as the denominator. Reduce the fraction to its lowest terms as needed. For example, 5 % is converted to a fraction in the following manner:

$$5\% = \frac{5}{100} = \frac{1}{20}$$

Study the following examples illustrating the conversion of percents to fractions.

EXAMPLES _____

Decimal				Fraction
14%	=	$\frac{14}{100}$	=	$\frac{7}{50}$
13%	=	$\frac{13}{100}$	=	$\frac{13}{100}$
45%	=	$\frac{45}{100}$	=	$\frac{9}{20}$
130%	=	$\frac{130}{100}$	=	$1\frac{3}{10}$

PRACTICE PROBLEMS

Convert the following percents to fractions. Write your answers in the blanks provided. Reduce the fraction to its lowest terms as needed.

1. 4% = _____   **2.** 36% = _____

3. 8% = _____   **4.** 20% = _____

5. 88% = _____   **6.** 12% = _____

Solutions: **1.** $\frac{1}{25}$; **2.** $\frac{9}{25}$; **3.** $\frac{2}{25}$; **4.** $\frac{1}{5}$; **5.** $\frac{22}{25}$; **6.** $\frac{3}{25}$

Frequently, a percent will appear with a decimal. To convert to a fraction, follow these steps.

STEPS

1. Change the percent to a decimal.

$$12.5\% = 0.125$$

2. Change the decimal to a fraction; reduce.

$$0.125 = \frac{125}{1,000} = \frac{1}{8}$$

PRACTICE PROBLEMS

Convert these percents to fractions, reducing your answers to lowest terms. Write your answers in the blanks provided.

1. 3.5% = _____   **2.** 53.7% = _____

3. 9.375% = _____   **4.** 37.5% = _____

5. 11.7% = _____   **6.** 8.5% = _____

Solutions: **1.** $\frac{7}{200}$; **2.** $\frac{537}{1,000}$; **3.** $\frac{3}{32}$; **4.** $\frac{3}{8}$; **5.** $\frac{117}{1,000}$; **6.** $\frac{17}{200}$

A percent containing a fraction, such as $8\frac{2}{3}\%$, can be converted to a fraction. Let's convert $8\frac{2}{3}\%$ to a fraction by following these steps.

STEPS

1. Drop the percent sign and write the percent over 100.

$$\frac{8\frac{2}{3}}{100}$$

2. Divide the numerator by the denominator. Remember to invert the divisor.

$$8\frac{2}{3} \div 100$$

$$\frac{\overset{13}{\cancel{26}}}{3} \times \frac{1}{\underset{50}{\cancel{100}}} = \frac{13}{150}$$

— PRACTICE PROBLEMS —

Convert these percents to fractions. Write your answers in the blanks provided.

1. $8\frac{1}{3}\%$ = _____   **2.** $7\frac{1}{2}\%$ = _____

3. $5\frac{3}{4}\%$ = _____   **4.** $\frac{1}{8}\%$ = _____

5. $16\frac{2}{3}\%$ = _____   **6.** $10\frac{1}{4}\%$ = _____

Solutions: **1.** $\frac{1}{12}$; **2.** $\frac{3}{40}$; **3.** $\frac{23}{400}$; **4.** $\frac{1}{800}$; **5.** $\frac{1}{6}$; **6.** $\frac{41}{400}$

6.5 Converting Common Fractions to Percents

One method for changing a common fraction to a percent is to first change the fraction to a decimal and then change the decimal to a percent. To convert a common fraction to a percent, use these steps.

STEPS

1. Convert the fraction to its decimal equivalent. For example, let's express the fraction $\frac{1}{5}$ as a decimal.

$$\frac{1}{5} = 0.20 \ (1 \div 5 = 0.20)$$

2. Convert the decimal to a percent as illustrated earlier in this chapter.

$$0.20 = 20\%$$

Study these examples illustrating the conversion of fractions to percents.

EXAMPLES _____

Fraction				Percent
$\frac{5}{8}$	=	0.625	=	62.5 %
$\frac{3}{6}$	=	0.5	=	50 %
$\frac{6}{8}$	=	0.75	=	75 %

Remember to convert fractions to decimals by dividing the numerator by the denominator or by using a table of decimal equivalents.

Some common decimal equivalents are listed below:

$\frac{1}{2} = 0.5$	$\frac{1}{5} = 0.2$	$\frac{3}{8} = 0.375$
$\frac{1}{3} = 0.3333$	$\frac{2}{5} = 0.4$	$\frac{5}{8} = 0.625$
$\frac{2}{3} = 0.667$	$\frac{3}{5} = 0.6$	$\frac{1}{10} = 0.1$
$\frac{1}{4} = 0.25$	$\frac{4}{5} = 0.8$	$\frac{5}{10} = 0.5$
$\frac{3}{4} = 0.75$	$\frac{1}{8} = 0.125$	$\frac{7}{10} = 0.7$

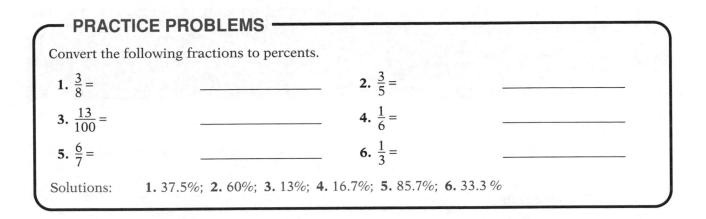

PRACTICE PROBLEMS

Convert the following fractions to percents.

1. $\frac{3}{8} =$ _____

2. $\frac{3}{5} =$ _____

3. $\frac{13}{100} =$ _____

4. $\frac{1}{6} =$ _____

5. $\frac{6}{7} =$ _____

6. $\frac{1}{3} =$ _____

Solutions: **1.** 37.5%; **2.** 60%; **3.** 13%; **4.** 16.7%; **5.** 85.7%; **6.** 33.3 %

Math Alert

Change the following percents (fourth column named Pct. Chg.) to decimals.

1. _____
2. _____
3. _____
4. _____
5. _____
6. _____
7. _____
8. _____
9. _____
10. _____

SPURTS IN VOLUME

Week's volume as a multiple of the issue's 10-month norm. All issues listed are traded on the New York Stock Exchange.

		Multiple	Close	Pct. Chg.
1.	Signal Apparel	12.359	$10\frac{3}{8}$	5.06
2.	United Inns	7.306	$3\frac{1}{4}$	4.00
3.	Neiman-Marcus	7.116	$15\frac{1}{8}$	−3.97
4.	Inspiration Res	5.924	$4\frac{1}{8}$	13.79
5.	Manhattan Natl	5.649	$7\frac{1}{4}$	0.00
6.	Nevada Power	5.346	$17\frac{1}{4}$	−10.39
7.	WMS Industries	5.331	$17\frac{5}{8}$	20.51
8.	Kemper	5.224	$29\frac{1}{4}$	−6.02
9.	Emerson Radio	5.118	$2\frac{1}{8}$	0.00
10.	Pansophic Sys	4.525	$15\frac{3}{4}$	1.61

Dallas Morning News, **September 15, 1994**

Study Guide

I.   Terminology

page 108   percent   Represents some part of a whole. Percent means hundredths
                                         or parts in 100.

II.   Working With Percents

page 108   *Converting percents to decimals:* Drop the percent symbol and divide the number
              by 100.

$$\text{Example: } 100\overline{)45.00}^{\,0.45} \qquad 45\% = 0.45$$

page 108   *Shortcut:* Drop the percent symbol and move the decimal point 2 places to the left.

Example: $45\% = 0.45_\times = 0.45$

page 109   *Converting decimals to percents:* Multiply the decimal number by 100 and add the
              percent sign.

Example: $0.572 = 0.572 \times 100 = 57.2 = 57.2\%$

page 110   *Converting percents to fractions:* Drop the percent sign and write a fraction, placing the
              percent as the numerator and 100 as the denominator. Reduce the fraction to its lowest
              terms as needed.

Example: $10\% = \dfrac{10}{100} = \dfrac{1}{10}$

page 112   *Converting common fractions to percents:* Convert the fraction to its decimal equivalent;
              then convert the decimal to a percent.

Example: $\dfrac{1}{5} = 0.20, \qquad (1 \div 5 = 0.20) = 20\%$

Study Skills

How to Take Tests

Your primary goal is to do well on a test; however, what you have learned while studying is what
is really important. What you have learned is what you will use in life. A test is important as a
temporary measurement of what you have learned—but will not demonstrate all that you know.

Tips

1. Write your name on the test and read the directions.
2. Budget your time. If you have 50 minutes to take the test, estimate the amount of time in
   relation to the number of questions and the importance assigned to those questions. Don't
   spend half your time on ten multiple-choice questions that are worth 10 points. Complete
   the questions that have the most points first.
3. Take all the time allowed.
4. Allow time to go over the test and correct any errors.

Assignment 1   Name_____ Date _____

Complete the following problems.

A. Convert the following percents to decimals.

1. 3%   = _____   **2.** 2.7%   = _____

3. 15.2%   = _____   **4.** 57%   = _____

5. 31.2%   = _____   **6.** 6.5%   = _____

7. 2.54%   = _____   **8.** 80%   = _____

9. 128%   = _____   **10.** 90%   = _____

11. 15%   = _____   **12.** 11.2%   = _____

13. 7%   = _____   **14.** 2.6%   = _____

15. 355%   = _____   **16.** 16.8%   = _____

17. 44%   = _____   **18.** 9.5%   = _____

19. 212%   = _____   **20.** 24%   = _____

B. Change the following decimals to percents.

21. 6.3   = _____   **22.** 0.861   = _____

23. 0.07   = _____   **24.** 4.00   = _____

25. 0.43   = _____   **26.** 11.8   = _____

27. 0.3   = _____   **28.** 6.05   = _____

29. 0.897   = _____   **30.** 0.62   = _____

31. 0.05   = _____   **32.** 0.037   = _____

33. 0.125   = _____   **34.** 0.37   = _____

35. 0.025   = _____   **36.** 0.0436   = _____

37. 0.60   = _____   **38.** 2.18   = _____

39. 0.4   = _____   **40.** 0.57   = _____

Assignment 2

Name_____ Date _____

Complete the following problems. Write your answers in the blanks provided.

A. Change the following percents to decimals and reduce to lowest terms.

1. 8%	= _____	**2.** 58%	= _____
3. 0.1%	= _____	**4.** 9.63%	= _____
5. 175%	= _____	**6.** $6\frac{2}{3}$%	= _____
7. 10.6%	= _____	**8.** 0.15%	= _____
9. 6%	= _____	**10.** 112%	= _____
11. 88%	= _____	**12.** 20%	= _____
13. 32.5%	= _____	**14.** 37.5%	= _____
15. 260%	= _____	**16.** 180%	= _____
17. 9.375%	= _____	**18.** 81.81%	= _____
19. 42.9%	= _____	**20.** 60%	= _____

B. Write the following fractions as percents. Round to the nearest tenth.

21. $2\frac{3}{4}$	= _____	**22.** $\frac{1}{4}$	= _____
23. $\frac{3}{8}$	= _____	**24.** $\frac{6}{10}$	= _____
25. $\frac{10}{12}$	= _____	**26.** $\frac{11}{12}$	= _____
27. $\frac{5}{8}$	= _____	**28.** $8\frac{5}{9}$	= _____
29. $\frac{6}{8}$	= _____	**30.** $1\frac{8}{9}$	= _____
31. $12\frac{5}{6}$	= _____	**32.** $\frac{7}{12}$	= _____
33. $3\frac{7}{9}$	= _____	**34.** $\frac{12}{25}$	= _____
35. $5\frac{2}{3}$	= _____	**36.** $7\frac{1}{9}$	= _____
37. $\frac{7}{50}$	= _____	**38.** $12\frac{1}{7}$	= _____
39. $\frac{7}{8}$	= _____	**40.** $\frac{4}{7}$	= _____

Assignment 3

Name_____ Date _____

Complete the following problems. Write your answers in the blanks provided.

A. Write the following decimals as percents.

1. 0.667	= _____		**2.** 0.375	= _____
3. 0.47	= _____		**4.** 0.391	= _____
5. 1.45	= _____		**6.** 0.05	= _____
7. 0.417	= _____		**8.** 5.01	= _____
9. 0.003	= _____		**10.** 0.516	= _____
11. 0.091	= _____		**12.** 2.16	= _____
13. 0.47	= _____		**14.** 0.28	= _____
15. 5.61	= _____		**16.** 0.935	= _____
17. 0.008	= _____		**18.** 10.13	= _____
19. 0.317	= _____		**20.** 0.4	= _____

B. Change the following percents to decimals.

21. 15.5%	= _____		**22.** 17.2%	= _____
23. 30%	= _____		**24.** 7.6%	= _____
25. 72%	= _____		**26.** 1.35%	= _____
27. $\frac{5}{8}$%	= _____		**28.** $6\frac{1}{4}$%	= _____
29. $\frac{7}{10}$%	= _____		**30.** 72.45%	= _____
31. 119%	= _____		**32.** 5.6%	= _____
33. 100%	= _____		**34.** 6.63%	= _____
35. $25\frac{1}{2}$%	= _____		**36.** 19.8%	= _____
37. 9%	= _____		**38.** 0.125%	= _____
39. $32\frac{1}{8}$%	= _____		**40.** $6\frac{1}{10}$%	= _____

Assignment 4

Name_____ Date _____

Complete the following word problems. Write your answers in the blanks provided.

1. Harriet Mendoza is analyzing changes in retail sales for her accessory shop. Write the changes shown in decimals to percents.

Month (Week)	Change	Percent
March (Week 1)	−0.036	_____
(Week 2)	+0.012	_____
(Week 3)	+0.0432	_____
(Week 4)	−0.0098	_____

2. A report, published in the local newspaper, showed that there was a decline in new car sales. The report revealed the following: Cadillac sales fell 1.6 percent; Buick sales were down 37.3 percent; Pontiac sales fell 29.16 percent; Lincoln and Mercury sales 41.6 percent and 8.6 percent, respectively; and Dodge sales slipped 6.3 percent. Change the percents to decimals.

    Cadillac _____   Buick _____   Pontiac _____

    Lincoln _____   Mercury _____   Dodge _____

3. The local newspaper showed that the cost of living increased by 8.7 percent. Write 8.7 percent as a decimal. _____

4. The following results were reported of games won for the American Baseball League (West): Minnesota, .606; Chicago, .545; Texas, .529; Oakland, .528; Kansas City, .514; Seattle, .500; and California, .496. Change the results shown in decimals to percents.

    Minnesota _____   Chicago _____   Texas _____

    Oakland _____   Kansas City _____   Seattle _____

    California _____

5. Based on the dollar amount of stocks sold, Akemi Saga can receive the following commission rates: 1.9%, 0.7%, 1.5%, 0.9%, 0.3%, 1.6%, and 1.1%. Write these percentages as decimals.

    1.9% = _____   0.7% = _____   1.5% = _____

    0.9% = _____   0.3% = _____   1.6% = _____

    1.1% = _____

6. Jay's Hallmark Gift Shop reported the following monthly sales increases. Convert the percents to decimals.

    January 4.45% _____   February 4.25% _____   March 1.17% _____

    April 6.17% _____   May 8.29% _____   June 5.27% _____

Assignment 5

Name_____ Date _____

Complete the following word problems. Write your answers in the blanks provided.

1. Cindy Tyler must convert the report on stock increases into decimals. The following are the amounts on the report:

0.75 _____	0.367 _____	1.01 _____
0.234 _____	0.033 _____	0.0025 _____
0.0085 _____	0.0425 _____	0.875 _____
0.011 _____		

   Convert these decimals to percents.

2. Jacy's Department Store advertised sales on these items for their annual Labor Day Sale: 20% off Men's Shoes; 35% off Ladies' Dresses; 15% off Children's Clothes; 40% off Bedding; 55% off Clearance Items; and 33% off all Men's Suits. Convert these percents to decimals.

20% _____	35% _____	15% _____
40% _____	55% _____	33% _____

3. National Bank advertises their interest rates as percentages. Convert the following percents to decimals.

3.25% _____	4.45% _____	1.25% _____
3.02% _____	2.7% _____	$5\frac{2}{3}$% _____

4. Jolly Amusement Park released their annual report on the market survey comparing the popular roller coasters. The report listed the survey in fractions. Convert the following fractions to percents. (Round to the nearest hundredth percent.)

JollyBones $\frac{1}{2}$ _____	Tornado $\frac{4}{5}$ _____	Popper $\frac{2}{9}$ _____
Shaker $\frac{3}{8}$ _____	Screamer $\frac{1}{5}$ _____	Hurricane $\frac{1}{6}$ _____

5. Smith Publishing reported their profits in various departments. Convert the following percents to decimals.

Mysteries, 23% _____	Classics, −8% _____
Biographies, −11% _____	Educational, 1% _____
Romance, 67.3% _____	

6. The Super Spa Health Club measures the body fat percentage of each of its members. The following are some of this week's measurements. Convert the percents to fractions. Reduce as necessary.

22.1% _____	33.33% _____	37.2% _____
17.5% _____	13% _____	25% _____
28.8% _____	20% _____	

CHAPTER 6 Introduction to Percents

Assignment 6   Name_____ Date _____

Complete the following word problems. Write your answers in the blanks provided.

1. Based on the dollar amount of insurance policies sold, Jason Patterson will receive the following commission rates: 0.8%; 2.1%; 1.3%; 0.75%; 2.8%; 1,166%; and 1.25%. Write these percents as decimals.

   0.8% _____   2.1% _____   1.3% _____

   0.75% _____   2.8% _____   1,166% _____

   1.25% _____

2. Gregory Scott is analyzing changes in retail sales for his hobby shop. Write the changes shown in decimals to percents.

   Stamps, −0.022 _____   Coins, +0.455 _____

   Trains, +0.001 _____   Doll Houses, −0.151 _____

   Model Cars, +0.221 _____   Model Planes, −0.137 _____

3. Winston's Department Store advertised the following sales in the local newspaper in time for Father's Day: 45% off men's short sleeve dress shirts; 50% off all ties; 30% off golf shirts; 37% off underwear, and 25% off dress slacks. Convert these percents to fractions and reduce to the lowest terms.

   45% _____   50% _____   30% _____

   37% _____   25% _____

4. The First State Bank published its weekly interest rates on savings accounts. Money Market, 5.25%; Passbook Savings, 3.12%; 60-Day CD, 4.2%; 90-Day CD, 4.65%, 6-month CD, 4.88%; and 1-Year CD, 6.10%. Change the given percents to decimals.

   Money Market _____   Passbook Savings _____   60-Day CD _____

   90-Day CD _____   6-Month CD _____   1-Year CD _____

5. Johnson Elementary School has 500 pupils. This year the enrollment of boys increased in each grade. The following fractions show the increase in each grade. Convert the fractions to percents. Round to the nearest tenth.

   First Grade, $\frac{2}{5}$ _____   Second Grade, $\frac{2}{3}$ _____   Third Grade, $\frac{4}{5}$ _____

   Fourth Grade, $\frac{6}{7}$ _____   Fifth Grade, $\frac{2}{7}$ _____   Sixth Grade, $\frac{3}{8}$ _____

6. The Speer Junior High enrollment decreased in the amount of girls. The following fractions show the decrease of girls in each grade. Convert the fractions to decimals and percents.

	Decimals	**Percents**
Seventh Grade, $\frac{4}{9}$	_____	_____
Eighth Grade, $\frac{1}{8}$	_____	_____
Ninth Grade, $\frac{1}{9}$	_____	_____

CHAPTER
Percentage, Base, and Rate

7

7

Percentage, Base, and Rate

The ability to compare numbers will be necessary in both your business and personal activities. For example, it is essential to know how to figure the percent of return on an investment to determine whether you should make the investment. Also, you might want to figure the dollar amount of your 5% pay raise.

Percents are commonly used to determine interest, sales, taxes, commissions, and discounts or to make other comparisons. In working with percent problems, you are usually asked to find the percentage, base, or rate.

Study the terminology and concepts presented in this chapter very carefully because these fundamentals are crucial to the development of your business math skills. Learning these concepts will help you solve business and personal math problems involving percents.

7.1 Terms Used in Percentage, Base, and Rate

The computations involved in working with percent problems require that you be able to identify the following elements: base, rate, and percentage. The **base** represents 100%, or that to which something is being compared. The **rate** is always a number followed by a percent sign or written as *percentage rate*. It is the percent one number is of another. The **percentage** is the number that is part of the base; for example, 26 of 198. Do not confuse percentage with rate (shown as %).

7.2 Finding the Percentage

Whenever you need to know the amount of a commission, discount, or finance charge, you will need to find the percentage. To find the percentage (or number) when the base and rate are known, change the rate (percent) to a decimal and multiply it by the base.

$$\text{Percentage} = \text{Base} \times \text{Rate}$$

An easy way to determine the formula for percentage is by studying this illustration.

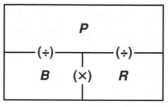

Locate the element in the square to be determined. In this case, locate the letter *P* for percentage. Cover the *P* with your finger. You can see that *R* and *B* remain. Think of this as *B* times *R* as in

$$P = B \times R$$

EXAMPLE _____

What is the percentage on a part of the cost of a $150 jacket that has a 30% discount? Written another way, you are finding the following.

What number is 30% of $150 or 30% of $150 is what?

Base is $150. Rate is 30%. Percentage is unknown.

$$\$150 \times 0.30 = \$45$$
$$\quad B \qquad R \qquad P$$

_____

■ **TIP** Notice how percentage problems are stated. Usually the base is preceded with the words *of* or *part of*. The rate is always the percent number or expressed with the word *percent* or its *symbol*. The percentage to be found is a number that is part of the base. The word *of* usually indicates multiplication.

_____

7.3 Percentage Shown with Decimals

To find a percentage when the rate is shown as a percent with or without a decimal, first change the percent to a decimal, then calculate the percentage.

EXAMPLE _____

What is 6.5% of 130?

STEPS

1. First identify the parts: Base is 130; Rate is 6.5%; Percentage is unknown.
2. Apply the formula: Percentage = Base × Rate.
3. Percentage = 130 × 6.5%.
4. Percentage = 130 × 0.065.

    (Change the percent to a decimal by dividing it by 100 or moving the decimal place 2 places to the left.)
5. Percentage = 8.45.

_____

■ **TIP** If you are using a calculator, use the percent key and skip step 4.

_____

7.4   Percentage Shown with Fractions

To find a percentage when the rate is shown in fractional form, first change the fractional form to a decimal, then calculate the percentage. For instance, suppose you had a gain of $6\frac{1}{4}\%$ on $400.

STEPS

---

1. Change the fractional part to a decimal, then move the decimal 2 places to the left, adding a zero.

$$6\frac{1}{4}\% = 6.25 \ (100 \div 4 = 0.25) = 0.06\underset{\curvearrowleft}{2}5$$

2. Apply the formula: Percentage = Base × Rate
                                     Base = $400
                                     Rate = 0.0625
                           Percentage = unknown

3. Percentage = $400 × 0.0625.

4. Percentage = $25.

7.5 Finding the Rate

Whenever you need to determine a percent as a return on an investment or as an increase or decrease in sales, you need to find the rate. You can find the rate or what percent one number is of another by dividing the percentage by the base. This formula is expressed as

Rate = Percentage ÷ Base

Study this illustration for determining the rate.

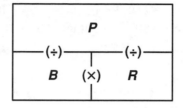

Locate the elements in the square to be calculated; in this case, locate the letter R. Cover the R with your finger. You can see that P and B remain and that P is shown over B as a fraction. Think of $\frac{P}{B}$ as in this formula:

$$R = \frac{P}{B} \text{ or } P \div B$$

Let's determine the rate or percent. Find what percent $15 is of $139 in the following example.

EXAMPLE _____

$15 is what percent of $139? or
↑   ↑
Percentage   Base

What percent of $139 is $15?

STEPS

1. First identify the parts:

    Base is $139.
    Percentage of number is $15.
    What is the rate?

2. Apply the formula: Rate = Percentage ÷ Base.

3. Rate = 15 ÷ 139 = 0.108 (Rounded).

4. Rate = 0.108 (Change decimal to percent).

5. Rate = 0.108 × 100 = 10.8 % or move the decimal 2 places to the right and add a percent sign.

■ **TIP**   Notice how rate problems are stated. Usually the base is preceded by the word *of*. The percentage is the number that is part of the base. You must determine the rate or percent because it is the element that is missing.

■ **TIP**   The rate is shown with a percent symbol; the base is preceded by the word *of*; the *percentage* is the number that is part of the base.

7.6 Finding the Base

Whenever you need to determine the selling price of an item or the total amount of sales for the month, you need to find the base. You can find the base when the percentage and rate are known. To figure the base, change the rate (percent) to a decimal and divide the percentage by the rate. This formula is expressed as

$$\text{Base} = \text{Percentage} \div \text{Rate}$$

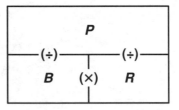

Locate the element in the square to be calculated; in this case, locate the letter *B* for base. Cover the *B* with your finger. You can see that *P* and *R* remain and that *P* is shown over *R* as a fraction. Think of $\frac{P}{R}$ as in the formula:

$$B = \frac{P}{R} \text{ or } P \div R$$

Let's determine the base in the following example:

■ **TIP**   If you are using a calculator, use the percent sign.

EXAMPLE _____

$1,325 is 9.5% of what amount?

STEPS

1. First identify the elements:

   Percentage is $1,325.
   Rate is 9.5%.
   What is the base?

2. Apply the formula: Base = Percentage ÷ Rate.

3. Base = 1,325 ÷ 0.095 (Change the rate to its decimal equivalent.)

4. Base = $13,947.37 (rounded)

■ **TIP** Notice how base problems are stated. The phrase, *what amount,* is preceded by *of*; therefore, you must find the base.

PRACTICE PROBLEMS

First identify the percentage and rate for these problems. Then find the base. Write your answers in the blanks provided. Round your answers to the nearest hundredth.

1. $249 is 6% of what amount?

Percentage = _____

Rate = _____

Base = _____

2. 3.5% of what amount is $320?

Percentage = _____

Rate = _____

Base = _____

Solutions: **1.** Percentage = $249, Rate = 6%, Base = $4,150;
             2. Percentage = $320, Rate = 3.5%, Base = $9,142.86

7.7 Identifying the Elements of Percent Problems

The key to working with the three formulas we have just discussed is being able to identify the elements: base, rate, and percentage. It is very important that you know which parts you have and which part is missing. Once you are able to identify the parts, you will be able to solve business math problems involving percents. Anytime there are two parts of a problem known, the third part can be found by using the base, rate, and percentage formulas.

It is easy to identify the rate because it has the percent symbol or the word "percent." However, sometimes it may be more difficult to distinguish between the base and percentage (or part). To help you distinguish between base and percentage, notice how the problems are written. For example, where the base refers to total sales, the percentage or part may refer to commission. Where the base refers to savings, the percentage or part may refer to the interest earned. Where the base refers to gross income, the percentage may refer to net income. Study the following table and become familiar with the terms.

Usually is the Base	Usually is the Part
Sales	Sales tax
Value of Bonds	Dividends
Retail Price	Discount
Old Salary	Raise
Value of real estate	Rents

PRACTICE PROBLEMS

Indicate *which formula* you would use for these problems.

1. Suppose the president of a community college claims that 30% of this year's student body went on to a four-year university. If 560 students went on to a four-year university, how many enrolled in the community college?

2. A real estate broker earns a commission of 3% on the total sale value of a house. If the house sells for $140,000, how is the commission calculated?

3. If your monthly budget is $2,150 and your food expense is $358, what percent of your monthly budget is spent for food?

Solutions: **1.** $\frac{P}{R} = B$;   **2.** $B \times R = P$;   **3.** $\frac{P}{B} = R$

Math Alert

Determine the percent of wins for the following teams. Before you find the percent, you must total the wins and losses to get the base number, then find the percent. For example, for Los Angeles, 142 is the base (81 + 61). Eighty-one is what percent of 142? The correct answer is 0.570. Percents of wins for baseball are shown using thousandths. Calculate the percent to the nearest tenth of a percent for each team in the National League (West).

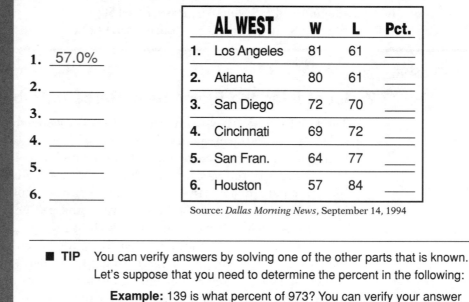

1. <u>57.0%</u>

2. _____

3. _____

4. _____

5. _____

6. _____

AL WEST	W	L	Pct.
1. Los Angeles	81	61	____
2. Atlanta	80	61	____
3. San Diego	72	70	____
4. Cincinnati	69	72	____
5. San Fran.	64	77	____
6. Houston	57	84	____

Source: *Dallas Morning News*, September 14, 1994

■ **TIP**   You can verify answers by solving one of the other parts that is known. Let's suppose that you need to determine the percent in the following:

Example: 139 is what percent of 973? You can verify your answer by multiplying the percent (14.3%) by the base (973).

Study Guide

I. Terminology

page 122	base	Represents 100% or that to which something is being compared.
page 122	rate	A number followed by a percent sign or written as a percent rate.
page 122	percentage	The number that is part of the base.

II. Working with Percentage, Rate, and Base

page 123 *Finding the Percentage Shown with Decimals:* The base and rate must be known; then change the rate (percent) to a decimal and multiply it by the base.

Formula: Percentage = Base × Rate ($P = B \times R$)

Example: What number is 30.3% of $150?
Base is $150.   Rate is 30.3%.   Percentage is unknown.
$150 × 0.303 = $45.45 (Percentage)

page 124 *Finding the Percentage Shown with Fractions:* The base and rate must be known; then change the fraction to a decimal by dividing the numerator by the denominator and adding the whole number; multiply the base by the rate.

Formula: Percentage = Base × Rate ($P = B \times R$)

Example: What number is $6\frac{1}{4}$% of $400?
Base is $400.   Rate is $6\frac{1}{4}$%.   Percentage is unknown.
$400 × 0.0625 = $25.00 (Percentage)

page 125 *Finding the Rate:* The percentage and the base must be known; divide the percentage by the base.

Formula: Rate = Percentage ÷ Base ($R = P \div B$)

Example: $15 is what percent of $139?
Base is $139.   Percentage is $15.   Rate is unknown.
$15 ÷ $139 = 0.108 or 10.8% (Rate)

page 126 *Finding the Base:* Percentage and rate must be known; change the rate (percent) to a decimal and divide the percentage by the rate.

Formula: Base = Percentage ÷ Rate ($B = P \div R$)

Example: $1,325 is 9.5% of what amount?
Percentage is $1,325.   Rate is 9.5%.   Base is unknown.
$1,325 ÷ 0.095 = $13,947.37 (rounded) (Base)

page 127 *Identifying the Elements of Percent Problems:* Rate is the percent; percentage is the part of the whole (base); base is the whole.

Assignment 1   Name_____ Date _____

Complete the following problems. Write your answers in the blanks provided.

A. Identify the parts of each problem. Do not calculate.

1. What number is 34% of 65?   P _____ R _____ B _____

2. 4 is what percent of 1,265?   P _____ R _____ B _____

3. 100% of what amount is 108.3?   P _____ R _____ B _____

4. What percent of 29 is 99?   P _____ R _____ B _____

5. 85.7 is 56% of what amount?   P _____ R _____ B _____

6. 20% of 16.25 is what number?   P _____ R _____ B _____

7. What number is 19% of 133?   P _____ R _____ B _____

8. 399 is 22% of what amount?   P _____ R _____ B _____

9. What percent of 9 is 6?   P _____ R _____ B _____

10. 20% of what amount is 126?   P _____ R _____ B _____

B. Find the percentage in the following problems. Round the answers to the nearest hundredth.

11. 0.6% of 125   = _____   12. 42% of $3,324   = _____

13. 3.3% of 631   = _____   14. 20% of 201   = _____

15. 3% of 96   = _____   16. 110% of 388   = _____

C. Find the percentage in the following problems. Round the answers to the nearest hundredth of a percent.

17. What percent of 40 is 19?   _____

18. 5.7 is what percent of 135?   _____

19. What percent of 75 is 13?   _____

20. 45 is what percent of 215?   _____

21. 31.27 is what percent of 313?   _____

D. Find the base — the total amount in these problems. Round the answers to the nearest hundredth.

22. 7 is 31% of what total amount?   _____

23. 125% of what total amount is 62?   _____

24. 44 is 27% of what total amount?   _____

25. 80% of what total amount is 60?   _____

26. 16 is 80% of what total amount?   _____

Assignment 2   Name_____ Date _____

Complete the following problems and write your answers in the blanks provided.

A. Determine the percentage in the following problems. Round the answers to the nearest hundredth.

    1. Find 9.6% of 160. _____

    2. What is 1.12% of 500? _____

    3. Find 45% of 600. _____

    4. Find 0.75% of 172. _____

    5. What is 27.4% of 80? _____

    6. 5.61% of 120 = _____

    7. 9% of 67 = _____

    8. 30% of 127 = _____

    9. 79% of 113 = _____

    10. $\frac{1}{4}$% of 800 = _____

B. Find the rate in the following problems. Round the answers to the nearest tenth of a percent.

    11. What percent of 885 is 99? _____

    12. 9.7 is what percent of 198? _____

    13. 61 is what percent of 462? _____

    14. What percent of 69 is 3.5? _____

    15. What percent of 12 is 7? _____

    16. 43 is what percent of 319? _____

    17. 82.5 is what percent of 102? _____

    18. 58 is what percent of 63? _____

    19. What percent of 150 is 22? _____

    20. 56 is what percent of 96? _____

C. Find the base in the following problems. Round the answers to the nearest hundredth.

    21. 9.1 is 32% of what total? _____

    22. 5.7% of what total is 92? _____

    23. 32.4% of what total is 120? _____

    24. 6% of what total is 48? _____

    25. $\frac{3}{4}$% of what total is 7? _____

    26. 30% of what total is 55? _____

    27. 4% of what total is 8? _____

    28. 2.9% of what total is 125? _____

    29. 3% of what total is 48? _____

    30. 27 is 32% of what total? _____

Assignment 3   Name_____ Date _____

Complete the following problems. Write your answers in the blanks provided.
Round the answers to the nearest hundredth for the percentage and base answers.
Round to the nearest tenth of a percent for rate answers.

1. 42% of $3,324 is what?   _____

2. 17.3% of $1,163 is what?   _____

3. $21 is what percent of $139?   _____

4. $28 is what percent of $275?   _____

5. $30 is 15% of what?   _____

6. $57 is 5% of what?   _____

7. 20% of $879 is what?   _____

8. $163 is what percent of $3,310?   _____

9. $16 is 19.5% of what?   _____

10. 7% of $125 is what?   _____

11. $546 is what percent of $1,961?   _____

12. $58 is 22% of what?   _____

13. 8% of $167 is what?   _____

14. 80% of 978 is what?   _____

15. $342 is what percent of $698?   _____

16. $68 is what percent of $275?   _____

17. $15 is 50% of what?   _____

18. $28 is 4% of what?   _____

19. 12% of 324 is what?   _____

20. What percent of 12,150 is 600?   _____

21. $3,000 is 36% of what?   _____

22. $12.24 is what percent of $98.90?   _____

Assignment 4

Name_____ Date _____

Determine the percentage, rate, or base in these problems. Round percentage and base answers to the nearest hundredth. Round rate answers to the nearest tenth of a percent.

1. Steve DeJong receives a commission of 3% of the total amount of his sales. He sold $3,895 worth of goods last week. How much was his commission?

   Percentage = _____

2. A purchase amounted to $145.50. If the sale discount is 20%, what is the amount of discount?

   Percentage = _____

3. Gloria Hutchins paid an electric bill of $182.30 last month. If this month's bill is 8% higher than her previous bill, what is the dollar difference between the two months' bills?

   Percentage = _____

4. You answered a test containing 60 questions. If you correctly answered 46 questions, what percent did you answer correctly?

   Rate = _____

5. You have been asked to pay a down payment of $180, which represents 8% of the purchase price of a microcomputer and accessories. What is the total purchase price?

   Base = _____

6. The grocery store delivered groceries totaling $120. The delivery charge was $9.95. What was the rate charged?

   Rate = _____

7. A real estate agent receives 8.25% of the selling price as her compensation for selling. What amount of money would she receive for selling a house for $76,400?

   Percentage = _____

8. A highway that is to be 54.4 miles long when finished has been 45% constructed. How many miles are built?

   Percentage = _____

9. Maurice paid a yearly insurance premium of $898.45 on his automobile. At the end of eight months, he cancelled his policy and sold his car. The insurance company charges him 67% of his yearly premium. What amount of money was returned to Maurice as the unexpired part of this yearly premium?

   Percentage = _____

10. Jody Martin, a salesperson in a specialty shop, is paid a salary of $500 per month, plus a $4\frac{1}{2}$% commission on all sales. Last month she sold $3,210 worth of goods. What were her total earnings for the month?

    Base = _____

11. A stereo that you would like to buy is priced at $895. If the sale discount is 25%, what is the amount of the discount?

    Percentage = _____

12. You must pay 8.25% sales tax on the purchase of computer software which costs $328. How much tax will you pay?

    Percentage = _____

Assignment 5   Name_____ Date _____

Determine the percentage, rate, or base in these problems. Round percentage and base answers to the nearest hundredth. Round rate answers to the nearest tenth of a percent.

1. Johnson County has a sales tax of 8.25%. Sally purchased a new computer for $2,338.47. How much sales tax will Sally pay on her purchase?

   _____

2. Ninety-two percent of one store's customers paid with credit cards. Fifty customers came in that day. How many customers paid for their purchases with credit cards?

   _____

3. Twenty-eight of 40 shareholders attended the annual stock meeting. What percent of the shareholders attended the meeting?

   _____

4. The new baseball stadium has a capacity of 47,333. If 37,665 fans attended a game, what percent of the seats were filled?

   _____

5. The financial officer for an accounting firm allows $2,500 for computer supplies in the annual budget. After 6 months, $1,239.34 has been spent on supplies. Is this figure within 50% of the annual budget?

   _____

6. Jules Mission earns a commission of $7\frac{3}{4}$% of his total sales. His sales were: Monday, $352.22; Tuesday, $558.98; Wednesday, $1,321.11; and Thursday, $447.73. How much did he earn as total commission for the week?

   _____

7. Amos worked during the summer selling subscriptions on a commission basis for a magazine publisher. His percent of commission was $33\frac{1}{3}$% of all sales. During the summer, he sold $8,998.43 worth of magazine subscriptions. What amount of money did he earn as total commission?

   _____

8. In her will, Henrietta Milestone left 30% of her estate to her husband, 20% to each of her three sons, and 5% to her niece. The balance was left to her favorite charity. The estate amounted to $239,339. What amount was left to charity?

   _____

9. Taylor Department Store employs 1,200 people. If 65% are sales staff, 14% are office staff and the remaining people are supervisory employees, how many people are supervisory employees?

   _____

10. Katie answered a test containing 75 questions. If she correctly answered 55 questions, what percent did she answer correctly?

    _____

Assignment 6

Name_____ Date _____

Determine the percentage, rate or base in these word problems. Round percentage and base answers to the nearest hundredth. Round rate answers to the nearest tenth of a percent.

1. Jacy Parks bought an IBM personal computer for $1,355. She made a 25% down payment. What was the amount of Jacy's down payment?

   _____

2. Harper's Bookstore ordered 125 accounting books but received 57 books. What percent of the order was missing?

   _____

3. Statewide Insurance pays its agents a 40% commission. Murray Townsend, an agent, earned $756 in commissions in one week. What were Murray's total sales for the week?

   _____

4. George Harlowe, a CPA, is reviewing the company's accounts payable. This week, the checks totaled $13,216. This represents 25% of the monthly accounts payable due. What is the company's total accounts payable?

   _____

5. Lee Pharmaceutical found 2% of all aspirin bottles were defective. Assuming 450 bottles were removed from production, how many total aspirin bottles were acceptable?

   _____

6. 15% of the 1,240 students at Southern State University are in the sophomore class. How many students are in the sophomore class?

   _____

7. At a benefit concert, the ticket booth estimated that 42% of the audience has traveled from out of state; 6,000 people attended the concert. How many people from within the state attended the concert?

   _____

8. Frank Tarleton earned $500 in commissions for selling life insurance. He receives a 20% commission. What were Frank's total sales?

   _____

9. Funtime Events ordered 200 round tablecloths. When they received the order, 23 tablecloths were stained. What percent of the order was stained?

   _____

10. Georgia has a 6.25% sales tax rate. Karen bought a computer and desk and paid $128 in sales tax. What was the cost of the computer and desk?

    _____

11. Thirty-six of 80 students made their oral presentations in record time. What percent of the students made their presentations?

    _____

12. Of the total budget of $157,510, 7.25% was expended in the first two weeks. How much has been expended?

    _____

Practical Math Applications

CHAPTER 8

Percent of Increase and Decrease

CHAPTER

8

Percent of Increase and Decrease

OBJECTIVES

After completing this chapter, you will be able to:

1. Calculate percent of increase and decrease.
2. Distinguish between increase and decrease problems.
3. Figure percentage distribution.

In business it is often necessary to compare current expenses, costs, sales, or profits with amounts from previous months or years. Businesses must analyze their statements and reports from any number of years to make comparisons and identify trends. These trends can be used as a basis for sound business decisions. For example, you might need to compare a particular department's sales for this year with last year's sales.

As an employee, you should be able to figure the percent of increase in your salary when you receive a monthly pay raise. As a consumer, you may want to figure the amount or percent of increase in the price of a product you intend to buy. It is very important to be able to figure these types of basic math problems because they occur often in business and in your daily personal life.

8.1 Terms Used in Percent of Increase and Decrease

An *increase* or *decrease* is the difference between two numbers being compared. There is an *increase* when the current amount is larger than the amount for the previous period. There is a *decrease* when the current amount is smaller than the amount for the previous period. A *rate of increase or decrease* is the percent obtained when the amount of increase or decrease is divided by the base or previous amount. A *percentage distribution* shows the percent each part is of the total.

8.2 Finding the Actual Increase

To find the actual increase, you must determine the amount of change that has occurred. You can find the amount of change by subtracting the smaller number from the larger number. Let's see how the amount of change is calculated in the following example.

EXAMPLE

Sales this year:	$800,120
Sales last year:	− 620,125
Amount of change:	$179,995

You can quickly determine that $179,995 is an increase because the amount from the previous period is smaller than the current amount.

8.3 Finding the Percent of Increase

Find the percent of increase by using this formula:

Percent of Increase = Amount of Increase ÷ Previous Amount

The formula for finding the percent of increase is familiar because the amount of increase and previous amount relate to the percentage and base in the formula

Rate = Percentage ÷ Base

29% = $179,995 ÷ $620,125

The steps used to find the percent of increase are the same steps used to find the rate.

Using the preceding example—sales this year, $800,120; sales last year, $620,125; amount of increase, $179,995—find the percent of increase by applying the formula: amount of increase ÷ previous amount or $R = \dfrac{P}{B}$.

$179,995 ÷ $620,125 = 0.29
100 × 0.29 = 29% increase

PRACTICE PROBLEMS

Find the actual increase and percent of increase for these problems. Write your answers in the blanks provided. Round your answers to the nearest tenth of a percent.

1. Last year's equipment budget: $120,588
   This year's equipment budget: $122,740

   Actual Increase = _____

   Percent = _____

2. Last year's office expenses: $23,640
   This year's office expenses: $35,570

   Actual Increase = _____

   Percent = _____

Solutions: 1. Actual Increase = $2,152; Percent = 1.8%
            2. Actual Increase = $11,930; Percent = 50.0%

■ TIP   When dividing in percent of increase or percent decrease problems, always divide by the base (last or previous year).

8.4 Finding the Actual Increase and the Total When the Percent is Given

In the following problem, the percent of increase is given, but the actual increase and the total earnings must be calculated.

EXAMPLE _____

Anita Dazai earns 5% more this year than last. If her earnings last year were $22,450, find her earnings for this year.

1. Before you can find the total earnings, identify the parts of the problem. In this example, the base is $22,450 and the rate is 5%.

2. Now calculate the actual increase by using the formula:

   $P = B \times R$
   $22,450 \times 5\%$
   $22,450 \times 0.05 = \$1,122.50$ actual increase

3. Calculate the total earnings for this year by adding the amount of increase to the base.

$$\begin{array}{ll} \$22,450.00 & \text{base} \\ +1,122.50 & \text{amount of increase} \\ \hline \$23,572.50 & \text{total earnings for this year} \end{array}$$

As a review, to determine the total earnings in this example, use these steps.

STEPS

1. Identify the parts of the problem.

2. Figure the actual increase by using the formula, $P = B \times R$.

3. Determine the total by adding the actual increase to the amount.

■ **TIP** A quick method for calculating total earnings is to add 100% to the rate and multiply the sum by the base.

Using the same example, follow these steps to calculate the total earnings.

STEPS

1. Add 100% to rate.

$$\begin{array}{r} 5\% \\ +100\% \\ \hline 105\% \end{array}$$

2. Multiply the sum by the base.

   $22,450 \times 105\% = \$22,450 \times 1.05 = \$23,572.50$ total earnings

---

8.5 Finding the Percent of Decrease

Study the following example and steps to determine how to figure the percent of decrease.

EXAMPLE

Sales this year: $127,430
Sales last year: $145,507

STEPS

1. Find the actual decrease by subtracting the current amount from the previous amount.

$145,507
$-127,430$
$18,077$ actual decrease

2. After calculating the actual decrease, use the following formula to find the percent (rate) of decrease:

Percent of Decrease = Actual Decrease ÷ Previous Amount

■ **TIP**   This formula is the same one used to find the percent of increase except the word "increase" has been replaced by the word "decrease."

You should recognize that this is the same formula as the one used to find the rate in problems presented in Chapter 7: $R = P \div B$. Last year's sales is the base of comparison, and the actual decrease is the percentage.

EXAMPLE

Let's find the percent of decrease by applying the formula, actual decrease ÷ previous amount.

$18,077 \div \$145,507 = 0.1242$
$\quad (P) \qquad\quad (B)$

$0.1242 \times 100 = 12.4\%$ decrease

8.6   Finding the Actual Decrease and the Total When the Percent is Given

In the following example, the percentage or the amount is calculated and is then subtracted from the base. Let's see how this works by finding the actual decrease in this example.

EXAMPLE _____

Greene College's enrollment of 8,425 declined by 8%. What was the total college enrollment after the decline?

STEPS

1. Determine the actual decrease by using the formula, $P = B \times R$.

$$8,425 \times 8\% = 8,425 \times 0.08 = 674 \text{ actual decrease}$$

2. To calculate the total enrollment after the decline, subtract the actual decrease from the base. For example,

$$
\begin{array}{r}
8,425 \\
-\ 674 \\
\hline
7,751 \\
\end{array}
\text{ total enrollment after 8\% decline}
$$

■ **TIP**   A quick method for calculating the *total amount after the decrease* is first to determine the complement of the percent. To find the complement of a percent, change it to its decimal equivalent and subtract from 1.00. To determine the total amount after the decrease, multiply the complement of the percent by the base.

Study these steps, which illustrate the complement method for calculating the total amount after the decrease using the preceding example.

STEPS

1. Change the percent to its decimal equivalent and subtract from 1.00.

$$\begin{array}{r} 1.00 \\ -0.08 \\ \hline 0.92 \end{array}$$

2. Multiply the base by the complement.

$$8,425 \times 0.92 = 7,751 \text{ after the 8\% decline}$$

■ **TIP** The following phrases might help you identify an increase or decrease problem.

Increase—"greater than," "more than," "an increase of," "were up," "exceeded."

Decrease—"less than," "a decrease of," "a reduction of," "marked down," "down from."

PRACTICE PROBLEMS

Solve these decrease problems. Write your answers in the blanks provided.

1. Base = $1,240; Rate = 9%

Actual Decrease = _____

Total Amount = _____

2. Base = $35,201; Rate = 30%

Actual Decrease = _____

Total Amount = _____

Solutions:　　**1.** Actual Decrease = $111.60; Total Amount = $1,128.40
　　　　　　　2. Actual Decrease = $10,560.30; Total Amount = $24,640.70

8.7 Determining Percentage Distribution

A percentage distribution shows the percent each component is of the total. In the following example, the sales amount for each department is a component of the total sales.

EXAMPLE

Departments	Sales	Percent
Hosiery	$2,937	_____
China	5,845	_____
Stationery	2,432	_____
Boys' Wear	6,079	_____
Total Sales	$17,293	100%

To determine the percent for each department, divide each department's sales amount by the total amount. This is the same basic formula you used to determine rate: $R = \dfrac{P}{B}$. Because you must determine a percent for each department, you should recognize that the base is $17,293 and the percentage or part is each department's amount. Study the example which follows.

Total each component

↓

Departments	Sales	Percent	
Hosiery	$2,937	17.0	$(2,937 \div 17,293 = 0.1698 = 17.0\%)$
China	5,845	33.8	$(5,845 \div 17,293 = 0.3379 = 33.8\%)$
Stationary	2,432	14.1	$(2,432 \div 17,293 = 0.1406 = 14.1\%)$
Boys' Wear	6,079	35.1	$(6,079 \div 17,293 = 0.3515 = 35.1\%)$
Total Sales	$17,293	100%	

↑

Divide each component by the total

NOTE: Many times you will have to adjust a percent a tenth of a percent to make it add to 100% because of rounding.

Being able to determine what percent each amount is of the total is important in your personal budgeting. To make comparisons in your own budget, you need to know what percent of your net income is spent for food, clothing, shelter, entertainment, and so on. After making these comparisons, you might discover areas in which you need to adjust your spending.

PRACTICE PROBLEMS

Determine what percent each amount is of the total in the following problems. Write your answers in the blanks provided. Round your answers to the nearest tenth of a percent.

1.

Departments	Sales	Percent
Dept. A1	$13,468	_____
Dept. B3	15,298	_____
Dept. C6	16,470	_____
Dept. F5	3,950	_____
Total Sales	$49,186	100%

2.

Departments	Sales	Percent
1 - Paint Supplies	$14,642	_____
2 - Automotive	12,403	_____
3 - Plumbing	5,202	_____
4 - Electrical	4,226	_____
Total Sales	$36,473	100%

Solutions

1. Dept. A1: 27.4%; Dept. B3: 31.1%; Dept. C6: 33.5%; Dept. F5: 8.0%;

2. Dept. 1: 40.1%; Dept. 2: 34.0%; Dept. 3: 14.3%; Dept. 4: 11.6%

Math Alert

A newspaper article on local car sales reported that:

Chevrolet registrations rose to 1,324 in August after a 10.6% increase in sales. GM's total sales fell 2.95% to 2,404 units. Despite the drop in sales, their market share was 26.4%; higher than last year's 26.1%.

Cadillac sales fell 1.6% to 240 units.

Buick sales were down 37.3% to 252 cars from 402 cars the previous year. Buick is still 17.4% above 1990 sales.

Ford auto sales were off 6.7% to 240 units.

Chrysler sales fell 23.11% to 782 cars, or 8.59% of the market, down from 10.73 percent a year ago.

1. Calculate the Chevrolet registrations before the 10.6 percent increase. _____

2. What were the original number of units before GM's total sales fell by 2.95 percent? _____

3. Buick's previous units were 402; this year's units are 252. How many units were decreased this year? _____

4. What were last year's number of units before Chrysler Corp's sales fell 23.11 percent? _____

CHAPTER 8 Percent of Increase and Decrease
Study Guide

I. Terminology

page 138   increase   When the current amount is larger than the amount for the previous period.

page 138   decrease   When the current amount is smaller than the amount for the previous period.

page 138   percentage distribution   The percent each part is of the total.

II. Percent of Increase and Decrease

page 138   *Finding the Percent of Increase:* Determine the amount of change that has occurred by subtracting the smaller number (last year's sales) from the larger number (this year's sales). The difference is the actual increase. Find the percent of increase by dividing the amount of increase by the previous amount.

Example: Sales this year : $800,120   Percent of increase = $179,995 ÷ $620,125
Sales last year :   − 620,125
Amount of change : $179,995   = 0.29 or 29%

page 140   *Finding the Actual Increase and the Total When the Percent is Given:* Identify the parts of the problem—percentage, base, and rate; figure the actual increase by using the formula, $P = B \times R$; determine the total by adding the actual increase to the base.

Example: Anita Dazai earns 5% more this year than last. If her earnings last year were $22,450, find her earnings for this year.

$22,450 × 0.05 = $1,122.50 actual increase
$22,450 + $1,122.50 = $23,572.50 this year's earnings

page 141   *Finding the Percent of Decrease:* Find the actual decrease by subtracting the current amount from the previous amount; then use the formula, Percent of Decrease = Actual Decrease ÷ Previous Amount.

Example: Sales last year : $145,507   Percent of decrease = $18,077 ÷ $145,507
Sales this year :   − 127,430
Amount of change : $18,077   = 0.124 or 12.4%

page 142   *Finding the Actual Decrease and the Total When the Percent is Given:* Determine the actual decrease by using the formula, $P = B \times R$; then subtract it from the base.

Example: Taylor College's enrollment of 8,425 declined 8%. What was the total college enrollment after the decline?

8,425 × 0.08 = 674
8,425 − 674 = 7,751 total enrollment after 8% decline

III. Determining Percentage Distribution

page 143   A percentage distribution shows the percent each component is of the total. To calculate percentage distribution, total each component; then divide each component part by the total.

Total each component
↓

Departments	Sales	Percent	
Hosiery	$2,937	17.0	(2,937 ÷ 17,293 = 0.1698 = 17.0%)
China	5,845	33.8	(5,845 ÷ 17,293 = 0.3379 = 33.8%)
Stationery	2,432	14.1	(2,432 ÷ 17,293 = 0.1406 = 14.1%)
Boys' Wear	6,079	35.1	(6,079 ÷ 17,293 = 0.3515 = 35.1%)
Total Sales	$17,293	100%	

↑
Divide each component by the total

Practical Math Applications

Assignment 1   Name_____ Date _____

Write your answers in the blanks provided. For the rate answers, round to the nearest tenth of a percent.

A. Find the actual increase and rate of increase in sales in the following problems.

	Sales Last Year	Sales Present Year	Actual Increase	Rate of Increase
1.	$64,919	$83,518	_____	_____
2.	$54,980	$73,876	_____	_____
3.	$28,899	$37,975	_____	_____
4.	$102,720	$115,176	_____	_____

B. Determine the actual decrease and rate of decrease in sales in the following problems.

	Sales Last Year	Sales Present Year	Actual Decrease	Rate of Decrease
5.	$73,078	$68,000	_____	_____
6.	$85,780	$67,800	_____	_____
7.	$84,799	$68,065	_____	_____
8.	$109,358	$99,875	_____	_____

C. Find the rate of increase in enrollments in four high schools from last year to the present year.

	Enrollments Last Year	Enrollments Present Year	Rate of Increase
9.	4,237	5,180	_____
10.	6,199	6,401	_____
11.	3,257	3,648	_____
12.	3,012	3,278	_____

Assignment 2

Name_____ Date _____

Write your answers in the blanks provided. For rate answers, round to the nearest tenth of a percent.

A. Find what percent each amount is of the total in the following problems.

	Departments	Sales	Percent
1.	Department A-1A	$3,781.24	_____
2.	Department A-1B	$5,043.36	_____
3.	Department A-1C	$2,909.47	_____
4.	Department A-1D	$6,188.43	_____
5.	Department A-1E	$7,010.12	_____
		$24,932.62	100.0%

	Departments	Sales	Percent
6.	Department 1	$2,919.28	_____
7.	Department 2	$4,012.36	_____
8.	Department 3	$1,982.91	_____
9.	Department 4	$2,354.10	_____
10.	Department 5	$2,887.70	_____
		$14,156.35	100.0%

B. Find what percent each amount is of the total in the following problems.

	Expenses	Amount	Percent
11.	Administrative	$500.00	_____
12.	Office Supplies	$127.00	_____
13.	Utilities	$250.00	_____
14.	Rent	$420.00	_____
15.	Depreciation	$131.00	_____
		$1,428.00	100.0%

Assignment 3   Name_____ Date _____

Write your answers in the blanks provided. For the rate answers, round to the nearest tenth of a percent.

A.   Find the actual increase and rate of increase in salaries in the following problems.

	Salaries Last Year	Salaries Present Year	Actual Increase	Rate of Increase
1.	$25,680	$27,250	_____	_____
2.	$24,370	$25,290	_____	_____
3.	$36,700	$38,520	_____	_____
4.	$32,475	$33,390	_____	_____
5.	$41,600	$42,750	_____	_____

B.   Determine the actual decrease and rate of decrease in gasoline prices in the following problems.

	Prices Last Year	Prices Present Year	Actual Decrease	Rate of Decrease
6.	$1.10	$1.07	_____	_____
7.	$1.12	$1.05	_____	_____
8.	$1.13	$1.09	_____	_____
9.	$1.11	$1.06	_____	_____
10.	$1.09	$1.06	_____	_____

C.   Find the the rate of increase or decrease in the following five elementary school enrollments.

	Enrollments Last Year	Enrollments Present Year	Rate of Increase(+) or Decrease (–)
11.	605	703	_____
12.	585	553	_____
13.	725	690	_____
14.	590	623	_____
15.	390	420	_____

Assignment 4   Name_____   Date _____

Solve the following increase and decrease problems. Carefully read the problems to determine whether they involve an increase of decrease. Round rate percents to nearest tenth of a percent.

1. Cheng Card & Gift Shop's total sales last year were $52,198. The shop sold 9% more this year than last. Find the total sales for this year.

   _____

2. Aaron Pike earns 6% more this year than last. If his earnings last year were $25,750, what are his earnings for this year?

   _____

3. Don Perkins' stock market investment of $5,320 has declined 5% in value. Determine the worth of the investment now.

   _____

4. Mario's Cheese Shop's sales have increased 9.8% from last year. His sales last year were $41,607. Find his sales for this year.

   _____

5. The Professional Women's Organization earned $2,630 last year on their fund-raising project. This year's fund is $2,420. Find the actual decrease and rate of decrease.

   _____

6. Fenton Art and Craft Center's inventory last year was valued at $82,970 which is 9.2% less than this year's. Determine the value of this year's inventory.

   _____

7. The printing costs in Chen Ding's department have increased from $1,340 to $1,608. Calculate the rate of increase.

   _____

8. Subscriptions for the *Weekly News* have increased from 1,911 to 2,472. Find the actual increase and the rate of increase.

   _____

9. The library processed 36,714 books this year but processed 32,980 books last year. Determine the actual increase and rate of increase.

   _____

10. You sold your golf clubs for $490, at a loss of 4%. How much did you lose?

    _____

11. Your company has laid off 15% of its 580 employees. How many were laid off? How many are currently employed?

    _____

12. Your stock dropped from $23.50 per share to $21.75. What was the rate of decrease?

    _____

13. CD-ROM sales have increased from 212 to 338. Find the rate of increase.

    _____

14. Enrollment at your school was 875 last year. This year's enrollment is 930. How many students are new to your school? What is the rate of increase?

    _____

Assignment 5

Name_____ Date _____

Complete the following word problems. Write your answers in the blanks provided. Carefully read the problems to determine whether they involve an increase or decrease. Round rate percents to nearest tenth percent.

1. Loretta Danzig, a secretary for Hoffman Electric, earns 6% more this year than last. If her earnings last year were $23,398, find her earnings for this year. _____

2. Volvo Motors raised the base price of its sedan by $2,400 to $34,400. What was the percent increase? _____

3. Last year, Brothers Hardware had $400,000 in sales. This year, Brothers' sales were up 25%. What are the sales for this year? _____.

4. Tom Batting sold a computer for $255 that originally cost him $1,200. What was Tom's percent of decrease based on the cost of the computer? _____

5. The price of a Southwest Airlines ticket to Houston, Texas decreased 12%. The original fare was $62. What is the price of the new fare? _____

6. Diane Reed had office supply sales of $3,285 this month. This was 78% of last month's sales. What was the amount of last month's sales? _____

7. The price of SVGA monitors at Computer Warehouse dropped from $299 to $245. What was the percent of decrease in price? _____

8. Paula Pruitt has invested $4,334 in the stock market. The investment has declined 7% in value. Determine the worth of the investment now. _____

9. An office furniture store has an inventory of $547,912.34. This is a 22% increase from last year's ending figure. Determine last year's amount of closing inventory. _____

10. Last year, Direct Marketing had $122,222 in sales. This year, Direct Marketing's sales were down by 9%. What were Direct Marketing's sales this year? _____

11. Rita paid $45 for a business English text at the college bookstore. The price has increased by 4% from last year. What was the old selling price of the business English text? _____

12. A cellular phone priced at $298 is marked down 5% to promote the new model. Find the amount the cellular phone was discounted. Calculate the new price. _____

13. Yolanda Hammond sold a portable building for $2,400. If she originally paid $1,400 for the building, find her percent of increase. _____

14. Robert is a department manager for Hoover's Hardware. Each quarter, he computes a breakdown of his sales. Given the following figures for sales, determine what percent each amount is of the total sales.

Power Tools	Sales	Percent
Sanders	$1,345.45	_____
Drills	986.77	_____
Saws	1,188.92	_____
Screwdrivers	1,444.65	_____
Routers	762.33	_____
Total Sales	**$5,728.12**	**100%**

Assignment 6

Name_____ Date _____

Complete the following word problems. Write your answers in the blanks provided. Carefully read the problems to determine whether they involve an increase or decrease. Round rate percents to nearest tenth percent.

1. Last year, Gaston Company had sales of $450,450. This year, sales were up 76%. What were the sales this year? _____

2. Samantha Clark receives an annual salary of $31,000 from J.C. Reddings, Inc. Today her supervisor informs her that she will be getting a $2,300 raise. What percent of her old salary is the $2,300 raise? _____

3. The price of a calculator at Justins Office Supply dropped from $49.95 to $38.95. What was the percent decrease in price? _____

4. This year, the enrollment for Eastfield Junior College was 5,323. This was a 14% increase from the enrollment last year. What was the Eastfield College enrollment last year? _____ (Round to the nearest unit.)

5. Jackie Morgan found a Chippendale chair in her attic. It was originally purchased for $2.50. The chair is now worth $1,250. What is the percent of increase? _____

6. The price of a Macintosh personal computer dropped from $1,800 to $1,200. What was the percent decrease? _____

7. This year, the price of a Hewlett Packard Deskjet printer rose to $359. This is an increase of 15% more than last year's price. What was the old selling price? _____ (Round to the nearest cent.)

8. Marble Creamery pays Ted Manson an annual salary of $46,000. Today, Ted's manager informs him that he will receive a $5,000 raise. What percent of Ted's salary is the $5,000 raise? _____

9. The price of an airline ticket to New York increased 12%. The ticket price is now $456. What is the old selling price? _____ (Round to the nearest cent.)

10. Last year, Kari Edwards earned $34,800, an increase of 13.4% over the previous year. What were Kari's earnings in the previous year? _____ (Round to the nearest cent.)

11. Pietro's Pizza ordered 300 pounds of flour this month. This was a 10% increase from the previous month's order. What was last month's order of flour? (Round to the nearest unit.) _____

12. Mooring Electric adjusted the payroll budget and decided each employee must take a 3% decrease in salary to cover increasing medical coverage costs. Sara Strauss earns an annual salary of $21,000. What will her new salary be after the decrease? _____

13. Last year, Victor Food Shoppes had $900,000 in sales. This year, Victor's sales were up 55%. What are the sales for this year?

    _____

14. Winns Department Store is calculating their end of year inventory. Given the following figures for each department's closing inventory, calculate the percent of inventory each department represents.

Departments	Inventory	Percent
Hosiery	34,112	_____
China	120,873	_____
Stationery	13,987	_____
Menswear	344,990	_____
Ladies Wear	876,654	_____
Bedding	203,836	_____
Shoes	176,348	_____
Childrens	485,291	_____
Total Inventory	**2,256,091**	**100.0%**

Practical Math Applications

Proficiency Quiz
R E V I E W

Name_____ Date_____

$$\frac{\text{Student's Score}}{\text{Maximum Score}} = \frac{\quad}{94} = \text{Grade}_____$$

A. Convert the following percents to decimals.

1. 0.0235% = _____ **2.** 63% = _____

3. 8.5% = _____ **4.** 7% = _____

5. 13.84% = _____ **6.** 21.4% = _____

7. 0.45% = _____ **8.** 73.4% = _____

9. 150% = _____ **10.** 0.1459% = _____

11. 1.7936% = _____ **12.** 24.95% = _____

13. 323.1% = _____ **14.** 85% = _____

B. Convert the following decimals to percents.

15. 0.41 = _____ **16.** 0.0651 = _____

17. 0.1115 = _____ **18.** 4.63 = _____

19. 0.728 = _____ **20.** 0.0538 = _____

21. 0.0001 = _____ **22.** 0.02 = _____

23. 3.65 = _____ **24.** 0.0999 = _____

25. 0.00045 = _____ **26.** 6.1504 = _____

27. 0.1965 = _____ **28.** 1.489 = _____

C. Change the following percents to fractions. Reduce each fraction to lowest terms as necessary.

29. 65% = _____ **30.** 50% = _____

31. 111% = _____ **32.** 5% = _____

33. 90% = _____ **34.** 150% = _____

35. 16% = _____ **36.** 45% = _____

37. 250% = _____ **38.** 15% = _____

D. Change the following fractions to percents.

39. $\frac{1}{5}$ = _____   **40.** $\frac{3}{8}$ = _____

41. $\frac{5}{10}$ = _____   **42.** $\frac{2}{3}$ = _____

43. $\frac{5}{8}$ = _____   **44.** $\frac{1}{8}$ = _____

45. $\frac{9}{10}$ = _____   **46.** $\frac{7}{8}$ = _____

47. $\frac{1}{3}$ = _____   **48.** $\frac{2}{5}$ = _____

E. Find the percentage in the following problems.

49. 80% of 300 = _____   **50.** 6% of 2 = _____

51. 15% of 50 = _____   **52.** 40% of 111 = _____

53. 25% of 100 = _____   **54.** 34% of $30.00 = _____

55. 61% of $15.00 = _____   **56.** 49% of $651.00 = _____

F. Find the rate in the following problems.

57. 20 is what percent of 100?   _____

58. 75 is what percent of 350?   _____

59. 90 is what percent of 175?   _____

60. 200 is what percent of 400?   _____

G. Determine the percentage, rate, or base for the following problems.

61. 25% of $360 = __?__.   _____

62. $86 is what percent of $2,500?   _____

63. $58 is 15% of __?__.   _____

64. 21% of $580 is __?__.   _____

65. $25 is 1% of __?__.   _____

66. $42 is what percent of $300?   _____

67. $200 is what percent of $1,000?   _____

68. $13 is 3% of __?__.   _____

69. 68% of $200 is __?__.   _____

70. $4 is what percent of $50?   _____

71. $18 is 50% of __?__.   _____

72. 95% of $800 is __?__.   _____

    Practical Math Applications

73. $75 is 10% of __?__.

74. $5 is 2% of __?__.

75. 33% of $105 is __?__.

76. $9 is what percent of $40?

77. 88% of $2 is __?__.

78. $14 is what percent of $28?

H. Find the rate of increase or decrease for the following problems.

	Sales Last Year	Sales Current Year	Rate of Increase or Decrease
79.	$245,321	$294,678	
80.	35,123	65,888	
81.	111,111	95,000	
82.	21,050	30,050	
83.	650,066	400,000	
84.	325,325	300,000	
85.	30,500	43,986	
86.	154,678	150,000	
87.	25,900	45,652	
88.	329,555	254,339	
89.	10,444	13,984	
90.	438,941	633,318	

I. Determine Percent Distribution for the following problems.

	Departments	Sales	Percent
91.	Furniture	$459	
92.	Lawn Care	685	
93.	Women's Wear	1,350	
94.	Children's	2,105	
		$4,599	100.0%

CHAPTER

Checking Accounts, ATMs, and Statement Reconciliation

9

Checking Accounts, ATMs, and Statement Reconciliation

OBJECTIVES

After completing this chapter, you will be able to:

1. Identify the parts of a check and deposit slip.
2. Identify and complete the parts of a check register and check stub.
3. Reconcile a bank statement.

Financial institutions such as banks, savings and loan associations, and credit unions offer many types of personal and business services to their customers. In this chapter, you will be introduced to one very important banking service—the checking account. Because almost all business financial transactions involve checks, it is important that you have a good understanding of the terminology and procedures associated with checking accounts.

9.1 Terms Used with Checking Accounts

Some of the basic terms related to checking accounts and statement reconciliation are defined in this section, while more specific terms are defined as each area is explained. When the term *bank* is used in this chapter, it also means credit unions and savings and loan institutions.

A *checking account* is an account opened at a banking institution that provides a method for payment of funds on deposit in a bank. A *checkbook* includes checks and a record of checks written and deposits made into an account. A *bank statement* is a listing of the checks paid, deposits made, charges against the account, and a final balance. *Reconciliation* is the process of adjusting the balance in a checkbook to the bank statement so that the balances agree.

9.2 Checking Account Services

Most banks provide a wide range of financial services related to checking accounts. For example:

1. Automatic savings transfer programs offer an easy way to add money to a savings account. You specify the amount to automatically transfer from your checking account to your savings account each month.

2. Loan payments can be made automatically from your checking account to the loan.
3. Direct deposit of monthly payments such as Social Security checks or payroll checks is available.
4. The depositor is insured up to $100,000 by the Federal Deposit Insurance Corporation (FDIC).
5. Bill paying by telephone or personal computer.
6. Interest on checking account balances.
7. Overdraft check protection assures that should you write a check over the amount in your account, the bank automatically deposits funds to your account to cover the overdraft amount. These amounts are set up as a loan; you make monthly payments until the loan is paid.
8. Service charges are waived when minimum amounts are kept in checking accounts.

9.3 Automated Teller Machines (ATMs)

One of the most popular services provided by most banks today is the Automated Teller Machine. These machines are located outside the bank, usually near the drive-in service, and allow you 24-hour access to your account. You are issued a plastic card, similar to a bank credit card, which contains a code number (called a PIN or personal identification number). The bank will assign the PIN number or you can choose your own. This number is *not* written on the card, nor should it be. The absence of the number is for your security to keep unauthorized persons from using the card.

To withdraw funds, as an example, the card is inserted into the ATM and the amount keyed (usually in multiples of $20) shows up on the screen. The code and amount are verified and the cash is paid to you. You also receive your bank card back along with a printed receipt. To make a deposit, the same procedure is followed except the customer inserts an envelope containing the money and/or checks along with a deposit slip into the ATM.

Checking and savings accounts are conveniently accessed by the ATMs. Usually there is no charge if you use the ATM at your bank. You may access your funds through commercial ATMs that charge a fee per transaction.

9.4 Checking Account

A checking account provides a convenient method of paying for purchases, services rendered, and other monetary obligations. Common items used in most checking accounts are:
1. Checkbooks which contain checks, deposit slips, and a check register or check stubs
2. Statement of Bank Account
3. Reconciliation Forms

Checks

A *check* authorizes the bank to pay a certain sum from the checking account to an individual or business known as a *payee.* The person who writes the check is the *drawer,* the drawer's bank is called the *drawee.*

The check below has been completed by Phillip Rochelle, the drawer. It includes the following parts:

a. Drawer's name and address
b. Check number
c. Date of check
d. Name of party to whom check is written (payee)
e. Amount of check written in numbers and words
f. Imprinted numbers to identify bank, Federal Reserve routing number, and drawer's account number
g. Drawer's signature
h. Bank's name and address

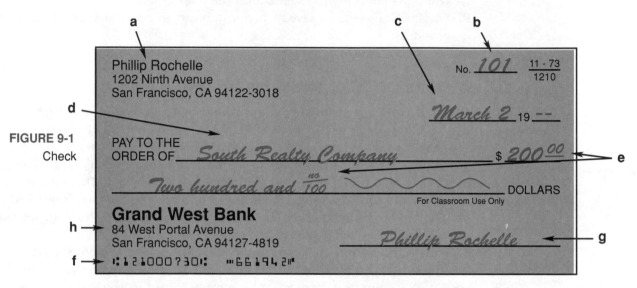

FIGURE 9-1
Check

The handwritten information on the check may also be completed on a typewriter or a computer except for the signature, which must be written by hand or a facsimile can be affixed with a signature machine.

Deposit Slip

A *deposit slip* is filled out when cash (currency or coin) and/or checks are deposited to a checking account. The form is filled out by the *depositor*. The deposit slip shown on the next page has been prepared by Phillip Rochelle, the depositor.

The deposit slip contains the following information:

a. Depositor's name and address, which are usually preprinted
b. Date of deposit
c. Depositor's signature for any cash received.
d. Bank's name and address
e. Bank's identification number and depositor's account number.
f. ABA (American Bankers Association) transit number
g. Amount of currency and coin
h. Amount of each check
i. Total amount deposited
j. Amount of cash depositor receives
k. Net deposit, which is actual amount being deposited into account

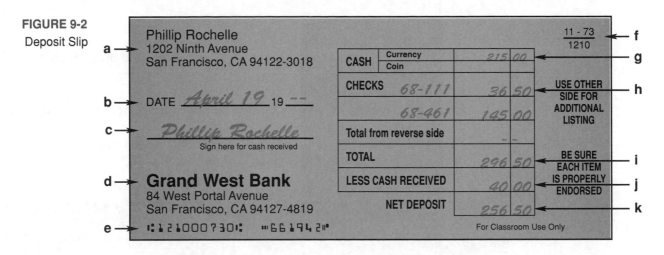

FIGURE 9-2
Deposit Slip

To help identify each deposited check, Mr. Rochelle listed each check's *American Bankers Association* (*ABA*) transit number on the deposit slip. The ABA number is usually located in the upper right corner of the check somewhere below the check number. It may appear as a fraction such as $\frac{11\text{-}73}{1210}$ (Mr. Rochelle's ABA number) or 11-73/1210. The first two numbers (or top numbers) identify the city or state in which the bank is located and the specific bank on which the check is drawn. The second number (or bottom number) identifies the Federal Reserve District where the check will be cleared and the routing number used by the Federal Reserve Bank. These numbers allow the check to be traced more easily should it become lost.

In order for Mr. Rochelle to deposit the two checks to his account, he must sign the backs of the checks. His signature is known as a *blank endorsement*. An example of a blank endorsement follows:

FIGURE 9-3
Blank Endorsement

Phillip Rochelle

A blank endorsement would allow anyone to cash the checks if the checks were lost or stolen. To ensure protection should the checks be lost or stolen before being deposited, the words "for deposit only" are written on the checks above the signature. This type of endorsement, known as a *restrictive endorsement*, limits the transactions that can be performed with the checks. If a check is endorsed "for deposit only," cash cannot be received from the bank; the check can only be deposited. The figure below illustrates this type of endorsement.

FIGURE 9-4
Restrictive Endorsement

For deposit only
Juanita Lopez

Check Register or Check Stubs

A check register or check stub is used to help the depositor keep track of checks written. The *check register* or *check stubs* are records of each check number and date, the amount of each check, deposits made, and the balance in the checking account. Examples of a check register and a check stub are shown in the figures on the next page. Both forms provide for the following information:

a. Date
b. Check number
c. Individual or business to whom the check is written and other descriptions
d. Amount of check
e. Amount of any deposit made since last check
f. *Balance brought forward*, which is the balance in the account after the last check was written
g. Balance known as a *running balance* because it is determined after each check has been subtracted or a deposit has been added to the account.

FIGURE 9-5
Check Register

DATE	CHECK NUMBER	DESCRIPTION OF TRANSACTION	√	AMOUNT OF DEPOSIT	AMOUNT OF CHECK	BALANCE BROUGHT FORWARD $ 1062 37
4/2/–	101	*South Realty Co.*			800 00	800 00
		Rent				262 37
4/4/–	102	*Butler Pharmacy*			51 88	51 88
		Prescription				210 49
4/8/–	103	*Darin Crow*			7 16	7 16
		Ticket				203 33
4/8/–		*Deposit*		50 00		50 00
						253 33

FIGURE 9-6
Check and Check Stub

Let's suppose you need to write a check to McNally Garage for car repair, dated April 20, in the amount of $27.50. Study the examples illustrating the steps involved in completing a check register and check stub.

STEPS

1. Write in the current date.

2. Write in the check number if not preprinted.

3. Write to whom the check is written or a description of transaction; indicate purpose of check.

4. Write in amount of check twice, once under "Amount of Check" and once under the balance column on the same line. Some registers have you write the amount only one time.

5. Write in amount of deposit, if made. Write the amount twice as in step 4.

6. Subtract amount of check from previous balance and list new balance.

7. Add deposit (if a deposit has been made) for new balance.

Practical Math Applications

FIGURE 9-7
Check Register

DATE	CHECK NUMBER	DESCRIPTION OF TRANSACTION	√	AMOUNT OF DEPOSIT	AMOUNT OF CHECK	BALANCE BROUGHT FORWARD $ 1,021 00	
4/20/–	121	McNally Garage			27 50	27	50
		Car repair				993	50

FIGURE 9-8
Check Stub

No. _121_ $ _27.50_

April 20 19 _--_

TO _McNally Garage_

FOR _Car repair_

	Dollars	Cents
Bal. Bro't. For'd.	1,021	00
Amt Deposited		
Total		
Amt. This Check	27	50
Bal. Car'd For'd.	993	50

Personal Check Writer

You can purchase a personal check writer. The device is a little larger than a personal checkbook and holds a supply of blank checks. When you want to write a check, you retrieve a check, insert it in the bottom, key the necessary information (date, payee, etc.) and the check writer types the check for you. The check writer will also write payments from memory for repeating monthly payments such as for an automobile loan or utility.

The check writer keeps a check register automatically and shows the running balance of your checking account at all times. The check writer can complete a bank reconciliation for you.

Some people forget to record their checking transactions and fail to keep an up-to-date balance. If you fail to manage your check register or check stubs correctly, you may be faced with an embarrassing and costly situation if a check (or checks) is returned indicating that there are no funds to cover it. Most banks offer printed checks that make a carbonless copy of the check. These types of checks assure that you do not forget that the check was written and the amount of the check.

■ **TIP**　Make it a practice to record all checking transactions, including cash withdrawals, as they occur.

In addition, your check-writing habits help establish your credit rating. A poor check-writing background can be a liability when you apply for a charge account or a loan.

PRACTICE PROBLEMS

Fill in the check stubs using the information given here.

1. Balance forward: $2,793.00
    Date: April 20, 19--
    Check number: 130
    To: R & R Tax Service
    For: consultation
    Amount of check: $175

2. Date: April 20, 19--
    Check number: 131
    To: Bruce's Office Supply
    For: folders
    Amount of check: $27.50
    Deposit: $360.90

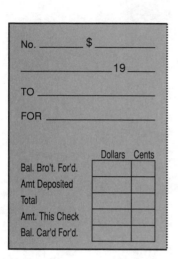

	Dollars	Cents
Bal. Bro't. For'd.		
Amt Deposited		
Total		
Amt. This Check		
Bal. Car'd For'd.		

Complete this check register by filling in the balance using the information given here.

3. Balance: $1,968.80
    Date: June 10, 19--
    Check no.:   321
    To: June's Deli
    For: lunch
    Amount: $9.95

4. Date: June 10, 19--
    Check no.:   322
    To: S & S Cleaners
    For: dry cleaning
    Amount: $20.40

5. Deposit: $814.00
    Date: June 10, 19--

DATE	CHECK NUMBER	DESCRIPTION OF TRANSACTION	√	AMOUNT OF DEPOSIT	AMOUNT OF CHECK	BALANCE BROUGHT FORWARD $	

Solutions:

1.

No. *130* $ *175.00*

April 20 19 *--*

TO *R&R Tax Service*

FOR *Consultation*

	Dollars	Cents
Bal. Bro't. For'd.	2,793	00
Amt Deposited		
Total	2,793	00
Amt. This Check	175	00
Bal. Car'd For'd.	2,618	00

2.

No. *131* $ *27.50*

April 20 19 *--*

TO *Bruce's Office Supply*

FOR *Folders*

	Dollars	Cents
Bal. Bro't. For'd.	2,618	00
Amt Deposited	360	90
Total	2,978	90
Amt. This Check	27	50
Bal. Car'd For'd.	2,951	40

3., 4., and **5.**

DATE	CHECK NUMBER	DESCRIPTION OF TRANSACTION	√	AMOUNT OF DEPOSIT	AMOUNT OF CHECK	BALANCE BROUGHT FORWARD $ 1,968 80
6/10/–	321	June's Deli			9 95	9 95
		Lunch				1,958 85
6/10/–	322	S & S Cleaners			20 40	20 40
		Dry cleaning				1,938 45
6/10/–		Deposit		814 00		814 00
						2,752 45

9.5 Statement of Bank Account

Usually each month the bank sends to each of its checking account customers a statement of his or her account. A *bank statement* is a list of all the account activity processed through the bank to date. The information on this statement helps the customer verify the check register or check stubs. Study the statement sent to Mr. Rochelle that is illustrated in the figure on the next page. It shows:

a. Customer's account number.

b. Total number of checks transacted and total amount of these checks during the month.

c. Total number of deposits made (credits) and total amount deposited during the month.

d. New balance.

e. Each check number, date paid by bank, and amount of check written. (Some statements do not show check numbers when cancelled checks are returned.)

f. Debits, such as bank charges, deducted from the account. (Specific bank charges will be explained later in this chapter.)

FIGURE 9-9
Bank Statement

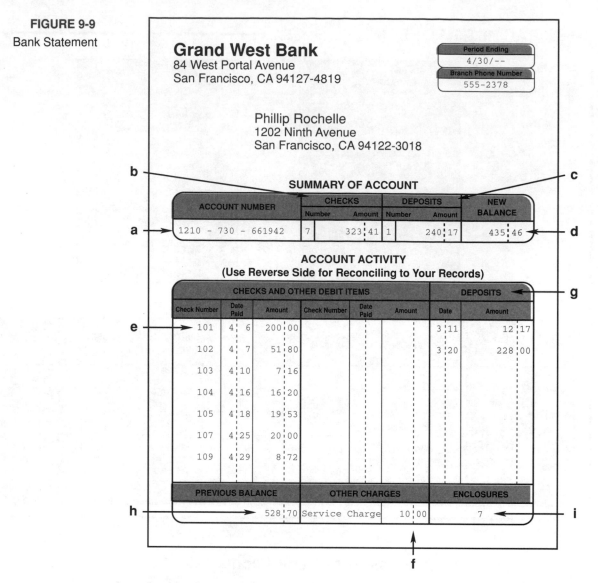

g. Date and amount of deposits made to the checking account.

h. Previous bank balance.

i. Enclosures are the total number of checks transacted that are being returned to the depositor. *Returned checks* (also known as *cancelled checks*) have been deducted from the depositor's checking account. However, there is a trend for banks not to return cancelled checks to the depositor when the checks are listed on the statement.

The statement balance may not agree with the depositor's checkbook balance. The balances may not agree for the following reasons:

1. *Outstanding checks.* These are checks that have been written and deducted by the depositor from the check register or check stubs but have not been received by the bank in time to be included on this month's bank statement.

2. *Outstanding deposits.* These are deposits that have been made by the depositor but have not been received by the bank in time to be included on this month's bank statement.

3. *Debits to the depositor's account.* These may include items such as *bank charges*, sometimes called *service charges* (SC), which are monthly fees charged to the depositor by the bank. These charges include fees for providing a checking account, for printing checks, and for collecting notes.

Other charges may be made if checks are drawn against an account when the account contains insufficient funds to cover the amount of the checks. These checks would be marked "not sufficient funds" (NSF) and would not be paid by the bank.

Other debits reflect money deducted from the checking account and automatically transferred to another account, such as a savings account.

4. *Credits to the depositor's account.* These may include items such as earned interest added to the depositor's checking account balance when a minimum balance is maintained or an amount collected by the bank and added to the customer's account. Although these items were deposited by the bank, they were not recorded by the depositor in the check register or on the check stubs.

5. *Errors made by either the depositor or the bank.* A common error made by the depositor occurs when a check is written but is not recorded or subtracted on the check register or on the check stub. Another typical error occurs when numbers are transposed; that is, reversed, such as $21.45 for $12.45. Another common error occurs when a deposit is made but the depositor forgets to record it on the check register or check stub.

For these reasons, it is important that the depositor compare the check register or check stub balance with the statement.

9.6   Statement Reconciliation

When the check register or check stub balance and the statement balance do not agree, the depositor must reconcile the balances. To *reconcile* the balances is to adjust each balance to the point that they agree. This procedure is usually completed on a *reconciliation form*. After the adjustments have been made, the balances should agree.

As an example, Ms. Mandy Rankin received her statement, and the balance was shown as $734.62. She noted the following items on the bank statement:

a. A service charge (marked SC) of $10.00.
b. A note for $330.40 (marked E) collected by the bank and deposited to Ms. Rankin's account.
c. An automatic transfer of $80.00 (marked S) to her savings account.

These items are shown in the figure at the top of the next page.

In comparing the bank statement with her check register, she noted that check No. 4563 in the amount of $34.35 had not been paid by the bank. Her final checkbook balance was $459.87.

The reconciliation form is usually printed on the back of the bank statement. The form contains the following parts (see the figure at the bottom of the next page):

a. Bank balance on statement.
b. Less outstanding checks.
c. Plus outstanding deposits.
d. Adjusted bank balance.

FIGURE 9-10

Ms. Rankin's Statement

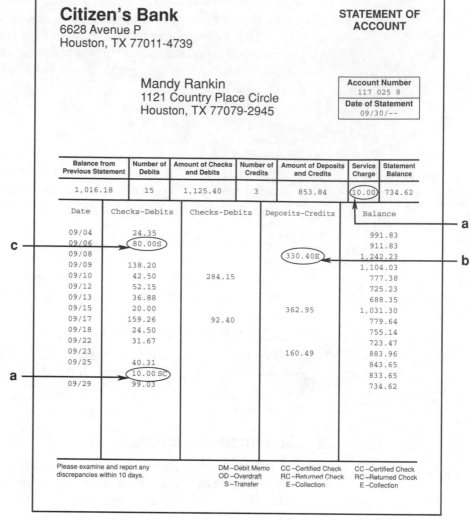

Citizen's Bank
6628 Avenue P
Houston, TX 77011-4739

STATEMENT OF ACCOUNT

Mandy Rankin
1121 Country Place Circle
Houston, TX 77079-2945

Account Number
117 025 8
Date of Statement
09/30/--

Balance from Previous Statement	Number of Debits	Amount of Checks and Debits	Number of Credits	Amount of Deposits and Credits	Service Charge	Statement Balance
1,016.18	15	1,125.40	3	853.84	10.00	734.62

Date	Checks-Debits	Checks-Debits	Deposits-Credits	Balance
09/04	24.35			991.83
09/06	80.00S			911.83
09/08			330.40E	1,242.23
09/09	138.20			1,104.03
09/10	42.50	284.15		777.38
09/12	52.15			725.23
09/13	36.88			688.35
09/15	20.00		362.95	1,031.30
09/17	159.26	92.40		779.64
09/18	24.50			755.14
09/22	31.67			723.47
09/23			160.49	883.96
09/25	40.31			843.65
	10.00 SC			833.65
09/29	99.03			734.62

Please examine and report any discrepancies within 10 days.

DM –Debit Memo
OD –Overdraft
S –Transfer

CC –Certified Check
RC –Returned Check
E –Collection

CC –Certified Check
RC –Returned Check
E –Collection

FIGURE 9-11

Reconciliation Form

Reconciliation of Bank Statement

a → Bank balance on statement — $734.62

b → Less outstanding checks — $34.35

$34.35

$700.27

c → Plus outstanding deposits

d → Adjusted bank balance — $700.27

e → Checkbook balance — $459.87

f → Less unrecorded debits — $80.00

$10.00 — $90.00

$369.87

g → Plus unrecorded credits — $330.40

$330.40

h → Adjusted checkbook balance — $700.27

e. Checkbook balance.

f. Less unrecorded bank charges (debits).

g. Plus unrecorded deposits (credits).

h. Adjusted checkbook balance.

Study the steps used in completing the form.

STEPS

1. Write in the ending bank statement balance.

2. Subtract outstanding checks and add unrecorded deposits.

3. Record adjusted bank balance.

4. Write in last checkbook (from check register or check stub) balance.

5. Subtract bank charges.

6. Add the note collected by the bank and deposited to Ms. Rankin's account.

7. Write in the adjusted checkbook balance.

After studying the figure at the bottom of the previous page, you can see that the balances agree because **d.** and **h.** are equal.

PRACTICE PROBLEMS

Complete the following reconciliation forms.

1. a. Bank statement balance: $97.80

   b. Outstanding checks: No. 1051, $32.18; No. 1062, $15.90

   c. Checkbook balance: $59.72

   d. Bank charges: $10.00

Reconciliation of Bank Statement

Bank balance on statement		_____
Less outstanding checks	_____	
	_____	
	_____	_____
Plus outstanding deposits	_____	
	_____	
Adjusted bank balance		_____
Checkbook balance		_____
Less unrecorded debits	_____	
	_____	
		_____
Plus unrecorded credits	_____	
	_____	_____
Adjusted checkbook balance		_____

2. a. Bank statement balance: $1,177.40
   b. Outstanding checks: No. 1055, $33.00; No. 1040, $17.29
   c. Plus outstanding deposits: $411.90; $55.11
   d. Checkbook balance: $508.02
   e. Bank charges: $10.00
   f. Unrecorded credits: $1,096.10 (note collected by bank)

Reconciliation of Bank Statement

Bank balance on statement

Less outstanding checks

Plus outstanding deposits

Adjusted bank balance

Checkbook balance

Less unrecorded debits

Plus unrecorded credits

Adjusted checkbook balance

Solutions: **1.**

Reconciliation of Bank Statement

Bank balance on statement		$97.80
Less outstanding checks	$32.18	
	$15.90	
		$48.08
		$49.72
Plus outstanding deposits		
Adjusted bank balance		$49.72
Checkbook balance		$59.72
Less unrecorded debits	$10.00	
		$10.00
		$49.72
Plus unrecorded credits		
Adjusted checkbook balance		$49.72

Solutions: **2.**

Reconciliation of Bank Statement

Bank balance on statement		$1,177.40
Less outstanding checks	$33.00	
	$17.29	
		$50.29
		$1,127.11
Plus outstanding deposits	$411.90	
	$55.11	$467.01
Adjusted bank balance		$1,594.12
Checkbook balance		$508.02
Less unrecorded debits	$10.00	
		$10.00
		$498.02
Plus unrecorded credits	$1,096.10	
		$1,096.10
Adjusted checkbook balance		$1,594.12

Math Alert

1. Call three banks, savings and loans, or credit unions. Ask the following questions at each institution:

   a. Do you provide automated teller machines for your customers?
   b. What ATM cards does your institution honor?
   c. What is the charge per transaction?
   d. What are the advantages of having an ATM card versus simply writing checks?

   Write a short paragraph explaining the information you gathered.

2. For each institution you called, compute the following charges.

	Institution Name	Rate per transaction	Number of transactions		Total charges
a.	_____	_____	× 12	=	_____
b.	_____	_____	× 25	=	_____
c.	_____	_____	× 40	=	_____

Study Guide

I. Terminology

page 158	checking account	A bank account that provides a method for payment from funds on deposit.
page 158	checkbook	Includes checks, a record of checks written, and deposits.
page 158	bank statement	A listing of all account transactions, and a final balance provided by the bank (usually monthly).
page 158	bank statement reconciliation	The process of adjusting the balance in a checkbook and that on the bank statement so that the balances agree.
page 158	automated savings transfer	An easy way to add money to your savings account from your checking account.
page 159	direct deposit	Payroll and other checks deposited directly into your account.
page 159	automated teller machine (ATM)	Machines that allow you 24-hour access to your bank account. They require an ID and/or a plastic card which contains a code number similar to a credit card.
page 159	check	Authorizes the bank to pay a certain sum from the checking account to an individual or business.
page 159	payee	The person or business to whom the check is written.
page 159	drawer	The person who writes the check.
page 159	drawee	The drawer's bank.
page 160	deposit slip	The form completed when money is added to a checking account.
page 160	depositor	The person filling out the deposit slip.
page 161	ABA number	The American Bankers Association transit number found on checks and deposit slips.
page 161	blank endorsement	The payee's signature written on the back left end of the check.
page 161	restrictive endorsement	When the check endorsement reads "for deposit only."
page 161	check register or check stub	Records of each check number, date, the amount of each check, deposits made, and balance in the checking account.
page 166	outstanding checks	Checks that have not yet been paid by (cleared) the bank.
page 166	cancelled checks	Checks that have been written and have been paid by the bank.
page 166	bank debits	Bank charges such as service charges.
page 166	service charge	A charge for handling your bank transactions.
page 167	bank credits	Credits to the depositor's account.
page 167	reconciliation form	A form (usually on the back of a bank statement) used to reconcile the statement.

II. The Checking Account

page 161 *Check Register and Check Stubs:* Write the: date; check number if not preprinted; description of transaction; purpose of check; amount of check; amount of deposit, and calculate a new balance.

page 167 *Statement Reconciliation:* Write in the ending bank statement balance; subtract outstanding checks and add unrecorded deposits and other debits; record adjusted bank balance; write in last checkbook balance; subtract bank charges; add credits if any; write in adjusted checkbook balance.

Assignment 1

Name_____ Date _____

Complete the following check register by filling in the cash balances for each transaction.

DATE	CHECK NUMBER	DESCRIPTION OF TRANSACTION	√	AMOUNT OF DEPOSIT		AMOUNT OF CHECK		BALANCE $ – 0 –	
6/15/–		Deposit		950	00			950	00
6/16/–	846	McCray Mgmt Corp Fee				350	00		
6/16/–	847	Money Saver Drugstore Prescriptions				72	40		
6/16/–	848	Citco Oil Co. Gasoline bill				36	90		
6/17/–	849	The Telephone Co. Phone bill				30	30		
6/20/–		Deposit		127	75				
6/22/–	850	Delhi Dept. Store Sweaters				42	20		
6/22/–	851	Bruce Johnston Mowing lawn				12	70		
6/23/–	852	Robert's Florist Roses				9	60		
6/23/–	853	Julie Anderson Child care				164	20		
6/24/–	854	Money Saver Drugstore Prescriptions				29	46		
6/24/–		Deposit		303	14				
6/25/–	855	XYZ Corporation Magazines				37	40		
6/25/–	856	Spring Valley Church Donation				75	00		
6/26/–	857	Our Neighborhood Grocery Groceries				82	40		
6/27/–		Deposit		303	14				
6/27/–	858	Triple A Cleaners Dry cleaning				27	60		
6/28/–	859	Delhi Dept. Store Hosiery				15	30		
6/29/–	860	Bruce Johnston Lawn care				25	00		
6/30/–	861	Julie Anderson Child care				164	20		

1.
2.
3.
4.
5.
6.
7.
8.
9.
10.
11.
12.
13.
14.
15.
16.
17.
18.
19.
20.

Assignment 2

Name_____ Date _____

Calculate the totals and balances on the check stubs. Carry each balance forward to the next numbered check stub.

No. *107* $ *33.10*

July 15, 19 --

TO *Zarif*

FOR *Lawn care*

	Dollars	Cents	
Bal. Bro't. For'd.	800	90	
Amt Deposited			
Total			**1.**
Amt. This Check			**2.**
Bal. Car'd For'd.			**3.**

No. *108* $ *6.55*

July 16, 19 --

TO *S. M. Cooper*

FOR *Weekly paper*

	Dollars	Cents	
Bal. Bro't. For'd.			**4.**
Amt Deposited			
Total			**5.**
Amt. This Check			**6.**
Bal. Car'd For'd.			**7.**

No. *109* $ *31.50*

July 18, 19 --

TO *The Dress-R*

FOR *Skirt*

	Dollars	Cents	
Bal. Bro't. For'd.			**8.**
Amt Deposited			
Total			**9.**
Amt. This Check			**10.**
Bal. Car'd For'd.			**11.**

No. *110* $ *13.45*

July 20, 19 --

TO *Sun Cleaners*

FOR *Dry cleaning*

	Dollars	Cents	
Bal. Bro't. For'd.			**12.**
Amt Deposited			
Total			**13.**
Amt. This Check			**14.**
Bal. Car'd For'd.			**15.**

No. *111* $ *22.70*

July 21, 19 --

TO *Dr. Garcia*

FOR *Eye exam*

	Dollars	Cents	
Bal. Bro't. For'd.			**16.**
Amt Deposited			
Total			**17.**
Amt. This Check			**18.**
Bal. Car'd For'd.			**19.**

No. *112* $ *97.20*

July 22, 19 --

TO *Trent's*

FOR *Tune-up*

	Dollars	Cents	
Bal. Bro't. For'd.			**20.**
Amt Deposited	750	00	
Total			**21.**
Amt. This Check			**22.**
Bal. Car'd For'd.			**23.**

No. *113* $ *72.60*

July 23, 19 --

TO *The Corner Drgst.*

FOR *Prescriptions*

	Dollars	Cents	
Bal. Bro't. For'd.			**24.**
Amt Deposited			
Total			**25.**
Amt. This Check			**26.**
Bal. Car'd For'd.			**27.**

No. *114* $ *113.00*

July 24, 19 --

TO *A. S. Auto*

FOR *Car payment*

	Dollars	Cents	
Bal. Bro't. For'd.			**28.**
Amt Deposited			
Total			**29.**
Amt. This Check			**30.**
Bal. Car'd For'd.			**31.**

No. *115* $ *28.59*

July 25, 19 --

TO *L. & B.*

FOR *Groceries*

	Dollars	Cents	
Bal. Bro't. For'd.			**32.**
Amt Deposited	100	00	
Total			**33.**
Amt. This Check			**34.**
Bal. Car'd For'd.			**35.**

Assignment 3

Name_____ Date _____

Interpret and reconcile the bank statement using the form provided.

Benito Moya received his bank statement from First Junction Bank. His check register balance was $792.37. The outstanding checks were:

273	21.47	**278**	31.15	**282**	62.00
275	18.22	**281**	2.19	**283**	75.00

A $300 deposit had not been recorded by the bank.

First Junction Bank
67 West Ridgeway, P. O. Box 9511
Los Alamos, NM 89544-9511

ACCOUNT NO.
92-118-20
STATEMENT DATE
12/17/--
PAGE NO.
1
CYCLE   SC CODE
12   04

BENITO MOYA
82 CLARK RD.
LOS ALAMOS, NM 87544

( )******( )   PERSONAL (PLUS)

CHECKING STATEMENT SUMMARY ·································
   PREVIOUS STATEMENT 11/19/--, BALANCE OF ············· 1,010.08
   1 DEPOSITS OR OTHER CREDITS TOTALING ····(1 ITEMS)   472.50
   6 CHECKS OR OTHER DEBITS TOTALING ····(6 ITEMS)   307.68
      SERVICE CHARGE AMOUNT ····························   7.00
CURRENT BALANCE AS OF 12/17/-- ·························· 1,167.90

MISCELLANEOUS DEBITS ····································
 DATE   AMOUNT   TRANSACTION DESCRIPTION
 12/17   7.00   REGULAR   SERVICE CHARGE

MISCELLANEOUS CREDITS ··········
 DATE   AMOUNT   TRANSACTION
 12/31   472.50

DAILY BALANCE SUMMARY ··········
 DATE···CHECK NO····AMOUNT
 11/18   272   29.49
 11/21   274   19.25
 11/26   276   38.15
 12/05   277   12.99
 12/12   279   92.00
 12/20   280   115.80

Reconciliation of Bank Statement

Bank balance on statement	_____	1.
Less outstanding checks	_____	
_____	_____	
_____	_____	
_____	_____	2.
	_____	3.
Plus outstanding deposits	_____	
	_____	4.
Adjusted bank balance	_____	5.
Checkbook balance	_____	6.
Less unrecorded debits	_____	7.
	_____	8.
	_____	9.
Plus unrecorded credits	_____	10.
	_____	11.
Adjusted checkbook balance	_____	12.

Assignment 4

Name_____ Date _____

Complete the following word problems. Write your answers in the blanks provided.

1. Marta Fuentes had a balance of $1,200.50 in her checking account. The bank issued her a credit of $505 and charged her $12 for new checks. There were no outstanding checks or deposits. What should her checkbook balance be?

    _____

2. Judy Martin had not checked off check number 112 for $42.50, number 115 for $13.75, and number 121 for $142.33. What was the total of her outstanding checks?

    _____

3. Heng-Che Fu had a checkbook balance of $2,500. The bank collected notes totaling $3,200.50 and charged a service charge of $15. What should her checkbook balance be?

    _____

4. After reconciling his bank statement, Mark Holmes found that he had not recorded a deposit of $93.25. His checkbook balance was $501.20. What is the balance after the checkbook was corrected?

    _____

5. On check stub No. 510, Dallas Johnson found he had transposed $32.10 when it should have been $23.10. His checkbook balance was $677.80. What should it have been?

    _____

6. The bank statement showed a balance of $4,772.12, and Carmen Cruz' checkbook balance showed $4,793.84. What is the difference between the two?

    _____

7. Amalia Herrera wrote a check for $39.50 but she recorded it in her checkbook register as $31.50. Will her running balance show more or less money than she actually has? _____ How much more or less? _____

8. Ruth Richard's bank statement showed a balance of $945.66. There were outstanding checks in the amount of $188.90. What should the adjusted bank balance be?

    _____

9. Eduardo Alvarez' adjusted checkbook balance was $1,899.90. The bank charged him $12 for service charge. What was his checkbook balance before he reconciled it?

    _____

10. Brad Nelson's bank statement balance was $5,600. He had outstanding checks amounting to $3,677.50 and an outstanding deposit of $290.90. What was his adjusted bank statement balance?

    _____

11. Chad had a bank statement balance of $2,900.10. He also showed $238 in out-standing checks and a $480 deposit in transit. The bank debited his account for $10 service charge and $20 NSF charge. His checkbook balance was $3,172.10. What was his check-book balance after he reconciled his bank statement?

    _____

Assignment 5

Name _____ Date _____

Complete the following word problems. Write your answers in the blanks provided.

1. Maria Gonzales showed a $1,181.11 checkbook balance. The bank statement showed a $2,111.46 balance and a $10.00 service charge. Outstanding checks total $940.35. What is Maria's corrected checkbook balance and does it balance to the bank statement?

   _____

2. Paul Hastings had a checkbook balance of $1,345.00. The bank collected a note of $423 and charged him $15 for service charge. What should his checkbook balance be?

   _____

3. Julie Marsh has a checkbook balance of $565.67. She has not recorded these checks: #123-$23.34; #124-$45.88; and #125-$76.05. Julie also needs to record a deposit of $125. What is her checkbook balance now?

   _____

4. On check stub No. 988, Hattie Kaufman found she had transposed $67.54 when it should have been $65.74. Her checkbook balance was $234.11. What is the corrected balance?

   _____

5. Richard Umberling has a checkbook balance of $23.19. The bank statement shows a balance of $128.66. The outstanding checks total $75.44. The bank collected a bill of $12.50. The service charge is $10.00. The difference in the balances is a deposit not recorded in the checkbook. What is the amount?

   _____

6. Penny Schwartz had a balance of $2,668.77 shown on her previous month's statement. The bank collected notes totaling $235.11 and charged her a $25.00 service charge. There were no outstanding checks or deposits. What should her reconciled checkbook balance be?

   _____

7. State National Bank sent Jeff Satterwhite his bank statement showing a $898.34 balance. Jeff's checkbook showed a $651.09 balance. The following checks were outstanding: #867-$23.98; #869-$11.11; #870-$2.98; and #871-$58.80. The monthly service charge was $12.50. The bank collected a note for $25.00. Jeff had forgotten to record a deposit in his checkbook for $137.88. What is the checkbook balance? _____ What is the bank balance? _____

8. Mary Wong had a balance of $1,754.03 in her checking account. The bank collected a bill of $623.12 and charged her $15 for service charge. There were no outstanding checks or deposits. What should her checkbook balance be?

   _____

9. Javier Sanchez had the following MasterCard deposits to his company account: 5/1-$55.76; 5/2-$67.99; 5/3-$128.11; 5/4-$298.04; and 5/5-$96.43. His beginning checkbook balance was $349.09. What is his current balance after these deposits?

   _____

10. On December 31, the checkbook balance of Huffman Company was $45,980.65. The bank statement showed $78,452.10. Checks outstanding totaled $12,878.03. The statement revealed an unrecorded deposit in Huffman's checking account of $19,617.27 as well as a service charge of $23.85. What is the bank statement balance after checks? _____ What is the corrected checkbook balance? _____

11. If a check was written for $92.01 and recorded in the check register as $29.01, the checkbook balance would be "off" by what amount?

    _____

Assignment 6

Name_____ Date _____

Complete the following word problems. Write your answers in the blanks provided.

1. On December 31, Bill's company checkbook showed a $587.98 balance. Bill's bank statement showed a $687.99 balance. Check No. 590 for $123.09 and check No. 592 for $33.44 were outstanding. A $200 deposit was in transit. The bank charged a $15 service charge. The statement showed $2.55 in interest earned and an unrecorded $150 deposit. Can you reconcile the statement? _____ If not, what is the difference in the balances? _____

2. On check stub No. 457, Gil Johnson found he had transposed $237.55 when it should have been $327.55. His checkbook balance was $505.22. What should it have been?

   _____

3. Tito Bates' bank statement showed a balance of $965.51. He had the following outstanding checks: #556-$23.09; #559-$86.44; #560-$1.25; and #561-$161.59. The bank collected a $45 note and charged a $13.50 service charge. Tito's balance in his checkbook was $661.64. What is the checkbook balance after reconciliation? _____

4. The Southside Company checkbook balance was $1,854.39. The bank charged the company $34 for service charges and collected a note for $505.22. The interest earned for the month was $7.86. What is the current checkbook balance after these items?

   _____

5. Reneé Terrace received her bank statement showing a balance of $1,475.98. The outstanding checks totaled $586.21. The deposits in transit were $758.03. She was charged no service charge. Her checkbook balance was $1,647.80. What is her adjusted bank statement balance? _____

6. Sassy Scissors, Inc. had a checkbook balance of $1,557.80. The statement showed a note collected of $100 and a service charge of $10. What is her adjusted checkbook balance? _____

7. Teresa Quitman wrote a check for $176.65, but she recorded it in her checkbook register as $716.65. What is the difference and is the checkbook balance affected and how?

   _____

8. Guy Yates found after reconciling his bank statement that he had not recorded a deposit of $225.15. His checkbook balance was $256.78. What is the balance after he corrected his checkbook? _____

9. Wanda received her bank statement that showed her balance to be $458.92. There were three outstanding checks: #111-$56.43, #115-$12.22; and #118-$39.30. The bank had collected a note for $125 and charged $15 for a service charge. Wanda had a checkbook balance of $190.97. She had forgotten to record a $50 deposit in her checkbook. What is the checkbook and bank balance after reconciliation? _____

10. Katherine Santos had a balance of $764.83 in her checking account. The bank collected a bill of $235.11 and charged her $13 for a service charge. The bank also credited her account with $5.45 in interest earned. There were no outstanding checks or deposits. What should her checkbook balance be?

    _____

11. If the checkbook balance is $511.40, and a deposit of $601.08 was not recorded, what is the correct balance in the checkbook?

    _____

Proficiency Quiz
R E V I E W

Name_____ Date_____

$$\frac{\text{Student's Score}}{\text{Maximum Score}} = \frac{}{40} = \text{Grade}\_\_\_\_\_$$

A. Complete the following partial check register by filling in the cash balance for each transaction.

DATE	CHECK NUMBER	DESCRIPTION OF TRANSACTION	√	AMOUNT OF DEPOSIT		AMOUNT OF CHECK		BALANCE	
								$ 5,952	80
4/01/–	1329	Segal Mfg. Co. Accts. Payable				310	15	310	15	**1.**
4/05/–		Deposit Professional fees		3,671	50			3,671	50	**2.**
4/10/–	1330	Dayton Eq. Co. Dental drills				75	90	75	90	**3.**
4/11/–	1331	Moore Dental Supplies Dental supplies				180	75	180	75	**4.**
4/12/–		Deposit Professional fees		2,190	90			2,190	90	**5.**

B. Calculate the totals and balances on the check stubs, carrying each balance forward to the next numbered check stub.

No. _1_ $ **65.39**
August 4, 19 _--_
TO _ABC Offfice_
FOR _Office supplies_

	Dollars	Cents	
Bal. Bro't. For'd.	620	50	
Amt Deposited	200	00	
Total			**6.**
Amt. This Check			**7.**
Bal. Car'd For'd.			**8.**

No. _2_ $ **25.00**
August 4, 19 _--_
TO _Right-Way Page_
FOR _Phone Expense_

	Dollars	Cents	
Bal. Bro't. For'd.			**9.**
Amt Deposited	—	—	
Total			**10.**
Amt. This Check			**11.**
Bal. Car'd For'd.			**12.**

No. _3_ $ **39.45**
August 5, 19 _--_
TO _Big Deals Furn._
FOR _Office Furniture_

	Dollars	Cents	
Bal. Bro't. For'd.			**13.**
Amt Deposited	—	—	
Total			**14.**
Amt. This Check			**15.**
Bal. Car'd For'd.			**16.**

No. _4_ $ **6.21**
August 6, 19 _--_
TO _US Post Office_
FOR _Postage Expense_

	Dollars	Cents	
Bal. Bro't. For'd.			**17.**
Amt Deposited	—	—	
Total			**18.**
Amt. This Check			**19.**
Bal. Car'd For'd.			**20.**

No. _5_ $ **88.89**
August 7, 19 _--_
TO _Mkt Auto Rental_
FOR _Leasing Expense_

	Dollars	Cents	
Bal. Bro't. For'd.			**21.**
Amt Deposited	451	39	
Total			**22.**
Amt. This Check			**23.**
Bal. Car'd For'd.			**24.**

No. _6_ $ **60.00**
August 7, 19 _--_
TO _Miller Clinic_
FOR _Medical Exp._

	Dollars	Cents	
Bal. Bro't. For'd.			**25.**
Amt Deposited	—	—	
Total			**26.**
Amt. This Check			**27.**
Bal. Car'd For'd.			**28.**

C. Using the information, reconcile the bank statement: Margaret Bolton received her bank statement from the National First Bank showing a balance of $3,811.42. Her check register balance was $3,265.21. The outstanding checks were:

No. 445 $32.11
447   10.95
448   98.10
449   300.50
450   211.05

The bank deducted a service charge of $12.50 and collected a note for $606.00. Margaret deposited $700 not recorded on her bank statement.

Reconciliation of Bank Statement

Bank balance on statement	_____	**29.**
Less outstanding checks	_____	
	_____	
	_____	
	_____	
	_____	_____ **30.**
		_____ **31.**
Plus outstanding deposits	_____	
	_____	_____ **32.**
Adjusted bank balance		_____ **33.**
Checkbook balance		_____ **34.**
Less unrecorded debits	_____	
	_____	_____ **35.**
Plus unrecorded credits	_____	
	_____	_____ **36.**
Adjusted checkbook balance		_____ **37.**

D. Complete the following word problems.

38. Marta Fuentes had a balance of $387.90 in her checking account. The bank collected a bill of $300 and charged her $15 for service charge. There were no outstanding checks or deposits. What should her checkbook balance be? _____

39. Grant Herrera had not checked off check numbers 255 for $87.90, 256 for $89, 257 for $13.25, and 258 for $25.25. What was the total of his outstanding checks? _____

40. On check stub No. 226, Tyler Sims found he had transposed numbers and wrote $21.25 when he should have written $12.25. His checkbook balance was $445.30. What should it have been. _____

CHAPTER

Purchasing and Pricing Merchandise

10

10

Purchasing and Pricing Merchandise

Buying, or "purchasing," as it is generally called in business, is an important function. A manufacturing center buys the materials needed for production, or a retail business buys the merchandise it needs for resale. The retail business attempts to purchase merchandise from manufacturers and wholesalers at the lowest possible cost.

10.1 Terms Used in Purchasing

Manufacturers purchase raw materials to make finished products or merchandise. Manufacturers sell their products or merchandise to other manufacturers, wholesalers, or other sales intermediaries. *Wholesalers* buy merchandise from manufacturers or other wholesalers and sell to *retailers*, who in turn sells directly to you, the consumer.

Just as most of you have probably received a *discount*—a reduction in the retail price of a product—businesses are also allowed discounts. One general kind of discount is a cash discount. A *cash discount* is extended to a business that pays its bill within a designated time period. The *list price* or suggested retail price is the price at which an item is usually sold to the public. *Net cost* or price is the actual amount to be paid by the buyer after the discount is deducted from the retail cost.

Different shipping methods or freight terms may appear on an invoice. You should become familiar with the following commonly used terms.

1. **C.O.D.**—cash on delivery. The purchaser pays cash for the merchandise on delivery.
2. **f.o.b. (shipping point)**—free on board. The buyer pays for shipping the merchandise to its destination.
3. **f.o.b. (destination).** The seller pays for shipping the merchandise to its destination.

10.2 Calculating Cash Discounts

Many businesses offer their customers a cash discount as an incentive to pay their invoices early. Cash discounts are advantageous to both the buyer and the seller. The type of cash discount a customer receives is usually shown under the heading "Terms." *Terms* refer to conditions or terms of payment. When the terms are shown as "2/10, n/30," it means that the customer receives a 2% cash discount if the invoice is paid within ten days of the invoice date. If the customer does not pay within ten days of the invoice date, the net amount without the discount is due within 30 days. Another type of discount is "3/10, 2/30, n/60." This discount means take a 3% discount if paid within ten days, 2% if paid within 30 days, or net amount due if paid within 60 days.

Other common payment terms that may appear on invoices are:

1. **n/EOM**—net due at end of month in which invoice is dated.
2. **n/30EOM**—net due 30 days after end of month in which the invoice is dated.
3. **n/30ROG**—net due 30 days after receipt of goods rather than date of invoice.
4. **C.O.D.**—or "terms cash"— cash is required when the merchandise is delivered. (This term has already been mentioned as a shipping term.)

To calculate the cash discount amount, multiply the total of the invoice by the percent of the cash discount allowed.

■ **TIP**   Remember to use the formula for determining percentage ($P = B \times R$). The total of the invoice is the base, and the percent of cash discount is the rate.

Use the following steps to find the cash discount and net amount due on this invoice if paid by March 25.

EXAMPLE _____

Invoice Date:	March 15
Terms:	2/10, n/30
Invoice Total:	$945.05

STEPS

1. Multiply the invoice total by the percent of the cash discount (2%) to obtain the cash discount.

$945.05
× 0.02
$18.9010 cash discount, rounded to $18.90

2. Subtract the cash discount from the invoice total to obtain the net amount due on the invoice.

$945.05
− 18.90
$926.15 net amount due on invoice

If the invoice is not paid within the 10 days, the full amount of $945.05 must be paid within 30 days after the date of the invoice, or payment is considered past due.

10.3 Using the Complement Method

An alternate method of determining the *net amount due* on an invoice is to multiply the original invoice amount by the complement. A ***complement*** is the difference between 100% and the discount rate. For example, a discount rate of 2% is given on an invoice of $425.72. Subtract the discount rate from 100%. In this case, 100% – 2% = 98%. Thus the amount due is 98% of the $425.72 as shown here:

$$\$425.72 \times 98\% = \$417.21$$

Your answers for some problems may vary by a few cents with the complement method.

PRACTICE PROBLEMS

Determine the cash discount and the net amount paid in the invoices for **1**, **2**, and **3**. Use the complement method to determine the net amount due on invoices **4**, **5**, and **6**. Write your answers in the blanks provided. Round all dollar amounts to the nearest penny.

Invoice Amount	Terms	Cash Discount	Net Amount
1. $392.10	1/10, n/30	_____	_____
2. $1,110.00	2/10, n/60	_____	_____
3. $15,896.00	2/10, n/45	_____	_____
4. $491.11	3/10, n/90		_____
5. $605.05	2/10, n/30		_____
6. $1,690.25	3/10, n/30		_____

Solutions:　**1.** $3.92; $388.18;　**2.** $22.20; $1,087.80;　**3.** $317.92; $15,578.08;
4. $476.38;　**5.** $592.95;　**6.** $1,639.54

10.4 Returned Merchandise

If merchandise is returned, the returned amount is deducted from the invoice amount before the cash discount is calculated. Study the following example where merchandise was returned.

EXAMPLE _____

Invoice Amount:　$3,688.00
Terms:　　　　　　2/10, n/30

A credit memorandum shows that the cost of the returned merchandise was $1,175.00. Follow these steps to verify the invoice amount subject to the returned goods and the discount. Then find the net amount due.

STEPS

1. Subtract the cost of returned merchandise (shown on the credit memorandum) from the invoice amount.

$$\begin{array}{r} \$3,688.00 \\ -\ 1,175.00 \\ \hline \$2,513.00 \end{array} \text{ basis for cash discount}$$

2. Obtain the discount amount by multiplying the amount subject to discount by 2%.

$$\begin{array}{r} \$2,513.00 \\ \times\ 0.02 \\ \hline \$50.26 \end{array} \text{ cash discount}$$

3. Subtract the discount amount from the invoice amount less the cost of returned merchandise.

$$\begin{array}{r} \$2,513.00 \\ -\ 50.26 \\ \hline \$2,462.74 \end{array} \text{ invoice net amount}$$

■ **TIP** Amount of returned merchandise is subtracted from the invoice amount before a cash discount is taken.

PRACTICE PROBLEMS

Determine the net amount due on these invoices assuming that the invoices have been paid within the discount time period and merchandise has been returned. Write your answers in the blanks provided.

Invoice Amount	Terms	Cost of Returned Merchandise	Net Amount
1. $4,689.50	2/10, n/30	$28.52	_____
2. $1,122.25	2/10, n/30	$87.95	_____
3. $990.85	3/10, n/45	$16.09	_____

Solutions: **1.** $4,567.76; **2.** $1,013.61; **3.** $945.52

10.5 Shipping and Insurance Charges

Cash discounts are not calculated on shipping or insurance charges. Therefore, to determine the total amount due on an invoice, you must add shipping or insurance charges *after* you have calculated the cash discount. If these charges have been added to the invoice already, they must be deducted from the total to determine the basis on which the discount is calculated. Study the following example.

EXAMPLE

An invoice of $455.20 has terms of 2/10, n/30 with a shipping charge of $19.50. Assume that the invoice is paid within the 10-day period. Let's see how the net amount due on the invoice is determined using these steps.

STEPS

1. Calculate the cash discount.

$$\begin{array}{r} \$455.20 \\ \times\ 0.02 \\ \hline \$9.10 \ \text{cash discount} \end{array}$$

2. Subtract the cash discount from the invoice amount.

$$\begin{array}{r} \$455.20 \\ -\ 9.10 \\ \hline \$446.10 \ \text{invoice amount minus discount} \end{array}$$

3. Add shipping charges.

$$\begin{array}{r} \$446.10 \\ +\ 19.50 \\ \hline \$465.60 \ \text{amount due on invoice} \end{array}$$

If the shipping charges have been added to the invoice amount, you would first subtract the shipping charges from the invoice amount, take the discount, and then add the shipping charges to the net invoice amount to obtain the total amount due on the invoice.

Remember that a cash discount is not allowed if an invoice is not paid within the stipulated time as defined by the terms. For example, an invoice amounting to $830.79 is dated April 29 and offers terms of 3/10, n/30 and has shipping charges of $22.29. Because the invoice was paid May 12, no cash discount is allowed.

$$\begin{array}{r} \$830.79 \\ +\ 22.29 \\ \hline \$853.08 \ \text{total amount due on invoice} \end{array}$$

PRACTICE PROBLEMS

Determine the cash discount and the net amount due on these invoices assuming that the invoices have been paid within the discount time period. Write your answers in the blanks provided.

Invoice Amount	Terms	Shipping Charges	Cash Discount	Net Amount Due
1. $1,622.00	2/10, n/60	$21.50	_____	_____
2. $176.00	2/15, n/30	$15.86	_____	_____
3. $685.79	2/10, n/30	$17.50	_____	_____

Solutions: **1.** $32.44; $1,611.06; **2.** $3.52; $188.34; **3.** $13.72; $689.57

10.6 Calculating Sales Tax

A *sales tax* is usually a specified percent charged when merchandise is sold to the customer. If you live in an area requiring a sales tax, you probably know that this tax is levied or set by your city, county, and/or state. A sales tax may be calculated on the basis of the unit of merchandise rather than the total purchase price. An example of sales tax calculated on the unit of merchandise is gasoline, which is taxed by the gallon.

Calculating a sales tax is simple because the process is basic multiplication and addition; however, the taxing agency usually provides a table with the tax already computed for you. Let's learn how it is calculated. Suppose the invoice total is $1,589.93. The sales tax is 6%; thus, the amount of tax can be figured using these steps.

STEPS

1. Multiply the invoice amount by the tax rate.

$$\begin{array}{r} \$1,589.93 \\ \times\ 0.06 \\ \hline \$95.40\ \text{tax} \end{array}$$

2. Add the tax to the invoice amount.

$$\begin{array}{r} \$1,589.93 \\ +\ 95.40 \\ \hline \$1,685.33\ \text{total due on invoice} \end{array}$$

10.7 Determining the Invoice Total

Working with invoices requires your close attention in determining or verifying the amounts. First, you calculate the total cost for all items. Then the sales tax, if any, is added to determine the invoice total. If a cash discount has been earned, it is deducted from the amount before the sales tax, then the sales tax is added to obtain the total due. When sending an invoice, use the following steps.

Sending an Invoice

Follow these steps when you are computing an invoice to send to a customer:

STEPS

1. Extend (multiply) each item to obtain total; add extended amounts.
2. Calculate and add sales tax.
3. Add any shipping and insurance charges to obtain a final total invoice.

Paying an Invoice

Follow these steps to verify an invoice when you are paying it:

STEPS

1. Extend (multiply) each item to obtain total; add extended amounts.
2. Calculate and subtract cash discounts (if earned) on the total extended amount.
3. Calculate and add sales tax.
4. Add shipping and insurance charges to obtain a final total invoice.

10.8 Terms Used in Pricing Merchandise

Selling price (sometimes called *retail price*) is the price at which merchandise is sold to retail customers. The difference between the cost of the merchandise to the retailer and the selling price is known as **markon** (sometimes called *markup*) or **gross profit Operating expenses** are the expenses incurred by a business, such as rent, utilities, wages and salaries, supplies, advertising, etc. **Net profit** is the amount left over after expenses have been paid. **Markdown** is a reduction from the regular retail or original selling price of merchandise.

10.9 Basic Pricing Equations

It will be helpful if you study three basic pricing equations showing the relationship of selling price, markon, and cost.

Formula 1: Selling price

Selling Price (SP) = Cost (C) + Markon (M)

Formula 2: Markon

Markon (M) = Selling Price (SP) – Cost (C)

Formula 3: Cost

Cost (C) = Selling Price (SP) – Markon (M)

Let's apply these formulas to the next examples to determine the dollar and cent figures.

PRACTICE PROBLEMS

Use the basic equations to determine the selling price, cost, or markon in these problems. Fill in the missing parts by writing your answers in the blanks provided.

	Cost	Selling Price	Markon
1.	$23.90	_____	$15.00
2.	$35.90	$73.40	_____
3.	_____	$568.95	$126.70
4.	$274.75	_____	$120.30
5.	$621.65	$730.50	_____
6.	_____	$44.79	$18.70

Solutions: **1.** SP = $38.90; **2.** M = $37.50; **3.** C = $442.25;
4. SP = $395.05; **5.** M = $108.85; **6.** C = $26.09

10.10 Determining Markdown Sale Price

Almost everyone wants to take advantage of special sales where merchandise has been marked down or reduced. Markdown is the difference between the original selling price and the reduced selling price. Markdown may be expressed as a percent, such as 20% off, or as a fraction, such as $\frac{1}{3}$ off. The original selling price of the merchandise is reduced for special sales.

■ **TIP**   The markdown is based on the original selling price. For example, if the markdown is 20%, the 20% is calculated on the original selling price.

You can use the steps given here to determine the sale price when the original selling price (*SP*) and markdown rate (*MD%*) are known.

EXAMPLE _____

An employee at The Crest Christmas Store has been asked to mark down the holiday decorations 40%. The decorations originally sold for $27.50.

STEPS

1. Obtain the amount of markdown by multiplying selling price by the markdown rate.

$27.50 × 0.40 = $11.00 markdown amount

2. Obtain the sale price by subtracting the markdown amount from the original selling price.

$27.50 − $11.00 = $16.50 marked down sale price

■ **TIP**   The complement method may be used to determine only the sale price.
For example:
1. 100% − 40% = 60% or 0.60 complement
2. $27.50 × 0.60 = $16.50

⎯ PRACTICE PROBLEMS ⎯

Calculate the sale price on these items. You may use the steps above or the complement method shown in the tip. Write your answers in the blanks provided.

Original Selling Price	Markdown Rate	Sale Price	Original Selling Price	Markdown Rate	Sale Price
1. $562.56	30%	_____	**2.** $364.50	25%	_____
3. $57	15%	_____	**4.** $142	30%	_____
5. $95.45	20%	_____	**6.** $102	33%	_____
7. $149	30%	_____	**8.** $1,750.50	40%	_____
9. $89.75	20%	_____	**10.** $36.54	15%	_____

Solutions: **1.** $393.79; **2.** $273.38; **3.** $48.45; **4.** $99.40; **5.** $76.36; **6.** $68.34; **7.** $104.30; **8.** $1,050.30; **9.** $71.80; **10.** $31.06

MATH ALERT

A. Assume you are paying the invoice below. Verify the invoice amounts to see if you obtain the same total shown. If you are having difficulty, review multiplication in Chapter 2 and the concepts presented in this chapter.

■ **TIP** List price means the cost of one particular item.

Oshima Business Supplies O_BS Invoice
8934 Madison Road
Columbus, OH 43219
(513) 555-9522

Sold To: Donnelley's Home Improvements   **Date:** January 18, 19--
      4323 Mt. Carmel-Tobasco Road   **Ship By:** Jettison Freight
      Columbus, OH 43219   **Terms:** 2/10, net 30

Quantity	Cat. No.	Description	List Price	Total
8	61320	Swingline staplers #80-A	10.95	87.60
5	3-A-102	Northwell desks, oak, 60 x 30	695.60	3,478.00
				3,565.60
		Cash discount 2% for early payment		71.31
				3,494.29
		Sales tax 4%		139.77
				3,634.06

B. Use the information given in the ad below to answer the questions which follow.

1. If you read this ad in the newspaper, could you calculate if the items were really marked down 20%-40%?
   A. What is 20% off of $19.95?
   B. What is 40% off of $45.95?
2. Is $19.99 a 50% markdown from the original selling price of $40.00?

Study Guide

I.   Terminology

page 182	manufacturer	A company that provides raw materials to make finished products or merchandise.
page 182	wholesaler	A company that purchases merchandise from manufacturers or other wholesalers for resale.
page 182	retailer	A company that sells directly to the consumer.
page 182	discount	A reduction in the retail price of a product.
page 182	list price	A suggested retail price at which an item is usually sold to the public.
page 182	cash discount	A discount on goods extended to a business that pays its invoice within a designated time period.
page 182	C.O.D	Cash on delivery. The purchaser pays cash for the merchandise on delivery.
page 182	f.o.b. (shipping point)	Free on board. The buyer pays for shipping the merchandise to its destination.
page 182	f.o.b. (destination)	Free on board. The seller pays for shipping the merchandise to its destination.
page 183	n/EOM	Net due at end of month in which invoice is dated.
page 183	n/30EOM	Net due 30 days after end of month in which invoice is dated.
page 183	n/30ROG	Net due 30 days after receipt of goods rather than date of invoice.
page 184	complement	The difference between 100% and the discount rate.
page 187	sales tax	A specified percent charged when merchandise is sold to the customer.
page 189	selling price	The price at which merchandise is sold to retail customers.
page 189	markon	The difference between the cost of the merchandise to the retailer and the selling price. Also called markup or gross profit.
page 189	operating expenses	The expenses incurred by a business, such as rent, utilities, wages, and salaries.
page 189	net profit	The amount left over after expenses have been paid.
page 189	markdown	A reduction from the regular retail or original selling price of merchandise.

II.   Cash Discounts

page 183   *Calculating Cash Discounts:* Multiply the invoice total by the percent of the cash discount to obtain the amount of the cash discount; subtract the cash discount from the invoice total to obtain the net amount due on the invoice.

Example: Invoice date: March 15
Terms: 2/10, n/30
Invoice total: $945.05

$945.05 × 0.02 = $18.90
$945.05 − $18.90 = $926.15 net amount due on invoice

Practical Math Applications

page 184 *Using the Complement Method:* Multiply the original invoice amount by the complement to compute the amount due on the invoice.

Example: Invoice date: March 15
Terms: 2/10, n/30
Invoice total: $945.05

100% − 2% = 98% or 0.98
$945.05 × 0.98 = $926.149 or $926.15 net amount due on invoice

III. Returned Merchandise

page 184 If merchandise has been returned, the returned amount is deducted from the invoice before the cash discount is calculated. A credit memorandum shows the cost of the returned merchandise. You should verify the net amount due by subtracting the cost of the returned merchandise from the invoice amount; obtain the discount amount by multiplying the amount subject to the discount by the percent of the discount; then subtract the discount amount from the invoice amount less the cost of returned merchandise.

Example: Invoice amount: $3,688.00
Terms: 2/10, n/30
Credit memo: $1,175.00

$3,688.00 − $1,175.00 = $2,513.00 subject to cash discount
$2,513.00 × 0.02 = $50.26 cash discount
$2,513.00 − $50.26 = $2,462.74 amount due on invoice

IV. Shipping Charges

page 186 Cash discounts are not calculated on shipping or insurance charges. Calculate the cash discount; subtract the cash discount from the invoice amount; add shipping charges.

Example: Invoice amount: $455.20
Terms: 2/10, n/30
Shipping charges: $19.50

$455.20 × 0.02 = $9.10 cash discount
$455.20 − $9.10 = $446.10 amount due on discounted invoice
$446.10 + $19.50 = $465.60 total due on invoice

V. Sales Tax

page 187 Multiply the total due on the invoice by the tax rate; add the tax to the invoice amount.

Example: Amount due on invoice: $1,589.93
Sales tax percent: 6%

$1,589.93 × 0.06 = $95.40 tax
$1,589.93 + $95.40 = $1,685.33 total due on invoice

VI. Determining the Invoice Total

page 188 *When Sending an Invoice to a Customer:* Extend (multiply) each item to obtain total; add extended amounts; calculate and add sales tax; add shipping and insurance charges (if any) to obtain invoice total.

page 188 *Verifying an Invoice When You Are Paying It:* Extend (multiply) each item to obtain total; add extended amounts; calculate and subtract cash discounts (if earned) on the total extended amount; calculate and add sales tax; add shipping and insurance charges (if any) to obtain a final total due on invoice.

VII. Basic Pricing Equations

page 189
    Formula 1: Selling Price
Selling Price (*SP*) = Cost (*C*) + Markon (*M*)
$119.00 (*SP*) = $99.95 (*C*) + $19.05 (*M*)

page 189
    Formula 2: Markon
Markon (*M*) = Selling Price (*SP*) – Cost (*C*)
$19.05 (*M*) = $119.00 (*SP*) – $99.95 (*C*)

page 189
    Formula 3: Cost
Cost (*C*) = Selling Price (*SP*) – Markon (*M*)
$99.95 (*C*) = $119.00 (*SP*) – $19.05 (*M*)

VIII. Determine Markdown Sale Price

page 190
Obtain the amount of markdown by multiplying the original selling price by the markdown rate; obtain the sale price by subtracting the markdown amount from the original selling price.

Example: Holiday decorations selling price: $27.50
              Markdown rate: 40%
              Marked down sale price: ?

$27.50 × 0.40 = $11.00
$27.50 – $11.00 = $16.50 marked down sale price

Assignment 1

Name _____ Date _____

Write your answers in the blanks provided.

A. Find the cash discount amount and the net amount due on these invoices. Assume each invoice is paid within the discount period. Round your answers to the nearest penny.

Invoice Amount	Terms	Cash Discount	Net Amount Due
$623.50	2/10, n/30	1. _____	2. _____
$4,635.65	3/10, n/60	3. _____	4. _____
$480	2/10, n/30	5. _____	6. _____
$385	1/10, n/30	7. _____	8. _____
$637	2/10, n/30	9. _____	10. _____
$4,114.75	2/10, n/30	11. _____	12. _____
$6,250.81	1/15, n/45	13. _____	14. _____
$2,044.17	2/10, n/30	15. _____	16. _____
$3,374.56	3/10, n/30	17. _____	18. _____
$4,501.08	2/10, n/30	19. _____	20. _____

B. Determine the cash discount and net amount due on these invoices. Round your answers to the nearest penny.

Invoice Amount	Terms	Paid Within Discount Period	Returned Merchandise	Shipping Charges	Cash Discount	Net Amount Due
$534.78	2/10, n/45	yes	none	$13.25	21. _____	22. _____
$638	1/10, n/30	yes	$33.50	none	23. _____	24. _____
$213.72	2/10, n/30	no	none	$15.50	25. _____	26. _____
$369	3/10, n/30	no	$57.40	none	27. _____	28. _____
$1,256.35	2/10, n/45	yes	none	$18.75	29. _____	30. _____
$471.08	1/10, n/30	no	$13.56	$11.82	31. _____	32. _____
$103.68	2/10, n/30	yes	none	none	33. _____	34. _____

Assignment 2

Name_____ Date _____

Write your answers in the blanks provided.

A. Find the sales tax and the net amount due on these invoices. Round your answers to the nearest penny.

Invoice Amount	Sales Tax Percent	Sales Tax Amount	Net Amount Due
$3,211.90	7%	1. _____	2. _____
$1,500.25	8%	3. _____	4. _____
$380.98	5.5%	5. _____	6. _____
$234.54	7.5%	7. _____	8. _____
$867.75	6%	9. _____	10. _____
$1,111.10	8.75%	11. _____	12. _____
$1,800.33	5%	13. _____	14. _____

B. Determine the sales tax and net amount due on these invoices. Round your answers to the nearest penny.

Invoice Amount	Sales Tax Percent	Sales Tax Amount	Shipping Charges	Net Amount Due
$1,443.22	5%	15. _____	$18.50	16. _____
$550.15	7%	17. _____	$7.75	18. _____
$775.30	8%	19. _____	$15.00	20. _____
$323.12	7.5%	21. _____	$4.50	22. _____
$9,000	7%	23. _____	$25.90	24. _____

C. Find the selling price, cost, or markon amount for the following items:

	Cost	Selling Price	Markon Amount
25.	_____	$52.98	$15
26.	$112.10	_____	$38.90
27.	$38	$45	_____
28.	$104.30	$118.30	_____
29.	$92.70	_____	$19.00
30.	_____	$136.80	$45.01

Assignment 3   Name_____ Date _____

Complete the following invoice by entering the amounts in the blanks provided.

ROCKFORD OFFICE SUPPLIES
4211 Beach Street, Beaumont, KY 42321

SOLD TO: Craig Glass & Mirror, Inc.
2366 East 114th Street
Beaumont, KY 42321

INVOICE NO.: 1255-2344
DATE: January 25, 19--

ORDER NO. 4588	SALES ASSOCIATE A. Miller	TERMS 3/10, net 30	DATE SHIPPED Jan. 15

QUANTITY	STOCK NO.	DESCRIPTION	UNIT PRICE	TOTAL	
6	L-2344	Gross, No. 2 pencil	$5.95	_____	1.
4	P-4665	Elect. pencil sharpener	$29.95	_____	2.
3	D-1233	Executive desk	$329.95	_____	3.
3	D-1234	Executive chair	$129.95	_____	4.
6	W-3324	Wastebasket	$12.95	_____	5.
5	C-5544	Computer desk	$388.00	_____	6.
5	C-5543	Computer hutch	$99.95	_____	7.
3	C-5542	Printer table	$149.00	_____	8.
5	D-1235	Computer chair	$128.50	_____	9.
5	C-5541	486 computer system	$2,955.00	_____	10.

INVOICE TOTAL	$_____	11.
BALANCE	$_____	12.
CASH DISCOUNT	$_____	13.
BALANCE	$_____	14.
6% SALES TAX	$_____	15.
BALANCE	$_____	16.
SHIPPING CHARGES	$_____89.00	
TOTAL AMOUNT DUE	$_____	17.

Assignment 4

Name_____ Date _____

Complete the following word problems. Write your answers in the blanks provided.

Trisha purchased 3 blouses at $32.50 each, 2 packages of 6 pairs of socks costing $7.50 per package, and 4 pair of hosiery costing $2.19 each. Sales tax was 6.5%.

1.   How much sales tax did Trisha pay? _____

2.   What was Trisha's total invoice including tax? _____

Bennett priced two computer systems at Riley's Computer Store. An 80486/50 processor, 4MB RAM, 360MB hard drive, 64K cache memory, and desktop case priced at $3,399; and another 80486/60 processor, with remaining features the same as the 486/50 priced at $3,699. Shipping costs would be the same for each system — $42.50. Sales tax would be 7%.

3.   What was the sales tax for the 80486/50? _____

4.   What was the total invoice price for the 80486/50? _____

5.   What was the sales tax for the 80486/60? _____

6.   What was the total invoice price for the 80486/60? _____

Bennett decided to purchase another 80486/60 computer. Before the computer was delivered, he decided to add more memory, costing $82.

7.   Recalculate the sales tax including the additional memory. _____

8.   Recalculate the total invoice including the additional memory. Shipping charges remain the same. _____

Bennett also purchased several software programs for the new system. They were: a desktop publishing program for $495, a database program for $419, a spreadsheet program for $399, a word processing program for $245, and a utility program for $114. Sales tax was 7%.

9.   What was the sales tax for all his software? _____

10.   What was the total invoice for the software? _____

When Bennett attempted to install the desktop publishing program on his computer, he found the disks were bad. Bennett decided to return the desktop publishing program to the store for a complete refund.

11.   What was the cost of the desktop publishing program? _____

12.   What was the amount of sales tax refunded? _____

13.   What was the new total invoice amount Bennett spent on all of the software after he returned the desktop publishing program? _____

Assignment 5

Name_____ Date _____

Complete the following problems. Write your answers in the blanks provided. Round answers to the nearest cent where needed.

1. Carlos Mendez ordered office equipment totaling $4,320. Sale terms were 3/10, n/30 FOB (Shipping point). The shipping charges were $23.00. If Carlos pays the invoice within the discount period, what does Carlos owe?

   _____

2. Cartland Office Supply received a $1,298.33 invoice dated 6/15/--. The $1,298.33 included $120 freight. Terms were 2/10, n/60. If Cartland pays the invoice on June 21, what will it pay? _____
   If Cartland pays the invoice on September 27, what will it pay? _____

3. Tate Tool and Die ordered new equipment for the machinists. There were 3 jigsaws @ $129.99 ea., 2 sanders @ $59.99 ea. and 4 circular saws @ $79.95 ea. The sales tax rate is 7.25%. What will Tate Tool and Die pay after the invoice is totaled including sales tax?

   _____

4. If Tate Tool and Die returns 1 jigsaw and 1 circular saw, what would the total invoice amount be including sales tax? _____

5. If the terms for the previous word problem (#4) were 3/10, n/30, what would the total invoice amount be including sales tax if paid within the discount time? _____

6. Macy's Office Supply wants to buy a new supply of desk chairs. Manufacturer A offers a 2/10, n/30 discount with FOB (shipping point) of $85.00. Manufacturer B offers a 1/10, n/30, with FOB (destination). The merchandise totals $1,697.06. Assuming you qualify for all discounts and want the lowest price, which manufacturer would you buy from? Why?

   _____

7. Jensen Supply received an invoice dated 8/15/--. The invoice had a $6,500 total that included $225 freight. Terms were 4/10, 3/30, n/60. Jensen pays the invoice on August 24. What amount does Jensen owe? _____
   If Jensen waits and pays it on August 27, what amount will Jensen owe? _____

8. In Exercise 7, the bookkeeper found that $45.56 tax was accidentally charged to Jensen Supply. What should the total be now for a 10-day discount after the tax is removed? _____ A 30-day discount? _____

9. Perkins Inc. ordered new office furnishings for the new president of the company. A purchase order was written for 1 desk @ $298.99, 1 chair @ $99.99, 2 side chairs @ $129.99 ea., 1 coffee table @ $109.00 and 2 lamps @ $69.99 ea. The sales tax rate is 8.25% and shipping is $125.00. The terms are 1/10, n/30. What will Perkins owe the furniture company assuming he pays within 10 days? _____

10. Another office furniture store offered a 4/10, n/30 discount for the same office furniture Perkins ordered previously in Exercise 9. The list price on each item was the same except the coffee table which was $139.99. The shipping is FOB destination. What would the total be for the invoice if Perkins orders from this office furniture store assuming he pays within 10 days? _____

11. The cost of a file cabinet is $75, and the selling price is $119. Calculate the amount of markon. _____

Assignment 6

Name_____ Date _____

**Complete the following problems. Write your answers in the blanks provided.
Round answers to the nearest cent where needed.**

1. Mrs. Southland received an invoice dated April 9, 19--, with terms 2/10, n/30, amounting to $732. She paid the bill on April 12. How much was the cash discount? _____
How much did Mrs. Southland pay?

_____

2. Peter Mays ordered office supplies for the month for his company. His order is as follows: 10 reams copy paper @ $3.50 each; 25 boxes of pencils @ $2.75 each; 1 case of Liquid Paper @ $32.49; 10 printer ribbons @ $7.79 each and 6 boxes of paper clips @ $1.25 each. The sales tax is 7.75%. What is the total of the invoice?

_____

3. If Peter returns 1 ream of copy paper and 3 printer ribbons, what is the new invoice total before sales tax?

_____

4. Quick Copy received a shipment of copier toner from Tyson, Inc. The bill of lading was marked FOB destination. Who paid the freight?

_____

5. How much would have to be paid on an invoice for $789 with terms 4/10, 3/30, 2/60, n/30 if the invoice is paid 7 days after the invoice date? _____
How much if the invoice is paid 31 days after the invoice date? _____
How much if the invoice is paid 75 days after the invoice date? _____

6. Jessica Hunt received an invoice in the amount of $1,298.30 that included $125 in freight and $76.80 in sales tax. The terms of the invoice were 1/10, n/30. How much does Jessica owe if she pays within the discount time?

_____

7. Gary Ewing orders books for Baily's Bookstore. This week he orders 4 cases of best-selling mysteries @ $19.95 ea.; 3 cases of romance novels @ $14.76 ea.; 1 case of western fiction @ $12.33 ea. and 10 cases of the best-selling autobiography @ $13.96 ea. There is no sales tax since this merchandise will be sold and tax collected at that time. What is the total of the invoice if FOB shipping point charges are $45?

_____

8. Gary finds 1 case of mysteries and 2 cases of autobiographies are damaged, and he must return them for credit; $3.50 in shipping will be refunded by the publisher. What is the total amount of the credit?

_____

9. Wing Manufacturing received an invoice dated 1/10/--. The invoice had a $4,231.11 total that included $100 freight. Terms were 3/10, 2/30, n/60. Wing Manufacturing pays the invoice on 2/11/--. What amount does Wing Manufacturing owe?

_____

10. Lee Chang ordered a new fan for her office. The price of the fan is $79.99. The shipping charge is $25 and the sales tax rate is 6.25%. The terms of the invoice are 1/10, n/30. If Lee pays within the discount time, what is the amount she will owe?

_____

Proficiency Quiz
R E V I E W

A. Find the cash discount and the net amount due on these invoices. Assume each is paid within the discount period. Be certain to add commas and dollar signs where needed.

Invoice Amount	Terms	Cash Discount Amount	Net Amount Due
$2,566	2/10, n/30	1. _____	2. _____
$3,150	2/10, n/30	3. _____	4. _____
$1,652	3/10, n/30	5. _____	6. _____
$2,989	2/10, n/30	7. _____	8. _____
$1,928	3/10, n/30	9. _____	10. _____
$3,000	2/10, n/30	11. _____	12. _____
$1,678	3/10, n/30	13. _____	14. _____
$2,223	3/10, n/30	15. _____	16. _____
$1,809	2/10, n/30	17. _____	18. _____
$2,050	3/10, n/30	19. _____	20. _____

B. Determine the cash discount and net amount due on these invoices. Assume each invoice is paid within the discount period. Add dollar signs.

Invoice Amount	Terms	Returned Merchandise	Shipping Charges	Cash Discount	Net Amount Due
$600.05	2/10, n/45	none	$10.25	21. _____	22. _____
$219.45	1/10, n/30	$32.50	none	23. _____	24. _____
$123.22	3/10, n/30	$10.55	$21.50	25. _____	26. _____
$751.98	2/10, n/30	none	$14.75	27. _____	28. _____
$398.65	3/10, n/45	$22.22	$16.45	29. _____	30. _____

C. Find the sales tax and the net amount due on these invoices. Add dollar signs, and insert commas where appropriate. Round answers to the nearest hundredth.

Invoice Amount	Sales Tax Percent	Sales Tax Amount	Shipping Charges	Net Amount Due
$254.87	5%	31. _____	$20.00	32. _____
$307.99	4%	33. _____	$45.45	34. _____
$1,989.99	7%	35. _____	$31.50	36. _____
$976.44	8%	37. _____	$15.95	38. _____
$195.20	6.5%	39. _____	$3.50	40. _____
$525.00	7.75%	41. _____	$16.60	42. _____

D. Calculate the markon rate based on the selling price. Round your answers to the nearest penny.

Cost		Selling Price		Markon Rate
$56.50	43. _____			$49.95
$35.90		$80.85	44. _____	
45. _____		$582.99		$219.90
$275.10	46. _____			$130.30
$625.25		$775.75	47. _____	

E. Complete the following invoice. Enter the amounts in the blanks provided. Add commas where necessary.

SMITH OFFICE SUPPLIES
2912 Connor Street, Dallas, TX 72225

SOLD TO: C and E Store
1245 North Expressway
Dallas, TX 75222

INVOICE NO.: 3455
DATE: April 5, 19--

ORDER NO. 1888	SALES ASSOCIATE B. Wells	TERMS 3/10, net 30	DATE SHIPPED Apr. 1, 19--

QUANTITY	STOCK NO.	DESCRIPTION	UNIT PRICE	TOTAL	
5	2952	Gross-file folders	$7.85	_____	48.
2	1919	Computer tables	$250.55	_____	49.
6	5544	Computer desks	$388.00	_____	50.

INVOICE TOTAL	$_____	51.
BALANCE	$_____	52.
CASH DISCOUNT	$_____	53.
BALANCE	$_____	54.
8% SALES TAX	$_____	55.
BALANCE	$_____	56.
SHIPPING CHARGES	$_____65.00	57.
TOTAL AMOUNT DUE	$_____	58.

CHAPTER 11

Gross Earnings and Payroll Deductions

DATE	EMPLOYEE NO.	REGULAR DAYS	REGULAR HOURS	OVERTIME	SICK TAKEN	VACATION TAKEN	PERSONAL LEAVE TAKEN
/84	000391	5	40	00.00	00.00	00.00	08.00

			CHECK NO.	NET AMOUNT	ACCRUED SICK	ACCRUED VACATION	ACCRUED PERSONAL LEAVE
			————	267.00	————	80.00	16.00

EARNINGS

EARNINGS	CURRENT	YEAR-TO-DATE
SS PAY	420.00	9,500

DEDUCTIONS

DEDUCTION	CURRENT	YEAR-TO-DATE
FICA	28.00	640.00
FED. TAX	53.00	1300.00
STATE TAX	12.00	283.00
CITY TAX	9.00	190.00
BEN.	26.00	
CU	25.00	

11

Gross Earnings and Payroll Deductions

Payroll is usually one of the largest operating expenses a company has. Employers must maintain accurate payroll records for three reasons. First, the company must keep accurate records because wages and salaries are income tax deductions for the employer. Second, data must be collected in order to compute earnings for each employee for each payroll period. Third, information must be provided to complete federal and state payroll reports that employers are required to keep by law. As an employee, you should understand how your employer makes the computations pertaining to employee paychecks. Let's learn the terminology used in computing gross earnings.

11.1 Terms Used in Computing Gross Earnings

Here are some common terms used in working with payroll information.

Compensation. Salary, wage, pay, or benefits received for the performance of a service.

Gross Earnings. The total amount of an employee's pay before deductions.

Fair Labor Standards Act. An act of law (sometimes called the Wage and Hour Law) establishing minimum wages and requiring employers whose firms are involved in interstate (from state to state) commerce (sale of goods) to pay their employees time and one-half for all hours worked in excess of 40 hours per week. The act also provides that certain employees (management and supervisory) are exempt from its regulations.

Time And-a-Half. One and one-half times an employee's hourly rate.

Double Time. Twice an employee's hourly rate.

Straight Time. Usually the first 40 hours worked per week. However, this number may vary from company to company. In this chapter, 40 hours will be used to represent straight time.

Overtime. All time worked in excess of straight time (over 40 hours for purposes of this chapter unless otherwise stated).

Salaried. A term applied to administrative or managerial personnel who are generally paid a salary and who are not paid for overtime.

Hourly Paid. Personnel who are paid wages by the number of hours worked.

Commission. Earnings paid an employee based on a percent of total sales.

Piecework Wage System. Paying an employee according to the number of units produced.

11.2 Calculating Gross Earnings

The calculations for gross earnings vary depending on how the employee is paid. Let's learn how to compute gross earnings using each method. In all computations dealing with money amounts, round up to the nearest penny.

Compensation by Hourly Wage: Straight Time

As previously mentioned, most employers consider 40 hours per week to be straight time. To compute Terrie McDowell's weekly salary (gross earnings) at $9.50 per hour, multiply the number of hours worked by the hourly rate.

EXAMPLE _____

40 × $9.50 = $380 per week (gross salary)

PRACTICE PROBLEMS

Compute the weekly gross earnings for the following employees. Write your answers in the blanks provided.

1. John Van Gilder: 40 hours @ $7.25 per hour   = _____

2. Jo Alfaro: 36 hours @ $9.90 per hour   = _____

Solutions: **1.** $290; **2.** $356.40

Compensation by Hourly Wage: Overtime

When a company is engaged in interstate commerce, it is required by the Fair Labor Standards Act to pay its employees time and one-half for all hours worked over 40.

EXAMPLE _____

Viona Lewis worked 48 hours operating a press for the ZIP Printing Company this week. Viona receives $12.75 per hour straight time, but what does she make per hour for her 8 hours overtime? Let's determine Viona's gross earnings by following these steps:

STEPS

1. Multiply straight time (40 hours) by hourly rate.

$$40 \times \$12.75 = \$510 \text{ straight time earnings}$$

2. Subtract 40 hours (straight time) from hours worked.

$$48 - 40 = 8 \text{ hours overtime}$$

3. Multiply overtime hours by time and one-half by hourly rate.

$$8 \times 1.5 \times \$12.75 = \$153 \text{ overtime earnings}$$

4. Add straight time and overtime earnings to determine gross earnings.

$$\$510 + \$153 = \$663 \text{ gross earnings}$$

PRACTICE PROBLEMS

Compute the weekly gross earnings for the following employees and write your answers in the blanks provided.

1. Tracy Saucedo worked 40 hours straight time and 8 hours overtime at an hourly rate of $9.95 per hour. Compute each of the following:

Straight time earnings = _____

Overtime earnings = _____

Gross Earnings = _____

2. Cindy Delano worked 40 hours straight time and 12 hours overtime at an hourly rate of $8.40 per hour. Compute each of the following:

Straight time earnings = _____

Overtime earnings = _____

Gross Earnings = _____

Solutions:
1. Straight time earnings = $398
   Overtime earnings = $119.40
   Gross earnings = $517.40
2. Straight time earnings = $336
   Overtime earnings = $151.20
   Gross earnings = $487.20

Compensation by Hourly Wage: Double Time

Some companies need workers on the job every day of the year. A utility company is an example. Many of these companies offer an added compensation to those employees who will work on Sundays and holidays. This compensation is in the form of double time; that is, double the worker's hourly wage.

EXAMPLE

Shawn Williams worked 52 hours in one week—8 hours each weekday, 8 hours on Saturday, and 4 hours on Sunday, which was New Year's Day (a holiday). His hourly rate is $7.05. Mr. Williams' gross earnings would be computed using the following steps.

STEPS

1. Compute straight time by multiplying straight time hours by the hourly rate.

   Straight time: 40 × $7.05 = $282

2. Compute overtime by multiplying the number of overtime hours by time and one-half, then by the hourly rate.

   Overtime: 8 × 1.5 × $7.05 = $84.60

3. Compute double time by multiplying the number of hours worked double time by 2, then by the hourly rate.

   Double time: 4 × 2 × $7.05 = $56.40

4. Determine total earnings (gross earnings) by adding straight time, overtime, and double time amounts.

   Total earnings this week = $423

PRACTICE PROBLEMS

Complete the following problems and write your answers in the blanks provided.

1. Garland Vanover worked 40 hours at $8.25 per hour straight time, 11 hours overtime, and 6 hours double time. Compute the following:

   Straight time earnings = _____   Overtime earnings = _____

   Double time earnings = _____   Gross earnings = _____

2. Mary Loy Harris worked 40 hours straight time, 9 hours on the July 4 holiday, and 8 hours overtime. Mary Lou's rate is $6.75 per hour. What were Mary Lou's gross earnings?

   Gross earnings = _____

   Solutions: **1.** Straight time earnings = $330
   Overtime earnings   = $136.13
   Double time earnings   = $99
   Gross earnings   = $565.13
   2. Gross earnings   = $472.50

Compensation by Salary

When an employee is salaried, pay is usually based on a yearly or monthly salary. Most salaried employees do not receive compensation for time worked over 40 hours.

EXAMPLE

Nadia Manez is an elementary school teacher working for the Lakemont Independent School District (LISD). Nadia signed a contract with LISD for an annual salary of $29,800, to be paid in 12 equal payments. To compute Nadia's monthly salary (gross earnings), divide the annual salary by 12 (months). Round the answer to the nearest whole dollar.

$29,800 ÷ 12 = $2,483 per month (gross earnings)

Compensation by Salary Plus Commission

Often sales personnel receive compensation in the form of a salary plus a percentage of their total sales, called a *commission*. Usually the commission is paid for sales over a set quota or amount. Follow these steps to determine this week's gross earnings for Gary Parsons.

EXAMPLE ——————————————————————

Gary's straight time salary is $500 per week. This week his total sales were $2,500; $75 was returned merchandise; his quota is set at $1,700; his commission is based on 6% of sales minus returned merchandise after he has met his quota.

STEPS

1. Determine the amount of actual sales by subtracting the amount of the returned merchandise from this week's total sales.

$$\$2,500 - \$75 = \$2,425 \text{ (actual sales)}$$

2. Determine the amount of sales above quota on which Gary will receive a commission by subtracting the quota from his actual sales.

$$\$2,425 - \$1,700 = \$725 \text{ (sales above quota)}$$

3. Multiply sales above quota by percent to determine Gary's commission.

$$\$725 \times 0.06 = \$43.50 \text{ (commission)}$$

4. Add weekly salary to commission to obtain this week's gross earnings.

$$\$43.50 + \$500 = \$543.50 \text{ (this week's gross earnings)}$$

— PRACTICE PROBLEMS ——————

Complete the following problems and write your answers in the blanks provided.

1. Rene Williams receives a salary of $325 per week plus 15% commission on all sales over $3,000. Rene's total sales last week were $4,150. There were no sales returned. What were her gross earnings for the week?

Gross earnings = _____

2. Carlos Villareal receives a salary of $300 per week plus a 15% commission on all sales over $3,000. Carlos' total sales last week were $5,600. A total of $110 of sales was returned. What were Carlos' gross earnings for the week?

Gross earnings = _____

Solutions: **1.** $497.50; **2.** $673.50

Compensation by Piecework

Compensation based on piecework is common among manufacturing companies, especially in the garment industry. The employee is paid for the number of pieces completed during the work day. Usually a quota is set that each employee must meet, but compensation is calculated on total pieces completed. To compute gross earnings, multiply the total pieces completed during the pay period by the pay rate per piece. Study this example.

EXAMPLE

Toby Jennings works for a wholesale drug warehouse. He is compensated based on the number of lines he fills from an order form. His job is to read the order, go to the proper shelf in the warehouse, and pull from that shelf the number of items ordered on that line. He is paid $0.04 for each line on an order form. Toby filled 15,193 line orders this pay period.

Multiply total pieces completed by pay rate per piece.

$$15,193 \times \$0.04 = \$607.72 \text{ gross earnings}$$

PRACTICE PROBLEMS

Complete the following piecework payroll for the Quick Garment Factory. Write your answers in the blanks provided.

1.   806 pieces @ $0.60 = _____
2.   1,509 pieces @ $0.35 = _____
3.   1,711 pieces @ $0.25 = _____
4.   1,600 pieces @ $0.37 = _____
5. 17,490 pieces @ $0.03 = _____
6.   692 pieces @ $0.11 = _____
7.   Total = _____

Solutions: **1.** $483.60; **2.** $528.15; **3.** $427.75; **4.** $592; **5.** $524.70; **6.** $76.12; **7.** $2,632.32

11.3 Payroll Deductions

Now that you have can compute gross earnings, you are ready to learn how to calculate the various deductions that appear on a pay stub. Some of the deductions are required by federal law; others may be deducted from your pay at your request.

11.4 Terms Used in Payroll Deductions

To understand the procedure for computing your paycheck, you must know the terminology related to the various deductions, forms, and laws involved in payroll deductions. Here are some of the terms you will use.

Federal Insurance Contributions Act (FICA). This law was passed in 1935 and provides for retirement funds after an employee reaches the age of 62; disability benefits for any employee who becomes disabled (and for his or her dependents), and a health insurance program after an employee reaches the age of 65. The funds (taxes) to support these programs are provided by workers through deductions withheld from their paychecks, with equal amounts also paid by employers. These taxes are shown on your payroll stub as Social Security and Medicare. In the examples and problems in this book, the tax rate for Social Security will be 6.2% on the first $61,200 (as of 1995) and Medicare at 1.45% for a combined FICA tax rate of 7.65%.

Federal Income Tax. The requirement that federal income tax be withheld from your paycheck came into being in 1943 with the passage of the Current Tax Payment Act. The act also requires that employers pay the tax withheld to the Internal Revenue Service and keep records of the names and addresses of persons employed, their earnings and withholdings, and the amounts and dates of payment. Employers must submit reports to the Internal Revenue Service on a quarterly (every three months) basis (Form 941) and to employees on an annual (yearly) basis (W-2 Form, called a Wage and Tax Statement). With few exceptions, employers must meet these requirements if they employ one or more persons.

Employee's Earnings Record. A record showing an employee's personal payroll information, yearly earnings, and deductions.

Employee's Withholding Allowance Certificate. This form specifies the number of withholding allowances claimed by an employee for tax purposes. Sometimes this certificate is called a W-4 Form.

Net Pay. The total amount of an employee's pay after deductions; that is, gross pay minus deductions.

Payroll Register. A summary record of payroll information for a particular pay period.

Accumulated Earnings. The accumulation or collection of earnings each pay period.

Let's learn how to compute your net pay by learning about the deductions that are withheld.

11.5 Payroll Deductions: FICA

The federal government has established by law the percentage of total earnings that will be withheld from your paycheck for FICA tax. Beginning in 1992, the percentage 7.65% was split—6.2% for Social Security (Old Age or Survivors Death Benefits) and 1.45% for Hospital Insurance (Medicare). For the first time, the Medicare portion had to be identified and reported separately. The Social Security wage base as of January, 1995 is $61,200. Medicare has no wage base (all wages are subject to the 1.45% tax). The maximum Social Security an employee could pay in 1995 would be $3,794 ($61,200 × 6.2%). Employers must match the amount of FICA tax deducted from each employee's wages. Self-employed persons pay at a rate the same as the total paid by the employer and employee on the same income.

The 1995 FICA rates will be used in the examples and problems in this chapter. To compute the amount of Social Security and Medicare to be withheld from an employee's gross earnings per pay period, use the following procedure.

STEPS

1. For Social Security tax, multiply gross earnings per pay period times 6.2% on the first $61,200 earned. Carry answers to 4 decimal places; then round to 2 places. If you use 2 or 3 places, your answer can be off a penny.

2. For Medicare tax, multiply gross earnings per pay period times 1.45%. Carry answers to 4 decimal places; then round to 2 places. If you use 2 or 3 places, your answer can be off a penny.

3. Social Security tax plus Medicare tax equals total FICA withheld.

— PRACTICE PROBLEMS —

Compute Social Security, Medicare, and total FICA tax withheld for each of the following employees and write your answers in the blanks provided. Remember to use a rate of 6.2% for Social Security and 1.45% for Medicare. Carry answers to 4 decimal places and round off to 2 places.

Employee	Weekly Gross Earnings	Social Security	Medicare	Total FICA Tax Withheld
1. Joan Cox	$552.60	_____	_____	_____
2. Dean Little	$660.50	_____	_____	_____
3. Joe House	$440.10	_____	_____	_____
4. Phoung Tai	$525.95	_____	_____	_____

Solutions: **1.** $34.26; $8.01; $42.27; **2.** $40.95; $9.58; $50.53; **3.** $27.29; $6.38; $33.67; **4.** $32.61; $7.63; $40.24

11.6 Federal Income Tax

The amount of federal income tax withheld is based on a person's total gross earnings, marital status, and number of allowances claimed. Each employee must complete a W-4 form which states the number of allowances claimed.

In 1987, the federal government encouraged every employee to complete a new W-4 form, which included a worksheet. This worksheet was designed to help employees calculate the number of withholding allowances that could be claimed.

Along with the new form, the tax reform law increased the required amount of federal income tax withholding from 80% to 90% of each employee's tax liability. Thus, computing the correct number of withholding allowances would assure that the employer would withhold the 90% required by law by year end. (Generally, if a minimum of 90% of an employee's tax liability is not withheld by year end, the employee is subject to a penalty.)

Study the worksheet and W-4 form in the figure below. The employee gives his or her name, address, Social Security number, marital status, and number of allowances claimed. In some instances, an employee may wish additional money withheld for federal income tax purposes, especially if he or she ends up paying additional taxes at the end of each year. Notice also there are instances where additional allowances may be claimed.

FIGURE 11-1

Worksheet and W-4 Form

Personal Allowances Worksheet

A Enter "1" for **yourself** if no one else can claim you as a dependent . . . . **A** ______

B Enter "1" if:
- You are single and have only one job; or
- You are married, have only one job, and your spouse does not work; or
- Your wages from a second job or your spouse's wages (or the total of both) are $1,000 or less.

. . **B** ______

C Enter "1" for your **spouse.** But, you may choose to enter -0- if you are married and have either a working spouse or more than one job (this may help you avoid having too little tax withheld) . . . . **C** ______

D Enter number of **dependents** (other than your spouse or yourself) you will claim on your tax return . . . . **D** ______

E Enter "1" if you will file as **head of household** on your tax return (see conditions under **Head of Household** above) . **E** ______

F Enter "1" if you have at least $1,500 of **child or dependent care expenses** for which you plan to claim a credit . . **F** ______

G Add lines A through F and enter total here. **Note:** This amount may be different from the number of exemptions you claim on your return ▶ **G** ______

For accuracy, do all worksheets that apply.
- If you plan to **itemize or claim adjustments to income** and want to reduce your withholding, see the Deductions and Adjustments Worksheet on page 2.
- If you are **single** and have **more than one job** and your combined earnings from all jobs exceed $30,000 OR if you are **married** and have a **working spouse or more than one job,** and the combined earnings from all jobs exceed $50,000, see the Two-Earner/Two-Job Worksheet on page 2 if you want to avoid having too little tax withheld.
- If **neither** of the above situations applies, **stop here** and enter the number from line G on line 5 of Form W-4 below.

Cut here and give the certificate to your employer. Keep the top portion for your records.

Form **W-4**
Department of the Treasury
Internal Revenue Service

Employee's Withholding Allowance Certificate

▶ **For Privacy Act and Paperwork Reduction Act Notice, see reverse.**

OMB No. 1545-0010

1995

1 Type or print your first name and middle initial | Last name | **2** Your social security number

Home address (number and street or rural route)

3 ☐ Single ☐ Married ☐ Married, but withhold at higher Single rate.
Note: *If married, but legally separated, or spouse is a nonresident alien, check the Single box.*

City or town, state, and ZIP code

4 If your last name differs from that on your social security card, check here and call 1-800-772-1213 for a new card . . . . . ▶ ☐

5 Total number of allowances you are claiming (from line G above or from the worksheets on page 2 if they apply) . **5** ______

6 Additional amount, if any, you want withheld from each paycheck . . . . **6** $ ______

7 I claim exemption from withholding for 1995 and I certify that I meet **BOTH** of the following conditions for exemption:
- Last year I had a right to a refund of **ALL** Federal income tax withheld because I had **NO** tax liability; **AND**
- This year I expect a refund of **ALL** Federal income tax withheld because I expect to have **NO** tax liability.

If you meet both conditions, enter "EXEMPT" here . . . . ▶ **7** ______

Under penalties of perjury, I certify that I am entitled to the number of withholding allowances claimed on this certificate or entitled to claim exempt status.

Employee's signature ▶ **Date ▶** , 19

8 Employer's name and address (Employer: Complete 8 and 10 only if sending to the IRS) | **9** Office code (optional) | **10** Employer identification number

Once the payroll clerk knows an employee's gross earnings, number of withholding allowances, and marital status, the amount of federal income tax to be withheld can be determined. In order to determine the tax to be withheld from an employee's gross earnings, most payroll clerks use the wage and bracket withholding tables contained in the Internal Revenue Service publication, Circular E, "Employer's Tax Guide." These tables are subdivided on the basis of marital status and cover payroll periods for monthly, semimonthly, biweekly, weekly, and daily payroll. Let's study how to determine the amount of tax to be withheld.

EXAMPLE

Let's assume Paula Russell's gross earnings were $334.25 per week and she had claimed one withholding allowances. Also assume that she is married. Follow these steps to use the federal income tax table on the next page.

STEPS

1. Beginning at the upper-left column entitled "At least," move down that column until you find the amount $330, then read across and note the number $340 in the "But less than" column. This entry would read Paula Russell made "at least" $330 "but less than" $340. Her gross earnings amount of $334.25 falls between these two numbers.

2. Continue moving across the same line until you find the number of withholding allowances claimed, 1. The amount of money shown in this column headed by the number 1 is the amount of federal income tax you are to withhold from gross earnings; $25.

Note there are no computations necessary to determine the amount of federal income tax to withhold—simply look up the amount on the proper tax table.

FIGURE 11-2
Withholding Tax Table

MARRIED Persons—**WEEKLY** Payroll Period

(For wages paid in 1995)

If the wages are–		And the number if withholding allowances claimed is—										
At least	But less than	0	1	2	3	4	5	6	7	8	9	10
		The amount of income tax to be withheld is—										
$0	$125	0	0	0	0	0	0	0	0	0	0	0
125	130	1	0	0	0	0	0	0	0	0	0	0
130	135	1	0	0	0	0	0	0	0	0	0	0
135	140	2	0	0	0	0	0	0	0	0	0	0
140	145	3	0	0	0	0	0	0	0	0	0	0
145	150	4	0	0	0	0	0	0	0	0	0	0
150	155	4	0	0	0	0	0	0	0	0	0	0
155	160	5	0	0	0	0	0	0	0	0	0	0
160	165	6	0	0	0	0	0	0	0	0	0	0
165	170	7	0	0	0	0	0	0	0	0	0	0
170	175	7	0	0	0	0	0	0	0	0	0	0
175	180	8	1	0	0	0	0	0	0	0	0	0
180	185	9	2	0	0	0	0	0	0	0	0	0
185	190	10	2	0	0	0	0	0	0	0	0	0
190	195	10	3	0	0	0	0	0	0	0	0	0
195	200	11	4	0	0	0	0	0	0	0	0	0
200	210	12	5	0	0	0	0	0	0	0	0	0
210	220	14	7	0	0	0	0	0	0	0	0	0
220	230	15	8	1	0	0	0	0	0	0	0	0
230	240	17	10	2	0	0	0	0	0	0	0	0
240	250	18	11	4	0	0	0	0	0	0	0	0
250	260	20	13	5	0	0	0	0	0	0	0	0
260	270	21	14	7	0	0	0	0	0	0	0	0
270	280	23	16	8	1	0	0	0	0	0	0	0
280	290	24	17	10	3	0	0	0	0	0	0	0
290	300	26	19	11	4	0	0	0	0	0	0	0
300	310	27	20	13	6	0	0	0	0	0	0	0
310	320	29	22	14	7	0	0	0	0	0	0	0
320	330	30	23	16	9	1	0	0	0	0	0	0
330	340	32	25	17	10	3	0	0	0	0	0	0
340	350	33	26	19	12	4	0	0	0	0	0	0
350	360	35	28	20	13	6	0	0	0	0	0	0
360	370	36	29	22	15	7	0	0	0	0	0	0
370	380	38	31	23	16	9	2	0	0	0	0	0
380	390	39	32	25	18	10	3	0	0	0	0	0
390	400	41	34	26	19	12	5	0	0	0	0	0
400	410	42	35	28	21	13	6	0	0	0	0	0
410	420	44	37	29	22	15	8	1	0	0	0	0
420	430	45	38	31	24	16	9	2	0	0	0	0
430	440	47	40	32	25	18	11	4	0	0	0	0
440	450	48	41	34	27	19	12	5	0	0	0	0
450	460	50	43	35	28	21	14	7	0	0	0	0
460	470	51	44	37	30	22	15	8	1	0	0	0
470	480	53	46	38	31	24	17	10	2	0	0	0
480	490	54	47	40	33	25	18	11	4	0	0	0
490	500	56	49	41	34	27	20	13	5	0	0	0
500	510	57	50	43	36	28	21	14	7	0	0	0
510	520	59	52	44	37	30	23	16	8	1	0	0
520	530	60	53	46	39	31	24	17	10	3	0	0
530	540	62	55	47	40	33	26	19	11	4	0	0
540	550	63	56	49	42	34	27	20	13	6	0	0
550	560	65	58	50	43	36	29	22	14	7	0	0
560	570	66	59	52	45	37	30	23	16	9	1	0
570	580	68	61	53	46	39	32	25	17	10	3	0
580	590	69	62	55	48	40	33	26	19	12	4	0
590	600	71	64	56	49	42	35	28	20	13	6	0
600	610	72	65	58	51	43	36	29	22	15	7	0
610	620	74	67	59	52	45	38	31	23	16	9	2
620	630	75	68	61	54	46	39	32	25	18	10	3
630	640	77	70	62	55	48	41	34	26	19	12	5
640	650	78	71	64	57	49	42	35	28	21	13	6
650	660	80	73	65	58	51	44	37	29	22	15	8
660	670	81	74	67	60	52	45	38	31	24	16	9
670	680	83	76	68	61	54	47	40	32	25	18	11
680	690	84	77	70	63	55	48	41	34	27	19	12
690	700	86	79	71	64	57	50	43	35	28	21	14
700	710	87	80	73	66	58	51	44	37	30	22	15
710	720	89	82	74	67	60	53	46	38	31	24	17
720	730	90	83	76	69	61	54	47	40	33	25	18
730	740	92	85	77	70	63	56	49	41	34	27	20

PRACTICE PROBLEMS

Using the tax table for weekly payroll for married persons on the previous page, look up the amount of federal income tax withheld for the following problems and write your answers in the blanks provided.

Gross Earnings	Number of Withholding Allowances	Federal Income Tax Withheld
1. $541.50	4	_____
2. $188.95	2	_____
3. $450.50	3	_____

Solutions: **1.** $34; **2.** $0; **3.** $28

11.7   Other Deductions: Computing Net Pay

An employee may have other deductions taken from his or her gross earnings. Examples are deductions for medical insurance, union dues, state income tax, savings bonds, and credit union savings or payments. It is the responsibility of the payroll clerk to keep accurate records of such deductions, and to have the necessary information on file authorizing these deductions. To aid in keeping this information, as well as information required by the federal government, an earnings record is usually kept for each employee. An example of a portion of this record for Courtney Langford is shown below.

Employee's Earnings Record

Name __Courtney Langford__   Employee No. __6__ Allowances __1__

Address __16 Jasmine Lane__   S.S. No. __900-16-9123__

_____Texarkana, TX 75503-6603__   Male _____ Female __X__

Married __X__ Single _____   Pay Rate __$9.00__ Per Hour __X__

Date Employed __1/2/--__   Per Day _____

Date Terminated _____   Job Class. __Typist II__

Date	Hrs. Worked Reg.	O.T.	Regular	Overtime	Gross Earnings	Federal In. Tax	Soc. Sec.	Medi-care	Med. Ins.	Union Dues	Total Ded.	Net Pay	Accum. Earnings
1/7	40	0	360.00	-0-	360.00	29.00	22.32	5.22	10.00	-0-	66.54	293.46	360.00
1/14	40	2	360.00	27.00	387.00	32.00	23.99	5.61	10.00	25.00	96.60	290.40	747.00

To complete an employee's earnings record for Courtney Langford for the week ending January 14, follow these steps. (Note that Courtney lives in Texas where there is no state income tax. Usually state income tax deductions are also listed on an employee's earnings record.) She also pays $10.00 medical insurance weekly and $25.00 for union dues every other week.

STEPS

1. Compute straight time (40 × $9 = $360).

2. Compute overtime (2 × 1.5 × $9 = $27).

3. Compute total earnings ($360 + $27 = $387).

4. Determine federal income tax withheld using the table on page 213 ($32).

5. Compute Social Security ($387 × 6.2% = $23.99). Carry answers to 4 decimal places and round to 2 places.

6. Compute Medicare ($387 × 1.45% = $5.61). Carry answers to 4 decimal places and round to 2 places.

7. Add all deductions ($32 + $23.99 + $5.61 + $10 + $25 = $96.60).

8. Compute net pay ($387 − $96.60 = $290.40).

11.8 Accumulated Earnings Column

The accumulated earnings column of the employee's earnings record is a listing of the gross wages the employee has earned to date. Each week the gross earnings amount is added to the previous gross earnings amount to show the current total earnings as of the end of that payroll period. This information is used by the employer to determine whether certain taxes must be paid by the employee and employer (you will learn more about employer's taxes in Chapter 12). Also, the employee's W-2 form is completed from this information and reported to the employee and the Internal Revenue Service. To complete accumulated earnings to date for the week ending January 14, you would add as follows:

$$
\begin{array}{rl}
\$360 & \text{(Accumulated earnings as of 1/7)} \\
+\,387 & \text{(This week's earnings)} \\
\hline
\$747 & \text{(Accumulated earnings to date)}
\end{array}
$$

— PRACTICE PROBLEM —

Complete the following employee's earnings record for Courtney Langford for the payroll period ending 2/14. Courtney worked 48 hours during that week. Write your payroll information in the blank space below Courtney's last pay period. Assume she pays $10 for medical insurance and no union dues this pay period. Remember when computing Social Security and Medicare, carry your answers to 4 decimal places and round to 2 places. Use the withholding tax table on page 213 to look up the federal income tax to be withheld.

(*continues on next page*)

Employee's Earnings Record

Name ___Courtney Langford___

Address ___16 Jasmine Lane___

___Texarkana, TX 75503-6603___

Married _X_ Single _____

Date Employed ___1/2/--___

Date Terminated _____

Employee No. ___6___ Allowances ___1___

S.S. No. ___900-16-9123___

Male _____ Female ___X___

Pay Rate ___$9.00___ Per Hour ___X___

Per Day _____

Job Class. ___Typist II___

Date	Reg.	O.T.	Regular	Overtime	Gross Earnings	Federal In. Tax	Soc. Sec.	Medi-care	Med. Ins.	Union Dues	Total Ded.	Net Pay	Accum. Earnings
2/7	40	2	360.00	27.00	387.00	32.00	23.99	5.61	10.00	25.00	96.60	290.40	2,214.00

Solution:

Date	Reg.	O.T.	Regular	Overtime	Gross Earnings	Federal In. Tax	Soc. Sec.	Medi-care	Med. Ins.	Union Dues	Total Ded.	Net Pay	Accum. Earnings
2/7	40	2	360.00	27.00	387.00	32.00	23.99	5.61	10.00	25.00	96.60	290.40	2,214.00
2/14	40	8	360.00	108.00	468.00	44.00	29.02	6.79	10.00	-0-	89.81	378.19	2,682.00

11.9 Completing a Payroll Register

After employees' earnings records are completed, the information is transferred to a payroll register. The payroll register is a summary record of the payroll information for a particular pay period. Study the payroll register below and compare its information with the employee's earnings record for Courtney Langford. Add Courtney's payroll information from the week ending 2/14 to the payroll register.

Payroll Register

Name	Allow.	Hrs. Wkd	Pay Rate	Reg.	O.T.	Gross Earnings	Federal In. Tax	Soc. Sec.	Medi-care	Med. Ins.	Union Dues	Total Ded.	Net Pay
Holman, J.	3	42	9.00	360.00	27.00	387.00	27.00	23.99	5.61	10.00	-0-	66.60	320.40
Langford, C.	1	48	9.00	360.00	108.00	468.00	44.00	29.02	6.79	10.00	-0-	89.81	378.19

Notice that all computations were completed on the employee's earnings record. After the information is transferred to the payroll register, you should total each column in the register, checking to make sure that:

1. Regular pay plus overtime pay equals gross earnings.
2. The total of all deduction columns equals the total of the Total Deductions column.
3. Gross earnings minus deductions equal the total of the Net Pay column.

This method of checking amounts is called **footing**. Your payroll register must balance; if not, check for errors in the transferring of amounts.

PRACTICE PROBLEM

Usually amounts entered on the payroll register are transferred from the employee's earnings record. For practice, however, compute the amounts on the following payroll register. All workers receive overtime pay for any hours worked in excess of 40 hours per week at a rate of time and one-half. Compute Social Security using 6.2% and Medicare using 1.45%. Use the tax table on page 213 for federal income tax withholding amounts (all workers are married.) Assume each employee pays $17.50 for medical insurance and $40 for union dues. Jones has 1 exemption; Hart, 3; Davis, 0; and Ely, 2. After completing the payroll register, be sure to check your totals by footing. Remember when computing Social Security and Medicare to carry your answers to 4 decimal places and round to 2 places. NOTE: Federal law requires that you deposit the exact amount you withhold from employee's pay. Social Security and Medicare columns should, therefore, be totaled rather than the totals obtained by calculating total gross earnings by the percentages for Social Security and Medicare.

Payroll Register

Name	Allow.	Hrs. Wkd	Pay Rate	Reg.	O.T.	Gross Earnings	Federal In. Tax	Soc. Sec.	Medi-care	Med. Ins.	Union Dues	Total Ded.	Net Pay
Jones, J.	1	44	8.70										
Hart, B.	3	40	6.60										
Davis, L.	0	48	7.40										
Ely, K.	2	42	8.60										
			Totals										

Solution:

Payroll Register

Name	Allow.	Hrs. Wkd	Pay Rate	Reg.	O.T.	Gross Earnings	Federal In. Tax	Soc. Sec.	Medi-care	Med. Ins.	Union Dues	Total Ded.	Net Pay
Jones, J.	1	44	8.70	348	52.20	400.20	35.00	24.81	5.80	17.50	40	123.11	277.09
Hart, B.	3	40	6.60	264	-0-	264.00	-0-	16.37	3.83	17.50	40	77.70	186.30
Davis, L.	0	48	7.40	296	88.80	384.80	39.00	23.86	5.58	17.50	40	125.94	258.86
Ely, K.	2	42	8.60	344	25.80	369.80	22.00	22.93	5.36	17.50	40	107.79	262.01
			Totals	1,252	166.80	1,418.80	96.00	87.97	20.57	70.00	160	434.54	984.26

MATH ALERT

1. You are the payroll clerk for Jason's Video Store. Compute the gross earnings for each employee during the pay period of February 19-26, 19--.

Employee	Hours Worked			Hourly Rate	Earnings		
	Straight	Overtime	Total		Straight	Overtime	Total
Holley, J.	40	6	46	$6.60			
Sanchez, L.	40	8	48	$5.50			
Stroner, B.	40	-0-	40	$4.25			
Barrett, G.	40	10.5	50.5	$6.00			

2. Compute the earnings for each employee of Devlin Manufacturing Company to complete the following payroll register. All workers receive overtime pay for any hours worked in excess of 40 hours per week at a rate of time and one-half. Compute Social Security using 6.2% and Medicare using 1.45%. Use the tax table on page 213 for federal income tax withholding (all employees are married). Assume each employee pays $21 per week for medical insurance and $12 union dues. Carry Social Security and Medicare calculations to 4 decimal places and then round to 2 places.

Payroll Register

Employee No.	Allow.	Hrs. Wkd	Pay Rate	Reg.	O.T.	Gross Earnings	Federal In. Tax	Soc. Sec.	Medi-care	Med. Ins.	Union Dues	Total Ded.	Net Pay
1	0	44	4.70										
2	2	40	5.60										
3	5	48	4.40										
4	2	42	5.60										
5	3	40	6.30										
6	0	40	7.75										
7	4	42	4.25										
8	1	41	6.95										
9	0	43	5.95										
10	2	44	4.50										
			Totals										

3. Juanita is a buyer for Fredericks Formal Fashions. She signed a year's contract for $28,000. She received her first monthly paycheck in the amount of $2,133.33. Juanita believes this amount is incorrect. Calculate (a) her monthly amount and (b) was her check overstated or understated, and (c) how much.

(a) Monthly gross amount: _____

(b) Overstated or understated: _____

(c) How much over- or understated: _____

Study Guide

I. Terminology

page 204	compensation	Salary, wage, pay, or benefits received for the performance of a service.
page 204	gross earnings	The total amount of an employee's pay before deductions.
page 204	Fair Labor Standards Act	An act of law (sometimes called the Wage and Hour Law) establishing minimum wages and requiring employers whose firms are involved in interstate (from state to state) commerce (sale of goods) to pay their employees time and one-half for all hours worked in excess of 40 hours per week.
page 204	time-and-a half	One and one-half times the employee's hourly rate.
page 204	double time	Twice the employee's hourly rate.
page 205	straight time	Usually considered to be the first 40 hours worked per week.
page 205	overtime	All the time worked in excess of straight time.
page 205	salaried	A term applied to administrative or managerial personnel who are generally paid a salary by the month and who are not usually paid overtime.
page 205	hourly paid	Personnel who are paid wages by the number of hours worked.
page 205	commission	Earnings paid an employee based on the percent of total sales.
page 205	piecework wage system	Paying an employee according to the number of units produced.
page 210	Federal Insurance Contributions Act (FICA)	A law passed in 1935 that provides for retirement funds after an employee reaches the age of 62, disability benefits for any employee who becomes disabled (and for his or her dependants), and a health insurance program after an employee reaches the age of 65. FICA is also called Social Security and is withheld from an employee's payroll at a rate of 6.2% of the first $61,200.
page 210	Medicare	A portion of FICA that is paid for old age or survivors benefits. Each employee pays 1.45% on total earnings.
page 210	federal income tax	A tax on each employee's income.
page 210	employee's earnings record	A record showing an employee's personal payroll information, yearly earnings, and deductions.
page 210	employee's withholding allowance certificate	A form that specifies the number of allowances claimed by the employee for tax purposes. Sometimes called a W-4 form.
page 210	net pay	The total amount of an employee's pay after deductions; that is, gross pay minus deductions.
page 210	payroll register	A summary record of the payroll information for a particular pay period.
page 210	accumulated earnings	The accumulation or total of earnings each payroll period.

II. Calculating Gross Earnings

page 205   *Calculating Straight Time Hourly Wage:* Multiply the hourly rate times the number of hours worked less than or equal to 40 (straight time) to get gross earnings.

Example: 40 × $9.50 (hourly rate) = $380 gross earnings

page 205   *Calculating Overtime:* Multiply straight time by hourly rate; subtract 40 fours from total hours worked to determine overtime hours; multiply overtime hours by time and one-half (1.5) by hourly rate; add straight time and overtime earnings to determine gross earnings.

Example: Number hours worked: 48
Hourly rate: $12.75

40 × $12.75 = $510 straight time earnings
48 − 40 = 8 hours overtime
8 × 1.5 × $12.75 = $153 overtime earnings
$510 + $153 = $663 gross earnings

page 206   *Calculating Double Time:* Calculate straight time and overtime the same as in the previous example; then calculate double time by multiplying the number of hours worked double time (you would have to be told how many) by 2 then by the hourly rate.

Example: Number hours worked: 48
Number of double time hours: 4
Hourly rate: $12.75

40 × $12.75 = $510 straight time earnings
48 − 40 = 8 hours overtime
8 × 1.5 × $12.75 = $153 overtime earnings
4 × 2 × $12.75 = $102
$510 + $153 + $102 = $765 gross earnings

page 207   *Calculating Compensation by Salary:* To calculate the monthly salary divide the annual salary by 12 months and round to the nearest dollar.

Example: $29,800 annual salary ÷ 12 months = $2,483 gross earnings per month

page 208   *Calculating Compensation by Salary Plus Commission:* Determine the amount of actual sales by subtracting the amount of the returned merchandise from this week's total sales; determine the amount of sales above quota on which employee will receive a commission by subtracting the quota from actual sales; multiply sales above quota by percent to determine commission; add weekly salary to commission to obtain gross earnings.

Example: Weekly salary: $500
Sales this week: $2,500
Sales quota $1,700
Returned merchandise: $75
Commission rate: 6%

$2,500 − $75 = $2,425 actual sales
$2,425 − $1,700 = $725 sales above quota
$725 × 0.06 = $43.50 commission
$500 + $43.50 = $543.50 gross earnings

page 209   *Calculating Compensation by Piecework:* Multiply total pieces completed times pay rate per piece.

Example: Pieces completed: 15,193
Rate per piece: $0.04

15,193 × $0.04 = $607.72 gross earnings

III. Calculating Deductions

page 210   *Calculating Social Security Deduction:* Multiply the employee's gross earnings (up to $61,200 per year) by 6.2% and round to the nearest penny.

Example: $550 gross earnings × 0.062 = $34.10 Social Security

page 210   *Calculating Medicare Tax Deduction:* Multiply the employee's gross earnings by 1.45% and round to the nearest penny.

Example: $550 gross earnings × 0.0145 = $7.98 Medicare tax

page 211   *Calculating Total FICA Tax Deduction:* Add Social Security and Medicare tax deductions.

Example: $34.10 + $7.98 = $42.08 FICA

page 213   *Determining Federal Income Tax:* Use the Circular E, "Employer's Tax Guide" and look up federal income tax based on pay period length, marital status, gross earnings, and the number of allowances each employee has declared.

page 214   *Calculating Net Pay:* Compute straight time, overtime, and gross earnings; look up federal income tax to be withheld from the Circular E booklet; compute Social Security; compute Medicare; add all deductions; subtract total deductions from gross earnings.

Example:

Compute straight time:	40 × $9 = $360
Compute overtime:	2 × 1.5 × $9 = $27
Compute total earnings:	$360 + $27 = $387
Look up federal income tax for $387 and 1 exemption:	$32
Compute Social Security:	$387 × 0.062 = $23.99
Compute Medicare:	$387 × 0.0145 = $5.61
Add all deductions:	$32 + $23.99 + $5.61 = $61.60
Compute net pay:	$387 − $61.60 = $325.40

page 215   *Calculating Accumulated Earnings:* Add this week's gross earnings to last week's accumulated earnings for the year.

Example: $2,333 accumulated earnings + $387 this week's earnings = $2,720 accumulated earnings to date.

Assignment 1

Name_____ Date _____

Complete the following problems. Write your answers in the blanks provided. Be sure to place commas and dollar signs where needed.

A. Calculate monthly earnings. Round the amounts to the nearest dollar.

1. Mary Nealy makes $56,800 annually   = _____ monthly

2. Bill Waddel makes $14,300 annually   = _____ monthly

3. Gary Ortiz makes $16,195   = _____ monthly

4. Ed Lair makes $22,600 annually   = _____ monthly

5. Desirie Lopez makes $12,600 annually   = _____ monthly

6. Seng Ounnarath makes $29,700 annually   = _____ monthly

7. Zann Nelson makes $42,100 annually   = _____ monthly

8. Jay Mercer makes $27,300 annually   = _____ monthly

9. Weldan Dittmer makes $19,500 annually   = _____ monthly

10. Lela walker makes $10,900 annually   = _____ monthly

B. Calculate straight time plus overtime.

Neal Sloan worked 40 hours straight time and 12 hours overtime at an hourly rate of $4.35. Compute the following:

11. Straight time earnings   = _____

12. Overtime earnings   = _____

13. Gross earnings   = _____

Lee Kashieta worked 40 hours straight time and 8 hours overtime at an hourly rate of $7.50. Compute the following:

14. Straight time earnings   = _____

15. Overtime earnings   = _____

16. Gross earnings   = _____

Rosa Martinez worked 40 hours straight time and 15 hours overtime at an hourly rate of $9.10. Compute the following:

17. Straight time earnings   = _____

18. Overtime earnings   = _____

19. Gross earnings   = _____

Letitia Alvarez worked 40 hours straight time and 20 hours overtime at an hourly rate of $10.50. Compute the following:

20. Straight time earnings   = _____

21. Overtime earnings   = _____

22. Gross earnings   = _____

Assignment 2

Name_____   Date _____

Complete the following problems. Write your answers in the blanks provided

A. Compute the earnings information for each employee of Gary Bartlett Management Services to complete the following payroll register. All workers receive overtime pay for any hours worked in excess of 40 hours per week at a rate of time and one-half. Compute Social Security tax using 6.2% and Medicare at 1.45%. Use the tax table on page 213 for federal income tax withholding (all employees are married). Assume each employee pays $10.50 for medical insurance and $5.50 for union dues. Carry Social Security and Medicare amounts to 4 decimal places, then round to the nearest penny.

Payroll Register

| Employee No. | Allow. | Hrs. Wkd | Pay Rate | Reg. | O.T. | Gross Earnings | Deductions | | | | | Total Ded. | Net Pay |
							Federal In. Tax	Soc. Sec.	Medi-care	Med. Ins.	Union Dues		
Allman, M.	1	49	7.70										
Bell, L.	0	50	6.60										
Cardino, J.	3	48	7.40										
Fiero, N.	2	52	6.60										
Ivy, S.	1	50	5.50										
Jones, I.	1	56	5.75										
Mulligan, J.	3	48	6.50										
Neels, N.	3	50	5.00										
Rodriguez, P.	5	61	6.00										
Beith, R.	2	49	7.25										
			Totals										

B. Compute gross pay in the following problems. Assume time and one-half for all hours over 40.

	Hours	Pay Rate	Straight Time	Overtime	Gross Earnings
1.	46	$6.95	_____	_____	_____
2.	51	$8.87	_____	_____	_____
3.	44	$9.50	_____	_____	_____
4.	62	$10.75	_____	_____	_____
5.	55	$8.75	_____	_____	_____
6.	49	$5.50	_____	_____	_____

Assignment 3

Name_____   Date _____

Complete the following problems. Write your answers in the blanks provided. Carry Social Security and Medicare answers to 4 decimal places, then round answers to the nearest percentage.

A.   Compute social security using 6.2% and medicare using 1.45%.

Gross Earnings	Social Security Tax Withheld	Medicare	Gross Earnings	Social Security Tax Withheld	Medicare
$408.11	1. _____	2. _____	$256.55	11. _____	12. _____
$516.10	3. _____	4. _____	$121.75	13. _____	14. _____
$332.90	5. _____	6. _____	$880.90	15. _____	16. _____
$708.15	7. _____	8. _____	$611.75	17. _____	18. _____
$90.60	9. _____	10. _____	$397.20	19. _____	20. _____

B.   Determine the amount of federal income tax to be withheld. (Use the tax table on page 213.)

	Gross Earnings	Allowances	Amount		Gross Earnings	Allowances	Amount
21.	$389.40	2	_____	31.	$110	3	_____
22.	$469.90	0	_____	32.	$125.18	0	_____
23.	$271.20	7	_____	33.	$342.40	1	_____
24.	$196.97	3	_____	34.	$60.95	0	_____
25.	$451.60	0	_____	35.	$169.21	2	_____
26.	$572.80	0	_____	36.	$340.40	3	_____
27.	$600.09	5	_____	37.	$501.01	4	_____
28.	$75.81	0	_____	38.	$609	8	_____
29.	$291.99	1	_____	39.	$251.50	2	_____
30.	$100.05	8	_____	40.	$400	4	_____

Assignment 4

Name_____ Date _____

Solve the following problems.

A. Calculate salary plus commission.

1. Johnny Haas receives a salary of $450 per week plus a 7% commission on all sales over $1,500. Johnny's total sales last week were $2,159. There were $471 sales returned. What were his gross earnings for the week?

    _____

2. Joyce Martin receives a salary of $600 per week plus a 2% commission on all sales over $45,000. Joyce's total sales last week were $51,000. There were $3,200 sales returned. What were her gross earnings for the week?

    _____

3. Larry Davidson receives a salary of $550 per week plus a 3% commission on all sales over $200. Larry's total sales last week were $811. There were no sales returned. What were his gross earnings for the week?

    _____

4. Don Ritchie receives a salary of $600 per week plus a 10% commission on all sales over $500. Don's total sales last week were $933. There were sales returns of $37. What were his gross earnings for the week?

    _____

5. Arlen Ortiz receives a salary of $590 per week plus a 5% commission on all sales over $650. Arlen's total sales last week were $1,880. There were no sales returns. What were his gross earnings for the week?

    _____

B. Calculate compensation by piecework.

6. Marie Miller works in a garment factory that makes shirts. Marie's job is sewing in sleeves. Marie receives $.45 for each pair of sleeves she completes. During this pay period, Marie completed 1,341 pairs of sleeves. What were Marie's gross earnings?

    _____

7. Elizabeth Graves works as a word processor for Millwood Mining Company in their word processing department. Elizabeth is paid by the number of typed lines per pay period. Her rate per line is $.04. This pay period Elizabeth completed 15,950 lines. What are Elizabeth's gross earnings for the period?

    _____

8. Danny Westwind works for a wholesale drug warehouse filling orders. He is paid $.04 for each line he fills from an order form. This pay period Danny filled 15,129 lines. What are Danny's earnings for the period?

    _____

9. Janet Johnson is employed by the Dewley Garment Manufacturing Company. Janet's job is sewing collars on blouses. She is paid $.30 for each collar she completes. This pay period Janet completed 1,625 collars. What are her gross earnings for the period?

    _____

10. Juan Martinez works on the assembly line at a lawn mower manufacturing plant. Juan's job is assembling the tires on the right side of each mower as it moves down the assembly line. Juan receives $.55 for each pair of tires assembled. This pay period Juan assembled 1,012 pairs of tires. What were Juan's gross earnings for the week?

    _____

Assignment 5

Name_____ Date _____

Solve these word problems. Write your answers in the blanks provided.
Round answers to the nearest cent.

1. Joseph Leeper worked a total of 59 hours in one week. Of these hours, he was paid for 8 at the regular overtime rate of 1.5 times his hourly wage and for 11 at the holiday rate of 2 times his hourly wage. Find his gross earnings for the week if his hourly pay is $12.65

   _____

2. Patricia Holmes has a salaried job. She earns $625 a week. In one week, she worked 60 hours. Find her gross earnings for the week.

   _____

3. For sewing collars on shirts, employees are paid $2.50 a shirt. Rita Mitz completes an average of 25 shirts a day. Find her average gross weekly earnings for a 5-day week.

   _____

4. Asa Clark, a car salesman, is paid 6% commission for all sales. If he needs a monthly income of $3,000, find the monthly sales volume he must meet.

   _____

5. Randy Taylor is a salesperson and is paid a salary of $325 plus 4% of all sales. Find his gross income if his sales are $7,650.

   _____

6. Keesha Denton works 49 hours in one week and earns $8.45 an hour with time and a half for overtime. Find her gross weekly pay.

   _____

7. William Griffith packages shrimp. He is paid $1 per package on all packages up to 100, $1.50 for packages from 101 to 150, and $1.75 for all packages over 150. Find his gross pay if he finished 275 packages in one pay period.

   _____

8. Stacy Breedlove, who sells medical supplies, works on 8% commission. If his sales for a week are $8,762.24, find his gross earnings.

   _____

9. Lisa Latham is a waitress earning $3.25 an hour plus a 15% gratuity automatically added to all food checks. If Lisa works 32 hours one week and sold $1,765.45 in food, how much food should her gross earnings be for the week?

   _____

10. Find the gross earnings if Inez Vasquez earns $325 plus 4% of all sales over $2,000 and the sales for a week are $6,983.34.

   _____

11. Sally Johansen is paid $500 plus 7.25% of the total sales volume. If she sold $2,400 in merchandise, find the gross earnings.

   _____

12. Peter Mullens worked a total of 57 hours in one week. Of these hours, he was paid for 8 at a holiday rate of 2 times his hourly wage and 9 hours for overtime rate of 1.5 times his hourly wage. Find his gross earnings for the week if his hourly pay is $15.00.

   _____

13. Corey is a migrant worker in south Texas. He worked last week picking lettuce. He is paid $2 a bushel. Last week he picked 60 bushels. What was his gross pay?

   _____

Assignment 6

Name_____ Date _____

Complete the following word problems. Write your answers in the blanks provided. Round answers to the nearest cent.

1. Hilton Sales pays its employees on a graduated commission scale: 4% on first $20,000 sales; 5% on sales above $20,000 to $30,000; and 6% on sales greater than $30,000. James Jones had $76,500 in sales. What commission did James earn?

_____

2. How much federal income tax must Taylor Publishing pay for Tim Johnson who earns $705 per week and claims two withholding allowances?

_____

3. How much Social Security tax must Taylor Publishing pay for Tim Johnson in problem 2 if the social security tax rate is 6.2%?

_____

4. Find the FICA tax at 6.2% and the Medicare tax at 1.45% for Kate Smith whose gross earnings are $421.97.

   FICA _____   Medicare _____

5. If Janita Hays has a gross income of $371.91 and total deductions of $76.54, find her net earnings.

_____

6. If Guy Morris had net earnings of $589.33 and total deductions of $129.01, find his gross earnings.

_____

7. Polly Maxwell has gross earnings of $341.87. She has a 4% retirement deduction and pays $31 for insurance. What is the total of these deductions?

_____

8. Deductions for Vera Mesa are as follows: withholding tax $52; FICA tax $32.89; retirement $12; insurance $31; and $6.50 union dues. Find the total deductions.

_____

9. Henry Talmadge earned $42,987 last year. What is his monthly income? Rounded to the nearest dollar?

_____

10. Portia Weems is paid on the following escalating piece rate: 1-100 (100), $1; 101-250 (150), $1.50; 251-400 (150), $2.25; and 401-up, $3. Find her gross earnings for completing 523 pieces.

_____

11. Keith O'Neil is paid 1.5 times his regular pay for all hours worked in a week exceeding 40. He worked 56.5 hours and earns $18.00 per hour. Calculate his gross pay.

_____

12. If Terry Cummings earns a salary $32,190 a year and is paid weekly, how much is his weekly paycheck before taxes are taken out?

_____

13. How much federal income tax must Joshua Bethel, who is married, pay if he earns $739 and carries 4 withholding allowances?

_____

Name_____ Date_____

$$\frac{\text{Student's Score}}{\text{Maximum Score}} = \frac{}{81} = \text{Grade}_____$$

A. Calculate monthly earnings. Round the amounts to the nearest dollar.

   1. Amy Shaw makes $39,500 annually = _____ monthly

   2. Chris Mace makes $21,300 annually = _____ monthly

   3. Tony Dawn makes $13,900 annually = _____ monthly

   4. Jerry Todd makes $18,450 annually = _____ monthly

   5. Sheila Smith makes $55,000 annually = _____ monthly

   6. Wilma Chess makes $29,560 annually = _____ monthly

B. Calculate straight time plus overtime. Overtime = Time and one-half.

   Theresa Minter worked 40 hours straight time and 10 hours overtime at an hourly rate of $8.50. Compute the following:

   7. Straight time earnings = _____

   8. Overtime earnings = _____

   9. Gross earnings = _____

C. Calculate straight time plus overtime plus double time.

   Carrie Harris worked 40 hours straight time, 9 hours overtime, and 12 hours double time. Her hourly rate is $9.25. Compute the following:

   10. Straight time earnings = _____

   11. Overtime earnings = _____

   12. Double time earnings = _____

   13. Gross earnings = _____

D. Maria Nolan receives a salary of $400 per week plus a 7% commission on all sales over $2,000. Maria's total sales were $6,422. There were $496 sales returned. Compute the following:

   14. Commission amount = _____

   15. Gross earnings = _____

   Seth Tyson earns a salary of $325 per week plus a 13% commission on all sales over $3,000. His total sales were $12,986. There were no sales returned. Compute the following:

   16. Commission amount = _____

   17. Gross earnings = _____

E. Calculate compensation by piecework. Maddie Marshall works in a garment factory that makes shorts. Maddie's job is sewing cuffs. Maddie receives $0.35 for each cuff completed. During this pay period, Maddie completed 845 cuffs.

18. Gross earnings = _____

F. Compute Social Security at 6.2% and Medicare using 1.45%. Carry answers to 4 decimal places and round answers to the nearest penny.

Gross Earnings	Social Security Tax Withheld	Medicare Withheld
$874.99	19. _____	20. _____
$620.44	21 _____	22. _____
$365.12	23. _____	24. _____
$281.73	25. _____	26. _____
$413.01	27. _____	28. _____
$1,390.45	29. _____	30. _____
$972.88	31. _____	32. _____

G. Complete the following employee's earnings record for Martha Vester. Use the tax table on page 213 of the text. Compute Social Security tax and Medicare. Martha's earnings to date are $4,150.67. Assume she pays $13.75 for medical insurance and $3.00 for union dues each week. Even though Martha is married, she claims only 1 allowance. Her rate of pay is $9.50 per hour. Any hours worked over 40 are paid at time and one-half.

Date	Hrs. Worked Reg.	Hrs. Worked O.T.	Regular	Overtime	Gross Earnings	Federal In. Tax	Soc. Sec.	Medi-care	Med. Ins.	Union Dues	Total Ded.	Net Pay	Accum. Earnings
4/1	40	12	33.	34.	35.	36.	37.	38.	39.	40.	41.	42.	43.
4/8	40	1	44.	45.	46.	47.	48.	49.	50.	51.	52.	53.	54.

H. Compute the payroll for the Magic Carpet Theater to complete the following payroll register. All workers receive overtime pay for any hours worked in excess of 40 hours per week at a rate of time and one-half. Use the tax table on page 213 of the text for federal income tax withholding. Assume each employee pays $12.50 per week for medical insurance. All workers are married.

						Payroll Register						
		Hrs. Pay		Earnings			Deductions					
Name	Allow.	Wkd.	Rate	Reg.	O.T.	Gross Earnings	Federal In. Tax	Soc. Sec.	Medi-care	Med. Ins.	Total Ded.	Net Pay
Bates, N	0	29	6.50	55.	56.	57.	58.	59.	60.	61.	62.	63.
Davis, B	2	45	5.75	64.	65.	66.	67.	68.	69.	70.	71.	72.
Stewart, R	1	42	8.00	73.	74.	75.	76.	77.	78.	79.	80.	81.

Practical Math Applications

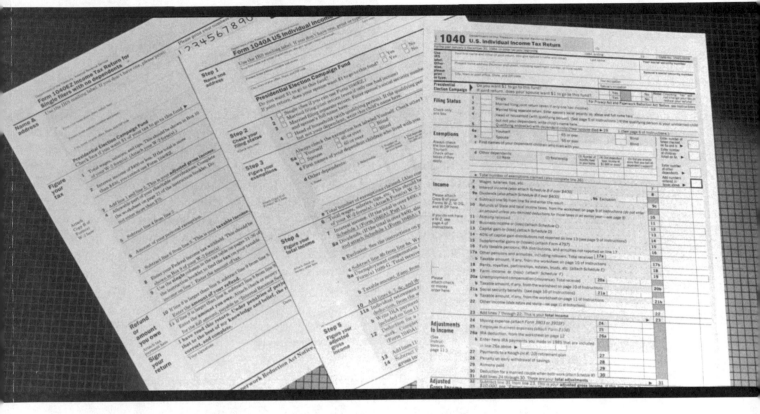

12
Taxes

The payment of taxes, whether they are local, state, or federal, is necessary so that your public officials will have money to run various programs, agencies, or public offices. Some taxes are imposed on both the employee and employer, whereas others are imposed only on the employer. The amount and type of taxes you are required to pay are determined by the various governing bodies in your city, state, or federal government, based on the passage of laws. You, as a voter, help determine these laws. Let's learn about some of these taxes and how they are computed.

12.1 Terms Used in Taxes

In this chapter, you will learn about property tax, state and federal unemployment taxes, and federal income tax. The terminology for these taxes will be covered as each is introduced. The first tax you will learn about is property tax; that is, a tax imposed on real estate such as land or buildings.

Property Tax

Property tax is a tax imposed on a property owner to help fund public services such as fire/police protection and schools. Some of the common terms used in calculating property tax are:

Market Value. The amount of money the property will bring in a competitive open market.

Assessed Rate. An arbitrary rate set by the taxing body.

Assessed Value. The amount of money for which property is listed on the books (tax records) for tax purposes.

Tax Rate. A certain percent based on the assessed value of the property. However, it is usually stated in dollar amounts, such as $1.80 per $100 of assessed value or $18 per $1,000 of assessed value.

Tax. The amount of money paid by the property owner.

Mill. One-thousandth of one cent (0.001¢) per dollar of assessed property value. When property tax levies are proposed, they are usually expressed in terms of mills. (i.e. 1.5 mill school levy)

12.2 Property Tax

Property tax is imposed by townships, cities, counties, states, or school districts on the owner of any type of real property (such as land and/or buildings) to raise funds for such areas as parks and school districts or for street repairs. This tax is *levied* (ordered to be paid) quarterly, semiannually, or annually. Property taxes are calculated on a value. This value is the amount for which property is listed in the tax records for tax purposes. It is called *assessed value* (as opposed to *market value*), the amount the property would bring in a competitive and open market. The tax usually is not shown as a percentage but is stated in dollar amounts, such as $1.80 per $100 of assessed value. It is determined by dividing the amount of funds needed by the assessed value.

Computing the Amount of Property Tax

Let's learn how to compute the amount of taxes Lee Jones, a farmer, must pay on his property.

The property in Bowie County is assessed at 50% of market value by the county, the governing body, with a tax rate of $6.95 per $1,000 valuation. Mr. Jones owns a farm with a market value of $100,000. To determine the amount of property tax he would be required to pay, you would follow these steps.

STEPS

1. Multiply the market value of the property by the assessed rate.

$$\$100,000 \times 50\% = \$50,000 \text{ assessed value}$$

2. Determine the number of $1,000s in the assessed value.

$$\$50,000 \div \$1,000 = 50$$

3. Multiply the number of $1,000s by the tax rate per $1,000 to determine the amount of tax due.

$$50 \times \$6.95 = \$347.50 \text{ tax due}$$

Determining the Property Tax Rate

Now, let's learn how governing authorities compute tax rates. Suppose Bowie County needs $50 million to complete repairs on all county roads. All the real property in the county has been assessed at $1,000,000,000. To compute the property tax rate, you would divide the amount of money needed by the total assessed value of property to obtain the tax rate.

Money needed: $50,000,000
Total assessed value: $1,000,000,000
$50,000,000 ÷ $1,000,000,000 = 0.05000 or 5% tax rate

12.3 State Unemployment Insurance Tax

State unemployment insurance tax is levied on an employer's payroll on the basis of the gross wages paid to the employees. This tax helps fund what the state pays as unemployment compensation. Like property taxes, these taxes are an expense to a business. They are paid by the employer only. The rate of the state unemployment tax varies considerably from state to state. However, most states adopt a base amount of earnings and a rate that varies, on which the employer must pay tax. An example is a base of at least $12,000 and rates that may be as much as 4.77 percent or as low a 0.1 percent. The employer would then pay state unemployment tax on each employee's first $12,000 earned. These taxes are usually paid quarterly; tax money must be received

30 days after the end of the quarter. For the purpose of computing problems in this chapter, let's assume a tax rate of 3% on the first $9,000. To compute state unemployment insurance tax, you would multiply gross wages subject to tax by the tax rate.

Carter Wholesale Drug Company had total wages of $219,300 for the first quarter. All wages were subject to state unemployment tax because no employee had earned the maximum of $9,000 by the end of the quarter.

$$\$219,300 \times 3\% = \$6,579 \text{ tax due}$$

The following employees' second quarter earnings are shown. A check mark has been placed beside those employees whose earnings have reached $9,000. Once the employee's earnings reach $9,000, the employer stops paying state unemployment tax. Therefore Carter Wholesale Drug Company no longer pays for Brown, Jones, and Green. Study these amounts.

Employee	Cumulative Earnings	
Brown, C.	$9,800	✔
Moore, T.	$8,890	
Jones, M.	$12,600	✔
Green, J.	$9,050	✔
Bixby, L.	$7,710	
Jordan, S.	$6,800	

The total gross wages at the end of the second quarter for Moore are $8,890; for Bixby, $7,710; and for Jordan, $6,800. What is the amount of state unemployment insurance tax their employer will pay on each of them? Add the three wages and multiply the total by the tax rate.

$$\$23,400 \times 3\% = \$702.00 \text{ due for all 3 employees}$$

PRACTICE PROBLEMS

Complete the following problems to compute state unemployment insurance tax; write your answers in the blanks provided. Assume a tax rate of 3% on the first $9,000.

1. $4,300 gross wages = _____ tax due

2. $12,000 gross wages = _____ tax due

Solutions: 1. $129 2. $270

Place a check mark beside each employee whose cumulative earnings have reached $9,000.

1. Don Mason, $16,000 gross earnings _____

2. T. Hall, $8,999 gross earnings _____

3. Ramon Ruiz, $9,050 gross earnings _____

Solutions: 1-3. Check marks should be placed beside Mason and Ruiz

12.4 Federal Unemployment Insurance Tax

Federal unemployment insurance tax is similar to state unemployment insurance tax and is levied on the employer only. This tax helps fund state employment compensation payments. It is set by Congress, and the rate may change from time to time. It is estimated quarterly. Generally, but there are many exceptions, this tax applies to every employer that, during the last or present year, pays wages of $1,500 or more in any calendar quarter or has one or more employees at any time in each of 20 calendar weeks. The employer must deposit the tax in a commercial bank or Federal Reserve bank, accompanied by a preprinted federal tax deposit card provided by the government. For the purposes of computing problems in this chapter, let's assume a tax rate of 0.8%. on the first $7,000 each employee earns. To compute federal unemployment tax, multiply gross earnings subject to tax by the tax rate.

Daniel Burton, a housing contractor, had gross wages of $95,000 for the first quarter. All wages were subject to federal unemployment tax.

$$\$95,000 \times 0.8\% = \$760.00 \text{ tax due}$$
(Notice that this is 0.8% or 0.008 and not 8%.)

When the employee earns over the maximum taxable amount, simply multiply the rate by the maximum similar to the computations for state unemployment insurance tax.

Mario Riveria earned $7,450 at the end of the second quarter. What is the amount of federal unemployment insurance tax Riveria's employer will pay?

Multiply the tax rate by the maximum amount.

$$\$7,000 \times 0.8\% = \$56 \text{ amount due}$$

PRACTICE PROBLEMS

Compute federal unemployment tax and write your answers in the blanks provided. Use 0.8% as the tax rate. Assume all earnings are subject to tax.

1. $42,000 gross earnings for Just Jeans, Inc. = _____ tax due

2. $58,942 gross earnings for Reyes & Sons = _____ tax due

Solutions: **1.** $336; **2.** $471.54

Place a check mark beside those employees whose gross earnings have exceeded the maximum ($7,000) and compute the amount of tax due.

1. Leon Branch, $15,800 _____

2. Betty Stienum, $7,200 _____

3. Ted O'Leary, $9,160 _____

Solutions: **1-3.** All names should have check marks beside them and each amount due is $56.

12.5 Who Should File Federal Income Tax

Your filing status, age, and gross income determine whether you have to file a tax return. Gross income usually means money, goods, and property you received that you must pay taxes on. It does not include nontaxable benefits, such as worker's compensation.

12.6 Social Security Number

Katherine M. Grayson is 18 years old and is in her first year of college. Katherine has worked part time as a receptionist for Morgan Reynolds, a local attorney, since her junior year in high school. When Katherine was hired by Reynolds and Associates, she was asked for her social security number, such as the one shown below. This number is her taxpayer's identification number, and it will appear on all tax forms that Katherine files.

FIGURE 12-1
Social Security Card

12.7 Federal Withholding and Income Forms

Employee's Withholding Allowance Certificate (W-4 Form)

Katherine was also asked to complete an Employee's Withholding Allowance Certificate (W-4 form). Refer to Chapter 11 to review the procedure for completing this form. Katherine claims 1 allowance for herself. Study the form she completed below.

FIGURE 12-2

Employee's Withholding Allowance Certificate (W-4 Form)

Form **W-4** Department of the Treasury Internal Revenue Service	**Employee's Withholding Allowance Certificate** ▶ For Privacy Act and Paperwork Reduction Act Notice, see reverse.	OMB No. 1545-0010 **1995**
1 Type or print your first name and middle initial Katherine M. Last name Grayson		2 Your social security number 900 43 4750
Home address (number and street or rural route) 4611 Valley Lane	3 ☒ Single ☐ Married ☐ Married, but withhold at higher Single rate. **Note:** If married, but legally separated, or spouse is a nonresident alien, check the Single box.	
City or town, state, and ZIP code Texarkana, AR 75501	4 If your last name differs from that on your social security card, check here and call 1-800-772-1213 for a new card . . . . . . ▶ ☐	

5 Total number of allowances you are claiming (from line G above or from the worksheets on page 2 if they apply) . **5** 1
6 Additional amount, if any, you want withheld from each paycheck . . . . **6** $
7 I claim exemption from withholding for 1995 and I certify that I meet **BOTH** of the following conditions for exemption:
 • Last year I had a right to a refund of **ALL** Federal income tax withheld because I had **NO** tax liability; **AND**
 • This year I expect a refund of **ALL** Federal income tax withheld because I expect to have **NO** tax liability.
If you meet both conditions, enter "EXEMPT" here . . . . ▶ **7**

Under penalties of perjury, I certify that I am entitled to the number of withholding allowances claimed on this certificate or entitled to claim exempt status.

Employee's signature ▶ *Katherine M. Grayson*   Date ▶ 4-30 , 19 95

8 Employer's name and address (Employer: Complete 8 and 10 only if sending to the IRS)	9 Office code (optional)	10 Employer identification number

PRACTICE PROBLEMS

Using your name and social security number, complete the following W-4 form, claiming 1 allowance for yourself.

Form W-4 Department of the Treasury Internal Revenue Service	Employee's Withholding Allowance Certificate ▶ For Privacy Act and Paperwork Reduction Act Notice, see reverse.	OMB No. 1545-0010 1995
1 Type or print your first name and middle initial Last name		2 Your social security number
Home address (number and street or rural route)	3 ☐ Single ☐ Married ☐ Married, but withhold at higher Single rate. Note: If married, but legally separated, or spouse is a nonresident alien, check the Single box.	
City or town, state, and ZIP code	4 If your last name differs from that on your social security card, check here and call 1-800-772-1213 for a new card . . . . ▶ ☐	

5 Total number of allowances you are claiming (from line G above or from the worksheets on page 2 if they apply) . | 5 |
6 Additional amount, if any, you want withheld from each paycheck . . . . | 6 | $
7 I claim exemption from withholding for 1995 and I certify that I meet **BOTH** of the following conditions for exemption:
   • Last year I had a right to a refund of **ALL** Federal income tax withheld because I had **NO** tax liability; **AND**
   • This year I expect a refund of **ALL** Federal income tax withheld because I expect to have **NO** tax liability.
   If you meet both conditions, enter "EXEMPT" here . . . . ▶ | 7 |

Under penalties of perjury, I certify that I am entitled to the number of withholding allowances claimed on this certificate or entitled to claim exempt status.

Employee's signature ▶

Date ▶   , 19

| 8 Employer's name and address (Employer: Complete 8 and 10 only if sending to the IRS) | 9 Office code
(optional) | 10 Employer identification number |

Solution: Answers will vary.

Wage and Tax Statement (W-2 Form)

At the end of the year, Katherine received a copy of her Wage and Tax Statement (W-2 form). This form shows Katherine's gross earnings ($8,950), total federal income tax withheld ($1,346), social security tax withheld ($554.90), and Medicare tax withheld ($129.78). Total state income tax withheld was $45 for the entire year. This Wage and Tax Statement must accompany her federal income tax return that she will file. Also, Katherine will retain a copy of her W-2 form for her personal files. Study the W-2 form she received (below).

The Wage and Tax Statement is completed by your employer. The employer would have this information compiled from the employer's records.

FIGURE 12-3
Wage and Tax Statement
(W-2 Form)

a Control number 22222 Void ☐	For Official Use Only ▶ OMB No. 1545-0008		
b Employer's identification number 75-5461978		1 Wages, tips, other compensation $8,950.00	2 Federal income tax withheld $1,346.00
c Employer's name, address, and ZIP code Morgan Reynolds 29 Professional Plaza Texarkana, AR 75501		3 Social security wages $8,950.00	4 Social security tax withheld $554.90
		5 Medicare wages and tips $8,950.00	6 Medicare tax withheld $129.78
		7 Social security tips	8 Allocated tips
d Employee's social security number 900-43-4750		9 Advance EIC payment	10 Dependent care benefits
e Employee's name (first, middle initial, last) Katherine M. Grayson 4611 Valley Lane Texarkana, AR 75501		11 Nonqualified plans	12 Benefits included in Box 1
		13 See Instrs. for Box 13	14 Other
f Employee's address and ZIP code		15 Statutory employee ☐ Deceased ☐ Pension plan ☐ Legal rep. ☐ 942 emp. ☐ Subtotal ☐ Deferred compensation ☐	

| 16 State   Employer's state I.D. No.
AR | 17 State wages, tips, etc.
$8,950.00 | 18 State income tax
$45.00 | 19 Locality name | 20 Local wages, tips, etc. | 21 Local income tax |

Annual 1099 Form

Katherine has a savings account at the National First Bank. At the end of the year, Katherine received from the bank a Form 1099-INT to notify her of the amount of interest ($752.05) her money earned during the year. Study the form she received (below).

Form 1099-INT is completed by your bank and sent to you. There is no form for you to complete.

FIGURE 12-4
Form 1099-INT

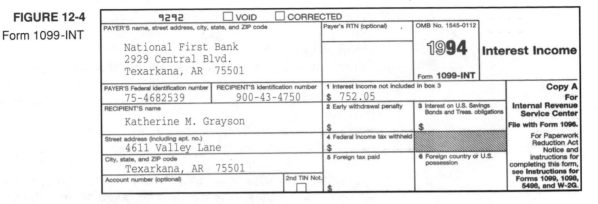

12.8 Federal Income Tax

The forms you just studied provide information necessary to file a federal income tax form. Let's study more about the computations and the forms you send to the Internal Revenue Service.*

The most common form used for filing personal income tax is the 1040 form; however, you may be able to file one of the shorter forms, Form 1040A or 1040EZ, instead. In this chapter, you will not complete these forms but will learn to compute the amount of tax due as if you were completing the forms. Let's learn some of the terminology associated with filing your federal income tax.

Exemptions. The taxpayer is allowed one exemption per dependent. You may subtract a deduction of $2450 from your taxable income in 1994 for each dependent.

Total Income. Includes all the income you receive such as wages, salaries, tips, interest income, dividend income, taxable refunds, credits or offsets of state and local income tax, alimony, business income, capital gain, pensions, IRA distributions, annuities, rents, royalties, partnerships, certain scholarships and fellowship grants, bartering income, endowments, prizes and awards, director's fees, estates, trusts, farm income, unemployment compensation, and, in some cases, social security benefits.

Adjustments to Income. Adjustments to your total income are current year moving expenses, IRA deduction, self-employment health insurance deduction, Keogh retirement plans, penalty for early withdrawal of savings, or alimony paid.

Adjusted Gross Income. Total income minus total adjustments.

Earned Income Credit. If you do not have any qualifying children, your earned income was under $9,000, your adjusted gross income was under $9,000, and you or your spouse were at least 25; or if you have one qualifying child and you earned less than $23,755 you may qualify for additional credit. You must complete schedule EIC.

---

*All information based on 1994 tax laws.

Itemized Deductions. Amounts subtracted from adjusted gross income before tax is computed. Deductions are listed on Schedule A. Here are some examples of deductions allowed in 1994.

(a) Medical and dental expenses that are more than 7.5% of your adjusted gross income.

(b) State and local income taxes.

(c) Real estate taxes that you paid on property you own that was not used for business but only if the taxes are based on the assessed value of the property.

(d) Personal property taxes but only if it is an annual tax based on value alone.

(e) Home mortgage interest.

(f) Contributions to tax exempt organizations that are religious, charitable, educational, scientific, literary in purpose or that work to prevent cruelty to children or animals, also contributions to federal, state or local governments.

(g) Each loss caused by theft, vandalism, fire, storm or similar causes, car, boat, and other accidents or similar causes that is more than 10% of adjusted gross income.

(h) Certain moving expenses incurred before 1994. Beginning in 1994, current year moving expenses will be taken as an adjustment to income on line 24, Form 1040.

(i) Periodic payments of alimony or separate maintenance under a court decree.

(j) Miscellaneous expenses: Unreimbursed employee business expenses (i.e. union dues), expenses of producing income, and most other miscellaneous expenses are deductible only to the extent the total amount of such expenses exceeds 2% of your adjusted gross income.

Standard Deduction. If you do not itemize your deductions, you may take a standard deduction in the following amounts (1994 figures given):

Single $3,800
Head of household 5,600
Married filing jointly or qualified widow(er) 6,350
Married filing separately 3,175

Additional amounts are allowed if you or your spouse is blind and/or over the age of 65.

Taxable Income. The amount of income subject to tax.

12.9 Tax Tables

The Internal Revenue Service provides tax tables that cover four categories of taxpayers: single, married filing jointly, married filing separately, and head of a household. Those who file declaring they are unmarried heads of households must meet qualifications having to do with support of children, parents, or relatives. Also, if you were married in 1994, had a child living with you, and lived apart from your spouse during the last 6 months of 1994, you may be able to file as head of household. If your spouse died in 1992 or 1993, and you had a dependent child living with you for all of 1994, you may file as a qualifying widow(er) with dependent child.

This table is read similar to the tax table you used in Chapter 11. Look at the tax table on pages 241 and 242 and its headings. In the upper left corner it reads "If line 37 (taxable income) is–"; line 37 is on the 1040 federal income tax return form. You are to use your taxable income from line 37.

If line 37 (taxable income) is— / And you are—

Columns: At least | But less than | Single | Married filing jointly * | Married filing separately | Head of a household — Your tax is—

5,000

At least	But less than	Single	Married filing jointly	Married filing separately	Head of a household
5,000	5,050	754	754	754	754
5,050	5,100	761	761	761	761
5,100	5,150	769	769	769	769
5,150	5,200	776	776	776	776
5,200	5,250	784	784	784	784
5,250	5,300	791	791	791	791
5,300	5,350	799	799	799	799
5,350	5,400	806	806	806	806
5,400	5,450	814	814	814	814
5,450	5,500	821	821	821	821
5,500	5,550	829	829	829	829
5,550	5,600	836	836	836	836
5,600	5,650	844	844	844	844
5,650	5,700	851	851	851	851
5,700	5,750	859	859	859	859
5,750	5,800	866	866	866	866
5,800	5,850	874	874	874	874
5,850	5,900	881	881	881	881
5,900	5,950	889	889	889	889
5,950	6,000	896	896	896	896

6,000

At least	But less than	Single	Married filing jointly	Married filing separately	Head of a household
6,000	6,050	904	904	904	904
6,050	6,100	911	911	911	911
6,100	6,150	919	919	919	919
6,150	6,200	926	926	926	926
6,200	6,250	934	934	934	934
6,250	6,300	941	941	941	941
6,300	6,350	949	949	949	949
6,350	6,400	956	956	956	956
6,400	6,450	964	964	964	964
6,450	6,500	971	971	971	971
6,500	6,550	979	979	979	979
6,550	6,600	986	986	986	986
6,600	6,650	994	994	994	994
6,650	6,700	1,001	1,001	1,001	1,001
6,700	6,750	1,009	1,009	1,009	1,009
6,750	6,800	1,016	1,016	1,016	1,016
6,800	6,850	1,024	1,024	1,024	1,024
6,850	6,900	1,031	1,031	1,031	1,031
6,900	6,950	1,039	1,039	1,039	1,039
6,950	7,000	1,046	1,046	1,046	1,046

7,000

At least	But less than	Single	Married filing jointly	Married filing separately	Head of a household
7,000	7,050	1,054	1,054	1,054	1,054
7,050	7,100	1,061	1,061	1,061	1,061
7,100	7,150	1,069	1,069	1,069	1,069
7,150	7,200	1,076	1,076	1,076	1,076
7,200	7,250	1,084	1,084	1,084	1,084
7,250	7,300	1,091	1,091	1,091	1,091
7,300	7,350	1,099	1,099	1,099	1,099
7,350	7,400	1,106	1,106	1,106	1,106
7,400	7,450	1,114	1,114	1,114	1,114
7,450	7,500	1,121	1,121	1,121	1,121
7,500	7,550	1,129	1,129	1,129	1,129
7,550	7,600	1,136	1,136	1,136	1,136
7,600	7,650	1,144	1,144	1,144	1,144
7,650	7,700	1,151	1,151	1,151	1,151
7,700	7,750	1,159	1,159	1,159	1,159
7,750	7,800	1,166	1,166	1,166	1,166
7,800	7,850	1,174	1,174	1,174	1,174
7,850	7,900	1,181	1,181	1,181	1,181
7,900	7,950	1,189	1,189	1,189	1,189
7,950	8,000	1,196	1,196	1,196	1,196

8,000

At least	But less than	Single	Married filing jointly	Married filing separately	Head of a household
8,000	8,050	1,204	1,204	1,204	1,204
8,050	8,100	1,211	1,211	1,211	1,211
8,100	8,150	1,219	1,219	1,219	1,219
8,150	8,200	1,226	1,226	1,226	1,226
8,200	8,250	1,234	1,234	1,234	1,234
8,250	8,300	1,241	1,241	1,241	1,241
8,300	8,350	1,249	1,249	1,249	1,249
8,350	8,400	1,256	1,256	1,256	1,256
8,400	8,450	1,264	1,264	1,264	1,264
8,450	8,500	1,271	1,271	1,271	1,271
8,500	8,550	1,279	1,279	1,279	1,279
8,550	8,600	1,286	1,286	1,286	1,286
8,600	8,650	1,294	1,294	1,294	1,294
8,650	8,700	1,301	1,301	1,301	1,301
8,700	8,750	1,309	1,309	1,309	1,309
8,750	8,800	1,316	1,316	1,316	1,316
8,800	8,850	1,324	1,324	1,324	1,324
8,850	8,900	1,331	1,331	1,331	1,331
8,900	8,950	1,339	1,339	1,339	1,339
8,950	9,000	1,346	1,346	1,346	1,346

9,000

At least	But less than	Single	Married filing jointly	Married filing separately	Head of a household
9,000	9,050	1,354	1,354	1,354	1,354
9,050	9,100	1,361	1,361	1,361	1,361
9,100	9,150	1,369	1,369	1,369	1,369
9,150	9,200	1,376	1,376	1,376	1,376
9,200	9,250	1,384	1,384	1,384	1,384
9,250	9,300	1,391	1,391	1,391	1,391
9,300	9,350	1,399	1,399	1,399	1,399
9,350	9,400	1,406	1,406	1,406	1,406
9,400	9,450	1,414	1,414	1,414	1,414
9,450	9,500	1,421	1,421	1,421	1,421
9,500	9,550	1,429	1,429	1,429	1,429
9,550	9,600	1,436	1,436	1,436	1,436
9,600	9,650	1,444	1,444	1,444	1,444
9,650	9,700	1,451	1,451	1,451	1,451
9,700	9,750	1,459	1,459	1,459	1,459
9,750	9,800	1,466	1,466	1,466	1,466
9,800	9,850	1,474	1,474	1,474	1,474
9,850	9,900	1,481	1,481	1,481	1,481
9,900	9,950	1,489	1,489	1,489	1,489
9,950	10,000	1,496	1,496	1,496	1,496

10,000

At least	But less than	Single	Married filing jointly	Married filing separately	Head of a household
10,000	10,050	1,504	1,504	1,504	1,504
10,050	10,100	1,511	1,511	1,511	1,511
10,100	10,150	1,519	1,519	1,519	1,519
10,150	10,200	1,526	1,526	1,526	1,526
10,200	10,250	1,534	1,534	1,534	1,534
10,250	10,300	1,541	1,541	1,541	1,541
10,300	10,350	1,549	1,549	1,549	1,549
10,350	10,400	1,556	1,556	1,556	1,556
10,400	10,450	1,564	1,564	1,564	1,564
10,450	10,500	1,571	1,571	1,571	1,571
10,500	10,550	1,579	1,579	1,579	1,579
10,550	10,600	1,586	1,586	1,586	1,586
10,600	10,650	1,594	1,594	1,594	1,594
10,650	10,700	1,601	1,601	1,601	1,601
10,700	10,750	1,609	1,609	1,609	1,609
10,750	10,800	1,616	1,616	1,616	1,616
10,800	10,850	1,624	1,624	1,624	1,624
10,850	10,900	1,631	1,631	1,631	1,631
10,900	10,950	1,639	1,639	1,639	1,639
10,950	11,000	1,646	1,646	1,646	1,646

11,000

At least	But less than	Single	Married filing jointly	Married filing separately	Head of a household
11,000	11,050	1,654	1,654	1,654	1,654
11,050	11,100	1,661	1,661	1,661	1,661
11,100	11,150	1,669	1,669	1,669	1,669
11,150	11,200	1,676	1,676	1,676	1,676
11,200	11,250	1,684	1,684	1,684	1,684
11,250	11,300	1,691	1,691	1,691	1,691
11,300	11,350	1,699	1,699	1,699	1,699
11,350	11,400	1,706	1,706	1,706	1,706
11,400	11,450	1,714	1,714	1,714	1,714
11,450	11,500	1,721	1,721	1,721	1,721
11,500	11,550	1,729	1,729	1,729	1,729
11,550	11,600	1,736	1,736	1,736	1,736
11,600	11,650	1,744	1,744	1,744	1,744
11,650	11,700	1,751	1,751	1,751	1,751
11,700	11,750	1,759	1,759	1,759	1,759
11,750	11,800	1,766	1,766	1,766	1,766
11,800	11,850	1,774	1,774	1,774	1,774
11,850	11,900	1,781	1,781	1,781	1,781
11,900	11,950	1,789	1,789	1,789	1,789
11,950	12,000	1,796	1,796	1,796	1,796

12,000

At least	But less than	Single	Married filing jointly	Married filing separately	Head of a household
12,000	12,050	1,804	1,804	1,804	1,804
12,050	12,100	1,811	1,811	1,811	1,811
12,100	12,150	1,819	1,819	1,819	1,819
12,150	12,200	1,826	1,826	1,826	1,826
12,200	12,250	1,834	1,834	1,834	1,834
12,250	12,300	1,841	1,841	1,841	1,841
12,300	12,350	1,849	1,849	1,849	1,849
12,350	12,400	1,856	1,856	1,856	1,856
12,400	12,450	1,864	1,864	1,864	1,864
12,450	12,500	1,871	1,871	1,871	1,871
12,500	12,550	1,879	1,879	1,879	1,879
12,550	12,600	1,886	1,886	1,886	1,886
12,600	12,650	1,894	1,894	1,894	1,894
12,650	12,700	1,901	1,901	1,901	1,901
12,700	12,750	1,909	1,909	1,909	1,909
12,750	12,800	1,916	1,916	1,916	1,916
12,800	12,850	1,924	1,924	1,924	1,924
12,850	12,900	1,931	1,931	1,931	1,931
12,900	12,950	1,939	1,939	1,939	1,939
12,950	13,000	1,946	1,946	1,946	1,946

13,000

At least	But less than	Single	Married filing jointly	Married filing separately	Head of a household
13,000	13,050	1,954	1,954	1,954	1,954
13,050	13,100	1,961	1,961	1,961	1,961
13,100	13,150	1,969	1,969	1,969	1,969
13,150	13,200	1,976	1,976	1,976	1,976
13,200	13,250	1,984	1,984	1,984	1,984
13,250	13,300	1,991	1,991	1,991	1,991
13,300	13,350	1,999	1,999	1,999	1,999
13,350	13,400	2,006	2,006	2,006	2,006
13,400	13,450	2,014	2,014	2,014	2,014
13,450	13,500	2,021	2,021	2,021	2,021
13,500	13,550	2,029	2,029	2,029	2,029
13,550	13,600	2,036	2,036	2,036	2,036
13,600	13,650	2,044	2,044	2,044	2,044
13,650	13,700	2,051	2,051	2,051	2,051
13,700	13,750	2,059	2,059	2,059	2,059
13,750	13,800	2,066	2,066	2,066	2,066
13,800	13,850	2,074	2,074	2,074	2,074
13,850	13,900	2,081	2,081	2,081	2,081
13,900	13,950	2,089	2,089	2,089	2,089
13,950	14,000	2,096	2,096	2,096	2,096

* This column must also be used by a qualifying widow(er).

Continued on next page

If line 37 (taxable income) is— At least	But less than	Single	Married filing jointly *	Married filing separately	Head of a household
			Your tax is—		
14,000					
14,000	14,050	2,104	2,104	2,104	2,104
14,050	14,100	2,111	2,111	2,111	2,111
14,100	14,150	2,119	2,119	2,119	2,119
14,150	14,200	2,126	2,126	2,126	2,126
14,200	14,250	2,134	2,134	2,134	2,134
14,250	14,300	2,141	2,141	2,141	2,141
14,300	14,350	2,149	2,149	2,149	2,149
14,350	14,400	2,156	2,156	2,156	2,156
14,400	14,450	2,164	2,164	2,164	2,164
14,450	14,500	2,171	2,171	2,171	2,171
14,500	14,550	2,179	2,179	2,179	2,179
14,550	14,600	2,186	2,186	2,186	2,186
14,600	14,650	2,194	2,194	2,194	2,194
14,650	14,700	2,201	2,201	2,201	2,201
14,700	14,750	2,209	2,209	2,209	2,209
14,750	14,800	2,216	2,216	2,216	2,216
14,800	14,850	2,224	2,224	2,224	2,224
14,850	14,900	2,231	2,231	2,231	2,231
14,900	14,950	2,239	2,239	2,239	2,239
14,950	15,000	2,246	2,246	2,246	2,246
15,000					
15,000	15,050	2,254	2,254	2,254	2,254
15,050	15,100	2,261	2,261	2,261	2,261
15,100	15,150	2,269	2,269	2,269	2,269
15,150	15,200	2,276	2,276	2,276	2,276
15,200	15,250	2,284	2,284	2,284	2,284
15,250	15,300	2,291	2,291	2,291	2,291
15,300	15,350	2,299	2,299	2,299	2,299
15,350	15,400	2,306	2,306	2,306	2,306
15,400	15,450	2,314	2,314	2,314	2,314
15,450	15,500	2,321	2,321	2,321	2,321
15,500	15,550	2,329	2,329	2,329	2,329
15,550	15,600	2,336	2,336	2,336	2,336
15,600	15,650	2,344	2,344	2,344	2,344
15,650	15,700	2,351	2,351	2,351	2,351
15,700	15,750	2,359	2,359	2,359	2,359
15,750	15,800	2,366	2,366	2,366	2,366
15,800	15,850	2,374	2,374	2,374	2,374
15,850	15,900	2,381	2,381	2,381	2,381
15,900	15,950	2,389	2,389	2,389	2,389
15,950	16,000	2,396	2,396	2,396	2,396
16,000					
16,000	16,050	2,404	2,404	2,404	2,404
16,050	16,100	2,411	2,411	2,411	2,411
16,100	16,150	2,419	2,419	2,419	2,419
16,150	16,200	2,426	2,426	2,426	2,426
16,200	16,250	2,434	2,434	2,434	2,434
16,250	16,300	2,441	2,441	2,441	2,441
16,300	16,350	2,449	2,449	2,449	2,449
16,350	16,400	2,456	2,456	2,456	2,456
16,400	16,450	2,464	2,464	2,464	2,464
16,450	16,500	2,471	2,471	2,471	2,471
16,500	16,550	2,479	2,479	2,479	2,479
16,550	16,600	2,486	2,486	2,486	2,486
16,600	16,650	2,494	2,494	2,494	2,494
16,650	16,700	2,501	2,501	2,501	2,501
16,700	16,750	2,509	2,509	2,509	2,509
16,750	16,800	2,516	2,516	2,516	2,516
16,800	16,850	2,524	2,524	2,524	2,524
16,850	16,900	2,531	2,531	2,531	2,531
16,900	16,950	2,539	2,539	2,539	2,539
16,950	17,000	2,546	2,546	2,546	2,546

If line 37 (taxable income) is— At least	But less than	Single	Married filing jointly *	Married filing separately	Head of a household
			Your tax is—		
17,000					
17,000	17,050	2,554	2,554	2,554	2,554
17,050	17,100	2,561	2,561	2,561	2,561
17,100	17,150	2,569	2,569	2,569	2,569
17,150	17,200	2,576	2,576	2,576	2,576
17,200	17,250	2,584	2,584	2,584	2,584
17,250	17,300	2,591	2,591	2,591	2,591
17,300	17,350	2,599	2,599	2,599	2,599
17,350	17,400	2,606	2,606	2,606	2,606
17,400	17,450	2,614	2,614	2,614	2,614
17,450	17,500	2,621	2,621	2,621	2,621
17,500	17,550	2,629	2,629	2,629	2,629
17,550	17,600	2,636	2,636	2,636	2,636
17,600	17,650	2,644	2,644	2,644	2,644
17,650	17,700	2,651	2,651	2,651	2,651
17,700	17,750	2,659	2,659	2,659	2,659
17,750	17,800	2,666	2,666	2,666	2,666
17,800	17,850	2,674	2,674	2,674	2,674
17,850	17,900	2,681	2,681	2,681	2,681
17,900	17,950	2,689	2,689	2,689	2,689
17,950	18,000	2,696	2,696	2,696	2,696
18,000					
18,000	18,050	2,704	2,704	2,704	2,704
18,050	18,100	2,711	2,711	2,711	2,711
18,100	18,150	2,719	2,719	2,719	2,719
18,150	18,200	2,726	2,726	2,726	2,726
18,200	18,250	2,734	2,734	2,734	2,734
18,250	18,300	2,741	2,741	2,741	2,741
18,300	18,350	2,749	2,749	2,749	2,749
18,350	18,400	2,756	2,756	2,756	2,756
18,400	18,450	2,764	2,764	2,764	2,764
18,450	18,500	2,771	2,771	2,771	2,771
18,500	18,550	2,779	2,779	2,779	2,779
18,550	18,600	2,786	2,786	2,786	2,786
18,600	18,650	2,794	2,794	2,794	2,794
18,650	18,700	2,801	2,801	2,801	2,801
18,700	18,750	2,809	2,809	2,809	2,809
18,750	18,800	2,816	2,816	2,816	2,816
18,800	18,850	2,824	2,824	2,824	2,824
18,850	18,900	2,831	2,831	2,831	2,831
18,900	18,950	2,839	2,839	2,839	2,839
18,950	19,000	2,846	2,846	2,846	2,846
19,000					
19,000	19,050	2,854	2,854	2,857	2,854
19,050	19,100	2,861	2,861	2,871	2,861
19,100	19,150	2,869	2,869	2,885	2,869
19,150	19,200	2,876	2,876	2,899	2,876
19,200	19,250	2,884	2,884	2,913	2,884
19,250	19,300	2,891	2,891	2,927	2,891
19,300	19,350	2,899	2,899	2,941	2,899
19,350	19,400	2,906	2,906	2,955	2,906
19,400	19,450	2,914	2,914	2,969	2,914
19,450	19,500	2,921	2,921	2,983	2,921
19,500	19,550	2,929	2,929	2,997	2,929
19,550	19,600	2,936	2,936	3,011	2,936
19,600	19,650	2,944	2,944	3,025	2,944
19,650	19,700	2,951	2,951	3,039	2,951
19,700	19,750	2,959	2,959	3,053	2,959
19,750	19,800	2,966	2,966	3,067	2,966
19,800	19,850	2,974	2,974	3,081	2,974
19,850	19,900	2,981	2,981	3,095	2,981
19,900	19,950	2,989	2,989	3,109	2,989
19,950	20,000	2,996	2,996	3,123	2,996

If line 37 (taxable income) is— At least	But less than	Single	Married filing jointly *	Married filing separately	Head of a household
			Your tax is—		
20,000					
20,000	20,050	3,004	3,004	3,137	3,004
20,050	20,100	3,011	3,011	3,151	3,011
20,100	20,150	3,019	3,019	3,165	3,019
20,150	20,200	3,026	3,026	3,179	3,026
20,200	20,250	3,034	3,034	3,193	3,034
20,250	20,300	3,041	3,041	3,207	3,041
20,300	20,350	3,049	3,049	3,221	3,049
20,350	20,400	3,056	3,056	3,235	3,056
20,400	20,450	3,064	3,064	3,249	3,064
20,450	20,500	3,071	3,071	3,263	3,071
20,500	20,550	3,079	3,079	3,277	3,079
20,550	20,600	3,086	3,086	3,291	3,086
20,600	20,650	3,094	3,094	3,305	3,094
20,650	20,700	3,101	3,101	3,319	3,101
20,700	20,750	3,109	3,109	3,333	3,109
20,750	20,800	3,116	3,116	3,347	3,116
20,800	20,850	3,124	3,124	3,361	3,124
20,850	20,900	3,131	3,131	3,375	3,131
20,900	20,950	3,139	3,139	3,389	3,139
20,950	21,000	3,146	3,146	3,403	3,146
21,000					
21,000	21,050	3,154	3,154	3,417	3,154
21,050	21,100	3,161	3,161	3,431	3,161
21,100	21,150	3,169	3,169	3,445	3,169
21,150	21,200	3,176	3,176	3,459	3,176
21,200	21,250	3,184	3,184	3,473	3,184
21,250	21,300	3,191	3,191	3,487	3,191
21,300	21,350	3,199	3,199	3,501	3,199
21,350	21,400	3,206	3,206	3,515	3,206
21,400	21,450	3,214	3,214	3,529	3,214
21,450	21,500	3,221	3,221	3,543	3,221
21,500	21,550	3,229	3,229	3,557	3,229
21,550	21,600	3,236	3,236	3,571	3,236
21,600	21,650	3,244	3,244	3,585	3,244
21,650	21,700	3,251	3,251	3,599	3,251
21,700	21,750	3,259	3,259	3,613	3,259
21,750	21,800	3,266	3,266	3,627	3,266
21,800	21,850	3,274	3,274	3,641	3,274
21,850	21,900	3,281	3,281	3,655	3,281
21,900	21,950	3,289	3,289	3,669	3,289
21,950	22,000	3,296	3,296	3,683	3,296
22,000					
22,000	22,050	3,304	3,304	3,697	3,304
22,050	22,100	3,311	3,311	3,711	3,311
22,100	22,150	3,319	3,319	3,725	3,319
22,150	22,200	3,326	3,326	3,739	3,326
22,200	22,250	3,334	3,334	3,753	3,334
22,250	22,300	3,341	3,341	3,767	3,341
22,300	22,350	3,349	3,349	3,781	3,349
22,350	22,400	3,356	3,356	3,795	3,356
22,400	22,450	3,364	3,364	3,809	3,364
22,450	22,500	3,371	3,371	3,823	3,371
22,500	22,550	3,379	3,379	3,837	3,379
22,550	22,600	3,386	3,386	3,851	3,386
22,600	22,650	3,394	3,394	3,865	3,394
22,650	22,700	3,401	3,401	3,879	3,401
22,700	22,750	3,409	3,409	3,893	3,409
22,750	22,800	3,420	3,416	3,907	3,416
22,800	22,850	3,434	3,424	3,921	3,424
22,850	22,900	3,448	3,431	3,935	3,431
22,900	22,950	3,462	3,439	3,949	3,439
22,950	23,000	3,476	3,446	3,963	3,446

* This column must also be used by a qualifying widow(er).

Continued on next page

EXAMPLE _____

Use the following steps to find the amount of federal income tax due for $17,525 taxable income, married filing jointly.

STEPS

1. Locate the $17,000 section on the tax table.

2. Beginning at the upper left column entitled "At least," move down that column until you find $17,500, then read across to note the amount $17,550 in the "But less than" column. This entry would read at least $17,500 but less than $17,550.

3. Notice at the top of the table the four categories just discussed. Continue moving across the "at least $17,500 but less than $17,550" line until you are under the column heading "married filing jointly." The amount of money shown in the column, $2,629, is the amount of federal income tax due for this year.

12.10 Computing Federal Income Tax

There are several versions of the federal tax form. You may complete the 1040EZ short form if you do not have enough deductions to itemize and if you had only wages, salaries, tips, taxable scholarships and fellowship grants, and you earned less than $400 in interest income, if you were single or married filing jointly and do not claim any dependents, were not 65 or older, or blind, and your taxable income is less than $100,000; if you did not receive any advance earned income credit payments; if you do not itemize deductions or claim any adjustments to income or tax credits; if your total wages were not over $60,600 if you had more than one employer. To complete the 1040EZ short form and compute taxable income, the following formula is used.

Taxable Income = Total Income – Standard Deduction – Exemptions

You would find your tax, based on your taxable income, in the tax table provided each year by the Internal Revenue Service. Here is a sample return based on the Form 1040EZ.

EXAMPLE _____

Bryan Phillips	$11,350
Plus interest income (savings account)	$780
Adjusted gross income	$12,130
Less standard deduction (single)	$3,800
Less exemptions (1)	$2,450
Taxable Income	$5,880
Income tax withheld	$1,005
Federal income tax due (single from tax table)	$881

If tax withheld is larger than tax due, you get a refund.

It tax withheld is smaller than tax due, you must pay.

Bryan received a $169 refund ($1,050 – $881 = $169).

JoAnne White is single and earned $19,500 in wages and $560 in interest from her savings account. JoAnne has no exemptions other than herself. Compute JoAnne's tax by completing the following blanks.

Total income (wages)	$19,500	
Plus interest income (savings account)	$560	
Adjusted gross income	_____	**(1.)**
Less standard deduction (single)	$3,800	
Less exemptions (1)	$2,450	
Taxable income	_____	**(2.)**
Income tax withheld	$3,245	
Federal income tax due (single)	_____	**(3.)**
Refund (if any)	_____	**(4.)**

Solutions: **1.** $20,060; **2.** $13,810; **3.** $2,074; **4.** $1,171

12.11 Form 1040

If you itemize your deductions (itemized deductions are listed on Schedule A), you would complete Form 1040 and compute your taxable income using the following formula.

$$\begin{array}{ccccccc} \text{Total} & - & \text{Adjustments} & = & \text{Adjusted} & - & \begin{array}{c}\text{Standard}\\\text{Deduction}\\\text{or}\\\text{Itemized}\\\text{Deductions}\end{array} & - & \text{Exemptions} & = & \text{Taxable} \\ \text{Income} & & \text{to Income} & & \text{Income} & & & & & & \text{Income} \end{array}$$

You would find your tax based on your taxable income from the tax tables provided by the Internal Revenue Service each year. Here is an example return based on Form 1040.

EXAMPLE _____

In 1994 Thomas Winters and his wife, Ann, filed a joint return. Thomas made $20,380 in wages, and Ann made $8,329 working part time. They have 1 child. Their itemized deductions are: $1,230, real estate tax; $480 personal property tax; $3,910, home mortgage interest; $5,600, medical expense; $2,800, charitable contributions. Thomas and Ann computed their tax as shown.

Thomas' income	$20,380
Plus Ann's income	$8,329
Total income	$28,709
Minus adjustments to income	– 0 –
Adjusted Gross Income	**$28,709**

Less itemized deductions (Schedule A):

Real estate tax	$1,230
Personal property tax	$480
Home mortgage interest	$3,910

Medical expense:

Total medical expenses	= $5,600	
Adjusted gross income times 7.5%	= −2,153 (rounded)	
Total deductible medical expense		$3,447

Charitable contributions	$2,800
Total deductions	**$11,867**
Less exemptions (3 × 2,450)	$7,350
Taxable income	$9,492
Federal income tax due (married filing jointly)	**$1,421**

■ **TIP**   When you itemize your deductions, (married filing jointly) if the amount is over $6,350, you use that figure. If your deductions are less than $6,350 you are allowed by law the $6,350 standard deduction anyway.

PRACTICE PROBLEMS

In 1994 Julia and Carlos Torres filed a joint return. Julia earned $18,300 in wages, and Carlos earned $17,500. They have 3 children. Julia contributes $800 a year to an IRA. Their itemized deductions are: $1,599, real estate tax; $399, personal property tax; $4,151, total medical expenses; $10,558, home mortgage interest; and $500, charitable contributions. Compute the amount of deductible medical expenses. Compute Julia and Carlos' tax.

Julia's income	$18,300	
Plus Carlos' income	$17,500	
Total income	_____	(1.)
Minus adjustments to income (employee IRA)	$800	
Adjusted gross income	_____	(2.)

Less itemized deductions (Schedule A):

Real estate tax	$1,599	
Personal property tax	$399	
Medical expense deductible	_____	(3.)
Home mortgage interest	$10,558	
Charitable contributions	$500	
Total deductions	_____	(4.)
Less exemptions (5 × $2,450)	_____	(5.)
Taxable income	_____	(6.)
Federal income tax due (married filing jointly)	_____	(7.)

Solutions: **1.** $35,800; **2.** $35,000; **3.** $1,526; **4.** $14,582
5. $12,250; **6.** $8,168; **7.** $1,226

MATH ALERT

So that you might actually experience working with a 1994 tax form, complete the 1040EZ form provided on the next page. Assume that you are single. Follow these steps:

A. Name & Address section.

Use your name, address, and social security number. Mark an "X" in the appropriate box indicating whether or not you wish to contribute $3 toward the Presidential Election Campaign.

B. Report your income section.

1. Assume you made $12,950 in total wages working part time while you were in school in 1994. Enter that figure in the appropriate boxes to the right.

2. Assume you earned $275 in interest income. Enter that figure in the appropriate boxes to the right.

3. Add line 1 and 2 to calculate your adjusted gross income.

4. Mark an "X" in the No box and enter your standard deduction and personal exemption of $6,250 in the appropriate boxes to the right.

5. Subtract line 4 (total exemptions and deductions) from Line 3 (adjusted gross income) to get your taxable income.

C. Figure you tax section.

6. Assume you had $1,349 in federal income tax withheld. You do not qualify for an earned income credit. Enter this figure in the appropriate boxes to the right.

7. Use the table on page 241 and look up your tax. Enter that figure in the appropriate boxes to the right.

D. Refund or amount you owe section.

Complete either line 8 or line 9 whichever is appropriate

8. If line 8 (taxes paid in) is larger than line 9 (tax owed), subtract line 9 from line 8. This is you refund.

9. If line 9 (tax owed) is larger than line 8 (tax paid in), subtract line 8 from line 9. This is the amount you owe. You would attach a check for your payment in full payable to the "Internal Revenue Service" and mail it with this form.

E. Sign your name section.

Read the statement that is bolded, then sign, date, and enter your occupation as "student" on the form.

Form
1040EZ

Department of the Treasury—Internal Revenue Service

Income Tax Return for Single and Joint Filers With No Dependents 1994 (P)

OMB No. 1545-0675

Use the IRS label (See page 12.) Otherwise, please print.

L
A
B
E
L

H
E
R
E

Print your name (first, initial, last)

If a joint return, print spouse's name (first, initial, last)

Home address (number and street). If you have a P.O. box, see page 12.   Apt. no.

City, town or post office, state and ZIP code. If you have a foreign address, see page 12.

Your social security number

Spouse's social security number

See instructions on back and in Form 1040EZ booklet.

Presidential Election Campaign (See page 12.)

Note: *Checking "Yes" will not change your tax or reduce your refund.*
Do you want $3 to go to this fund?   ▶
If a joint return, does your spouse want $3 to go to this fund? ▶

Yes   No

Income

Attach Copy B of Form(s) W-2 here. Enclose, but do not attach, any payment with your return.

Note: *You must check Yes or No.*

Dollars   Cents

1 Total wages, salaries, and tips. This should be shown in box 1 of your W-2 form(s). Attach your W-2 form(s).   **1**

2 Taxable interest income of $400 or less. If the total is over $400, you cannot use Form 1040EZ.   **2**

3 Add lines 1 and 2. This is your **adjusted gross income.** If less than $9,000, see page 15 to find out if you can claim the earned income credit on line 7.   **3**

4 Can your parents (or someone else) claim you on their return?
☐ **Yes.** Do worksheet on back; enter amount from line G here.   ☐ **No.** If **single,** enter 6,250.00. If **married,** enter 11,250.00. For an explanation of these amounts, see back of form.   **4**

5 Subtract line 4 from line 3. If line 4 is larger than line 3, enter 0. This is your **taxable income.**   ▶ **5**

Payments and tax

6 Enter your Federal income tax withheld from box 2 of your W-2 form(s).   **6**

7 **Earned income credit** (see page 15). Enter type and amount of nontaxable earned income below.

| Type | | |

7

8 Add lines 6 and 7 (don't include nontaxable earned income). These are your **total payments.**   **8**

9 **Tax.** Use the amount on **line 5** to find your tax in the tax table on pages 28–32 of the booklet. Then, enter the tax from the table on this line.   **9**

Refund or amount you owe

10 If line 8 is larger than line 9, subtract line 9 from line 8. This is your **refund.**   **10**

11 If line 9 is larger than line 8, subtract line 8 from line 9. This is the **amount you owe.** See page 20 for details on how to pay and what to write on your payment.   **11**

Sign your return

Keep a copy of this form for your records.

I have read this return. Under penalties of perjury, I declare that to the best of my knowledge and belief, the return is true, correct, and accurately lists all amounts and sources of income I received during the tax year.

| Your signature | Spouse's signature if joint return |
| Date | Your occupation | Date | Spouse's occupation |

For IRS Use Only — Please do not write in boxes below.

For Privacy Act and Paperwork Reduction Act Notice, see page 4.   Cat. No. 12616G   Form 1040EZ (1994)

Study Guide

I. Terminology

page 232	property tax	A tax imposed on a property owner to help fund public services such as fire, police protection, and schools.
page 232	market value	The amount of money the property will bring in a competitive and open market.
page 232	assessed rate	An arbitrary rate set by the taxing body.
page 232	assessed value	The amount of money for which property is listed on the books (tax records) for tax purposes.
page 233	tax rate	A certain percentage based on the assessed value of the property.
page 233	tax	The amount of money paid by the property owner.
page 237	employee's withholding allowance certificate	W-4 form for an employee to claim the number of exemptions he/she wishes to claim.
page 238	wage and tax statement	W-2 form sent to the employee and the IRS each year showing the amount of wages, federal income tax, social security, and medicare taxes withheld.
page 239	Form 1099-INT	A form used by your bank or other savings institution and the IRS reporting the amount of interest income you earned.
page 239	exemptions	Each dependent in your household including yourself.
page 239	adjustments to income	Adjustments to your total income for such items as IRAs, self-employment tax, Keogh retirement plans, to name a few.
page 239	adjusted gross income	Total income minus total adjustments.
page 240	itemized deductions	Deductions allowed on your income tax form as set by law.
page 240	standard deduction	A deduction allowed if you choose not to itemize your deductions.
page 240	taxable income	The amount of income subject to tax.
page 240	tax tables	Tables provided by the IRS for you or your tax preparer to determine the amount of income tax you owe.

II. Property Tax

page 233   *Computing the amount of property tax:* Multiply the market value of the property by the assessed rate; determine the number of $1,000s in the assessed value; multiply the number of $1,000s by the tax rate per $1,000 to determine the amount of the tax due.

Example: Property assessed: 50% of market value
Tax rate: $6.95 per $1,000 valuation
Market value of property: $100,000

$100,000 × 50% = $50,000 assessed value

$50,000 ÷ $1,000 = 50

50 × $6.95 = $347.50 tax due

page 234   *Determining the tax rate:* Divide the amount of money needed by the total assessed value of property to obtain the tax rate.

Example: Money needed: $50,000,000
Total assessed value: $1,000,000,000

$50,000,000 ÷ $1,000,000,000 = 0.0500 or 5% tax rate

II. State Unemployment Insurance Tax (SUTA)

page 235   *Computing the amount of state unemployment insurance tax due:* Multiply the percent tax rate set by governing authorities by the taxable wages.

Example: Taxable Wages: $219,300
         SUTA rate: 3% $219,300 × 0.03 = $6,579 tax due

III. Federal Unemployment Insurance Tax (FUTA)

page 236   *Computing the amount of federal unemployment insurance tax due:* Multiply the percent tax rate set by governing authorities by the taxable wages.

Example: Taxable Wages: $219,300 $219,300 × 0.008 = $1,754 tax due
         FUTA rate: 0.8%   (rounded to the nearest dollar)

IV. Federal Income Tax 1994 Tax Information

page 243   *Computing income tax if filing the Form 1040EZ:* Add all sources of wages; subtract $3,800 standard deduction; subtract $2,450 exemption to obtain taxable income; look up tax owed from tax table; subtract tax owed from amount of tax paid in to arrive at tax due or refund due.

Example:	
Total Wages	$11,350
Plus interest income	780
Adjusted gross income	$12,130
Less standard deduction (single)	3,800
Less exemptions (1)	2,450
Taxable income	$5,880
Income tax withheld	1,050
Less federal income tax due	881
Refund (if any)	$169

page 244   *Computing income tax if filing the Form 1040:* Add all sources of income then subtract any adjustments to income to arrrive at adjusted gross income; list deductions (calculate medical expenses at 7.5% of adjusted gross income); multiply the number of exemptions by $2,450; subtract from adjusted gross income itemized deductions and total exemptions to determine taxable income. Look up taxable income on tax table. If you paid in more than you owed your get a refund; if you did not pay enough you must deduct the amount you paid in from the amount owed and pay the difference.

Example: A married couple file a joint return. The husband earned $17,500 and the wife earned $18,200 in wages. They have 3 children. The wife contributes $800 a year to an IRA for retirement. Their itemized deductions are: $1,599 real estate tax; $399 personal property tax; $4,151 medical expenses; $10,558 home mortgage interest; and $500 charitable contributions. Here is how their tax would be computed:

Husband's income	$17,500
Wife's income	$18,300
Total income	$35,800
Minus adjustments to income	$800
Adjusted gross income	$35,000
Less itemized deductions (Schedule A)	
Real estate tax	$1,599
Personal property tax	$399
Medical expenses deductible (Total expenses less 7.5% of adjusted gross income)	$1,526
Home mortgage interest	$10,558
Charitable contributions	$500
Total deductions	$14,582
Less exemptions (5 × $2,450)	$12,250
Taxable income	$8,168
Federal income tax due (married filing jointly)	$1,226

Assignment 1

Name_____ Date _____

Complete the following problems. Write your answers in the blanks provided.

A. Compute the amount of property tax due.

	Market Value	Assessed Value	Tax Rate		Tax Due
1.	$51,000	80%	$1.50 per $1,000	=	_____
2.	$65,000	60%	$2.05 per $1,000	=	_____
3.	$150,000	50%	$3.95 per $10,000	=	_____
4.	$49,000	65%	$2.50 per $1,000	=	_____
5.	$900,000	80%	$50 per $10,000	=	_____
6.	$75,000	80%	$1.75 per $1,000	=	_____
7.	$108,000	50%	$4 per $10,000	=	_____
8.	$86,000	65%	$2.75 per $1,000	=	_____
9.	$1,000,000	80%	$1.95 per $10,000	=	_____
10.	$92,000	70%	$1.55 per $1,000	=	_____

B. Compute the property tax rate. Carry your answers to 4 places, and round to the nearest whole percent.

	Money Needed	Assessed Value		Tax Rate
11.	$800,000	$10,000,000	=	_____
12.	$62,000	$900,000	=	_____
13.	$45,000	$775,000	=	_____
14.	$208,000	$1,000,000	=	_____
15.	$139,000	$3,500,000	=	_____
16.	$80,000	$800,000	=	_____
17.	$149,000	$21,000,000	=	_____
18.	$500,000	$1,000,000	=	_____
19.	$200,000	$3,000,000	=	_____
20.	$1 billion	$6 billion	=	_____

Assignment 2

Name_____   Date _____

Complete the following problems. Write your answers in the blanks provided.

A. Compute the state unemployment tax due. Use 2.8% up to and including $9,000.

1. $5,600 gross wages   = _____ tax due   6. $7,715 gross wages   = _____ tax due

2. $8,150 gross wages   = _____ tax due   7. $10,250 gross wages = _____ tax due

3. $8,990 gross wages   = _____ tax due   8. $9,850 gross wages   = _____ tax due

4. $9,000 gross wages   = _____ tax due   9. $15,950 gross wages = _____ tax due

5. $5,142 gross wages   = _____ tax due   10. $21,800 gross wages = _____ tax due

B. Compute the federal unemployment tax due. Use 0.8% up to and including $7,000.

11. $5,990 gross wages   = _____ tax due   16. $5,000 gross wages   = _____ tax due

12. $9,560 gross wages   = _____ tax due   17. $4,225 gross wages   = _____ tax due

13. $10,200 gross wages = _____ tax due   18. $3,333 gross wages   = _____ tax due

14. $6,495 gross wages   = _____ tax due   19. $2,059 gross wages   = _____ tax due

15. $7,800 gross wages   = _____ tax due   20. $15,615 gross wages = _____ tax due

C. Using the income tax table on page 240, determine the amount of tax due for a
    couple married filing jointly.

Taxable Income	Tax Due		Taxable Income	Tax Due
21. $6,775	_____		26. $10,990	_____
22. $12,950	_____		27. $11,335	_____
23. $22,050	_____		28. $7,805	_____
24. $8,800	_____		29. $18,510	_____
25. $9,211	_____		30. $21,215	_____

Assignment 3   Name_____ Date _____

**Complete the following problems. Write your answers in the blanks provided.
Use the tax table on pages 240 and 241. If a refund is due, indicate the amount
with the word, "Refund" beside it.**

Compute federal income tax: Form 1040EZ.

Mary Bailey is single. Her total wages for the year were $26,000. She claims no dependents other
than herself. Her W-2 form shows she had $2,995 in federal income tax withheld. How much more
would Mary have to remit with her tax return?

1. Taxable income: _____   **2.** Tax due: _____

Rachel Greenwood is single. Her total wages for the year were $20,800. She claims no depend-
ents other than herself. How much total tax will she have to pay?

3. Taxable income: _____   **4.** Tax due: _____

Rosa Martinez is single. Her total wages were $19,500 and her interest earned from a savings
account was $350. She claimed herself as a dependent. How much total tax should Rosa have paid?

5. Taxable income: _____   **6.** Tax due: _____

Danny Roberts is a senior in high school and works part time at a local grocery store chain. Danny's
total wages were $10,710, and he had $1,150 in federal income tax withheld during the year. He
claims 0 exemptions because his father claims him. What should Danny pay in federal income tax?

7. Taxable income: _____   **8.** Tax due: _____

Jill Watkins earned $15,900 from a job she was laid off from in June. Jill began working for another
company and earned $3,000 from July to December. She earned $380 interest from a savings
account. Jill is single and claims herself as a dependent. Jill had $1,950 in federal income tax
withheld from her first job and $940 from her second job. What additional tax will Jill have to pay?

9. Taxable income: _____   **10.** Tax due: _____

Kyle O'Nealy is single and earned $18,500 at Hillside Manufacturing Company. He had paid $1,946
in federal taxes. How much tax does Kyle owe?

11. Taxable income: _____   **12.** Tax due: _____

Berry Ellis is single and earned $14,500 as a secretary for a law firm. Berry also worked for her
father on weekends and earned $5,950. How much additional tax should Berry pay if she had
already paid in $2,580?

13. Taxable income: _____   **14.** Tax due: _____

Les Pittman is single and earned $13,600 as a machine operator. He paid in $1,565 in taxes.
How much additional tax does he owe?

15. Taxable income: _____   **16.** Tax due: _____

Assignment 4

Name_____ Date _____

Complete the following problems. Write your answers in the blanks provided. Use the tax table on pages 240 and 241. If a refund is due, indicate the amount due with the word "Refund" beside it.

Compute federal income tax: Form 1040. Assume all itemized deductions are fully deductible.

Roy Jenkins and his wife, Alice, are filing a joint return. Roy's salary is $25,000 a year, and Alice does not work outside the home. They have 1 child, age 8. Their itemized deductions are: $1,110, real estate tax; $440, personal property tax; $2550, home mortgage interest; and $2,100, charitable contributions. How much tax should Roy and Alice pay?

1. Taxable income: _____   **2.** Tax due: _____

Beth is divorced, and her earnings are $28,000. She is filing as head of the household. She claims 4 dependents including herself. Her itemized deductions are: $3,250, real estate tax; $310, personal property tax; $2,500, home mortgage interest; and $600, charitable contributions. How much tax should Beth pay?

3. Taxable income: _____   **4.** Tax due: _____

Terry Thompson earned $22,000, and his wife, Theresa, earned $12,000 working part time. Their interest income is $2,500. They have 1 child. Their itemized deductions are: $1,700, real estate tax; $350, personal property tax; $12,000 home mortgage interest; and $1,200, charitable contributions. How much tax should they pay if they file a joint return?

5. Taxable income: _____   **6.** Tax due: _____

Tony Ventura and his wife, Margaret, are filing a joint return. They had $5,400 in federal income tax withheld during the year. They have a total income of $41,000 and have 2 children. their itemized deductions are: $2,500, real estate tax; $10,900, home mortgage interest; and $5,400, charitable contributions. How much additional tax should they pay?

7. Taxable income: _____   **8.** Tax due: _____

Manuel Ortega earned $30,000. His wife, Katrina, earned $11,000. They are filing a joint return and claim 5 dependents, including themselves. The Ortegas have an additional deduction for the alimony Manuel pays his first wife. Therefore, the adjustment to income figure is $6,000. Their itemized deductions are: $600, real estate tax; $9,000, home mortgage interest; $650, personal property tax; and $5,000, charitable contributions. What is their total taxable income? What is the total tax they should pay?

9. Taxable income: _____   **10.** Tax due: _____

CHAPTER
Insurance
13

CHAPTER

13
Insurance

Businesses and individuals may purchase protection against financial loss due to fire, theft, or accident. This protection is called insurance. Insurance is tailored to meet the individual needs of a person or a business. There are many types of insurance available. In this chapter you will learn about hospitalization, auto, life, and property insurance.

13.1 Terms Used in Insurance

Certain terms are basic to all types of insurance. These general terms are discussed here; the more specific terminology relating to each type of insurance will be discussed as each area is covered.

Insurance. Protection against loss.

Face Value. The amount of insurance purchased.

Premium. Payment to the insurance company for the purchase of an insurance policy.

Term. The time for which the insurance policy is in effect.

Insured. Individual or company receiving the insurance protection.

Policy. A written contract between the insurance company and the insured that explains the benefits of the protection purchased.

Claim. A form filed to request payment for losses covered in an insurance policy.

Beneficiary. Individual who is designated to receive the value of a life insurance policy in the event of the insured's death.

Carrier. The insurance company that "underwrites" or "insures."

Policyholder. The person or business that purchases insurance.

13.2 Health Insurance

Hospitalization insurance is one of the most common types of health insurance. An individual may purchase insurance that will provide protection against incurred costs due to various types of illnesses or accidents. Most employers provide, as a fringe benefit, some type of health insurance coverage. A company may receive cheaper rates than an individual because it can provide coverage for its workers under group plans available through many insurance companies. Some companies pay a portion of the premiums, some pay all, while others don't pay any of the premiums.

Because of the increasing cost of health care, plans such as a Health Maintenance Organization (HMO) or various Preferred Provider Organization (PPO) networks have become available. These plans provide a list of preferred health care providers and hospitals that participate in the plan. If the insured selects a doctor or hospital from this preferred list, the network organization pays a higher percentage of the cost of the health care than if the insured used their own doctor or hospital. PPO's are usually owned by local hospitals and are designed to offer coverage at a more reasonable cost.

Participation by the employee in a group coverage plan is usually voluntary, and an individual may be offered the choice of varying amounts of coverage. The amount of premium an employee pays depends on the amount of coverage he or she selects. The insurance company providing the coverage determines the amount of the premiums. This cost depends on the type of coverage desired, whether there is a deductible amount, and, if so, how much.

EXAMPLE

Assume that Riley Manufacturing Company has contracted with Health Care Insurance Company for a group plan. Riley has agreed to pay $75 of each employee's premium per month. The employee is offered two types of coverage: 100% coverage (Plan A) or $100 deductible (Plan B). Under Plan A, the insurance covers 100% of the medical costs the employee incurs; under Plan B, the employee pays the first $100 of medical expenses and the insurance covers the remainder. Under both plans monthly premiums are calculated the same way. The costs of these plans to individual employees are computed as follows:

Plan A: Annual premium, 100% coverage

Individual: $1,950

Family: Additional $75 monthly

To determine the monthly premium follow these steps.

STEPS

1. Divide individual annual premium by 12 months.

$$\$1,950 \div 12 = \$162.50 \text{ per month}$$

2. For family monthly premium, add $75 to individual monthly premium ($162.50).

$$\$162.50 + \$75 = \$237.50 \text{ per month}$$

3. Subtract amount of premium company will pay ($75).

Individual: $162.50 − $75 = $87.50 per month

Family: $237.50 − $75 = $162.50 per month

Plan B: Annual premium, $100 deductible

Individual: $1,600

Family: Additional $75 monthly for family coverage

To determine the monthly premium, follow the steps as in Plan A:

Individual: $1,600 ÷ 12 = $133.33 per month

Family: $133.33 + $75 = $208.33 per month

Individual: $133.33 – $75 = $58.33 per month

Family: $208.33 – $75 = $133.33 per month

The employee must pay the first $100 of medical expenses and the insurance covers the rest.

⎯ PRACTICE PROBLEMS ⎯⎯⎯⎯⎯⎯⎯⎯⎯

Determine the monthly premium for individual and family coverage and write your answers in the blanks provided. Round all answers to the nearest penny.

Plan A: Individual annual premium $1,410 (additional $80 a month for family coverage). The company pays nothing toward the individual's insurance premium.

1. Individual monthly premium   = _____

2. Family monthly premium   = _____

Solutions: **1.** $117.50; **2.** $197.50

Plan B: Individual annual premium $980 (additional $80 a month for family coverage) with company paying $45 of monthly premium for each employee.

3. Individual monthly premium   = _____

4. Family monthly premium   = _____

Solutions: **3.** $36.67; **4.** $116.67

13.3 Automobile Insurance

Almost all states require by law that individuals and businesses carry some type of automobile insurance. Many kinds of coverage are available. The cost of coverage depends on the driver's classification according to age, sex, marital status, driver training experience, geographic location, distance traveled to and from work, and whether the car is used for pleasure or business. If a driver has a record of several accidents and/or speeding tickets, insurance is difficult to get and the rates are much higher because of the increased risk for the insurance company.

Premiums are determined from tables used by the insurance companies for each of the geographic areas, and rates vary from state to state.

13.4 Automobile Liability Insurance

Automobile liability insurance covers damages for bodily injury or property damage for which any covered person becomes legally responsible because of an auto accident. The formula, 25/50/10, is an example of liability coverage. It means there will be $25,000 maximum payment for bodily injury to any one person in the accident, $50,000 total maximum payment for bodily injuries to all persons in the accident, and $10,000 maximum payment for property damage caused by the accident. An automobile policy is usually issued for one year at a time, and the exact amount of the premium is calculated. Depending on the agreement with the insurance company, premiums may be paid monthly, quarterly, semiannually, or annually. When an accident occurs, the insured, when determined at fault, usually must pay any expenses not covered by the insurance policy. The following is an example of payment of liability coverage.

EXAMPLE

Martin runs into Landers' automobile, causing Landers and Eileen (a passenger in Landers' car) injuries. A court suit awards Landers $28,000 for personal bodily injuries and Eileen, $22,000. The damage to Landers' car amounts to $8,000. Martin has 25/50/10 liability insurance. To determine how Martin and his insurance company pay for the liabilities (injuries and damages), you would follow these steps.

STEPS

1. Bodily injury:

Landers =	$28,000
Eileen = +	22,000
Total =	$50,000

   (Maximum paid to any one person is $25,000. The insurance company will pay $25,000 of Landers' $28,000 worth of expenses and all of Eileen's expenses, which are $22,000. The maximum payment to all persons is $50,000. The total of covered expenses, $47,000, falls within this total coverage figure.)

   $25,000 + $22,000 = $47,000 insurance pays

   $28,000 − $25,000 = $3,000 Martin pays

2. Property damage: Landers' car = $8,000

   (The maximum paid for any one automobile accident is $10,000. Landers' damages fall within the maximum amount; therefore, the $8,000 will be paid in full.)

   $8,000 insurance pays in full

Determine the amount of money the insurance company and Allen Grayson will pay. Write your answers in the blanks provided.

Grayson runs into Hargrove's automobile, injuring Hargrove and Mayfield (a passenger in Hargrove's car). Grayson has 25/50/10 liability coverage. Hargrove is awarded $26,000 for personal bodily injuries, Mayfield is awarded $20,000. It will cost $9,200 to repair Hargrove's automobile.

1. How much will Grayson's insurance company pay?

   Personal bodily injuries   _____

   Property damage   _____

2. How much will Grayson pay?   _____

Solutions: 1. Personal bodily injury = $45,000, Property damage = $9,200,
                  Insurance pays = $54,200

               2. Grayson pays = $1,000

13.5 Life Insurance

Life insurance is an agreement providing for the payment of a stipulated sum to a beneficiary upon the death of the insured person. A person may purchase any amount of life insurance coverage desired . The basic purpose is to give financial protection to the survivors after the insured's death. Upon the insured's death, the insurance company pays the full amount of the policy. Some of the most common types of life insurance are:

Whole Life. A premium is paid annually until the insured's death, at which time the face value of the policy is paid to the designated beneficiary. A whole life policy builds a cash value as premiums are paid. (A beneficiary is the person who is designated to receive the face value of the policy upon the death of the insured.)

Term Insurance. Provides coverage for a specified period of time. It is the least expensive type of life insurance. Upon the expiration of the time period, the coverage ends. If the death of the insured occurs during the term of the policy, the beneficiary receives the face value of the policy.

Limited Payment Life Insurance. Provides coverage for life, but allows payment of premiums for a specified period of time, usually during the wage-earning years. The face value of the policy is payable to the beneficiary upon the death of the insured.

Endowment Insurance. Provides coverage during a specified period of time. If the insured dies during the specified period of time, the beneficiary receives the face value of the policy. If the insured does not die at the end of the specified time, he or she receives a predetermined cash value.

Universal Life Insurance. Universal life insurance came into existence around 1980 to provide low-cost term insurance, with the balance of each premium going into an account that would earn a minimum amount of interest–usually 4 to 5 percent. The face value of the policy and the premium payment can vary. As the account earns money, premiums can be paid from the amount.

Variable Life Insurance. Variable life insurance is similar to universal life insurance. It offers more investment options and thus creates the possibility of higher rates of return on your money. The policy value and interest earned can fluctuate based on investment choices. Usually, the insured pays a set premium.

Variable Universal Life Insurance. This insurance guarantees a certain death benefit yet still allows the selection of investment choices. This policy may also be known as flexible-premium variable life insurance.

Premiums are usually set as a number of dollars per $1,000 insurance coverage desired based on age, health of the insured, and type of insurance. If the purchaser of a policy wants $100,000 worth of coverage, the rate might be quoted at $3.75 per $1,000 worth of coverage. To determine the annual premium amount, use the following formula

$$\frac{Amount\ of\ Coverage}{\$1,000} \times Cost\ per\ \$1,000 = Annual\ Premium$$

EXAMPLE

To determine the annual premium for $200,000 worth of coverage quoted at $3.75 per $1,000, follow these steps.

STEPS

1. Divided the amount of coverage by $1,000 to determine the number of $1,000s in $200,000.

$$\$200,000 \div \$1,000 = 200$$

2. Multiply the number of $1,000s by the cost of the coverage per $1,000.

$$200 \times \$3.75 = \$750\ \text{cost of 1 year's premium}$$

■ **TIP**   You can quickly divide the amount of coverage by 1,000 by moving the decimal 3 places to the left in the amount of coverage. Your answer is the number of 1,000s. Simply multiply this number by the rate per $1,000 to obtain the annual premium.

PRACTICE PROBLEMS

Determine the amount of annual premiums due in the following problems. Write your answers in the blanks provided.

1. $\dfrac{\$600,000}{\$1,000} \times \$3.50 = $ _____

2. $\dfrac{\$250,000}{\$1,000} \times \$1.95 = $ _____

3. $\dfrac{\$100,000}{\$1,000} \times \$2.50 = $ _____

4. $\dfrac{\$500,000}{\$1,000} \times \$3.10 = $ _____

Solutions: **1.** $2,100; **2.** $487.50; **3.** $250; **4.** $1,550

13.6 Property Insurance

Property insurance provides coverage of property against damage or loss. One of the most common types of property insurance is fire insurance. Premiums may be based on the nature of the risk involved, the amount of coverage purchased, the length of time, and the location of the property. Extended coverage may be purchased to cover more than just the damage caused by fire. Examples of extended coverage include damage caused by smoke, water, or chemicals; damage done by firemen breaking into the property or by measures taken to contain the fire; riots; civil commotion; hurricanes; wind; or hail.

Fire insurance premiums are usually quoted based on the number of dollars per $1,000 of insurance desired. Renters may also obtain property insurance to cover damages to their household furnishings in a rented apartment or home. Computations are the same as for life insurance. Use the following formula to determine the annual premium amount.

$$\frac{Amount\ of\ Coverage}{\$1,000} \times Cost\ per\ \$1,000 = Annual\ Premium$$

EXAMPLE _____

Let's assume Jan and Michael Thompson decide to purchase a fire insurance policy in the amount of $50,000 to cover the household furnishings in their apartment. The Royal Insurance Company quoted to them a cost of $4.95 per $1,000 for the coverage. Their annual premium would be computed using the following steps:

$$\frac{\$50,000}{\$1,000} \times \$4.95 = Annual\ Premium$$

STEPS

1. Divide the amount of coverage by $1,000 to determine the number of 1,000s in $50,000.

$$\$50,000 \div \$1,000 = 50$$

2. Multiply 50 (number of $1,000s in $50,000) by cost per $1,000 ($4.95).

$$50 \times \$4.95 = \$247.50\ cost\ of\ 1\ year's\ premium$$

― **PRACTICE PROBLEMS** ―――――――――――――

Compute the annual premiums for each of the following fire insurance policies and write your answers in the blanks provided.

1. $70,000 coverage on a mobile home; coverage quoted at $6.50 per $1,000.

   Annual premium = _____

2. $300,000 coverage on a building; coverage quoted at $3.75 per $1,000.

   Annual premium = _____

Solutions: 1. $455; 2. $1,125

13.7 Coinsurance: Coverage Equal to 80% of Property Value

If a company purchased property insurance such as fire insurance for the full value of the property, the premiums would be extremely expensive. Because of this expense, coinsurance may be used to help cut the costs of the premiums. A *coinsurance* policy allows the insured to bear part of the loss when there is damage due to fire. The insured agrees to purchase a policy for a stated percentage, usually 80% of the value of the property. Under an 80% coinsurance policy, property valued at $100,000 would be insured for $80,000. As long as the insured carries insurance covering 80% of the value of the property, the insurance company will pay for all fire damages up to $80,000. Study the formula:

$$\frac{Amount\ of\ Coverage}{Amount\ of\ Insurance\ Required} \times Loss = Recovery$$

EXAMPLE

Julie Craine is a cosmetologist. Julie purchased an 80% coinsurance policy to insure her place of business for $76,000; the business was valued at $95,000. The building caught fire and suffered $65,000 damage. To determine the amount of money Julie will collect from insurance, you would follow these steps.

Coverage	= $76,000
Loss	= $65,000
Value	= $95,000

$$\frac{\$76,000}{\$76,000} \times \$65,000 \quad = Recovery$$

STEPS

1. Multiply 80% by the value of the property.

    0.80 × $95,000 = $76,000 (amount of insurance required)

2. Divide the amount of the coverage Julie carried by the amount of insurance required.

    $76,000 ÷ $76,000 = 1 or 100% recovery

3. Multiply 1 by Julie's amount of loss.

    1 × $65,000 = $65,000 recovery

■ **TIP**   If the insured carries up to 80% of the value of the property, he or she will recover all losses up to that 80% amount; therefore, no calculations are necessary if the amount of coverage is more than the damage.

— PRACTICE PROBLEMS —

Compute the amount paid by the insurance company on an 80% coinsurance policy for the following properties and write your answers in the blanks provided.

Coverage	Value	Loss	Insurance Paid
1. $120,000	$150,000	$110,000	_____
2. $288,000	$360,000	$225,000	_____
3. $108,540	$135,675	$100,000	_____

Solutions: **1.** $110,000; **2.** $225,000; **3.** $100,000

13.8 Coinsurance: Coverage not Equal to 80% of Property Value

If the insured carries less than 80% of the value of the property, the insurance company will pay damages up to whatever fractional part of the actual coverage is of the required insurance. Study this example:

EXAMPLE _____

Cindy Haygood's home was valued at $165,000, and she purchased a coinsurance fire insurance policy. She covered the home for only $100,000. A fire occurred in Cindy's home, and the damage amounted to $90,000. You would determine how much the insurance company would pay by following these steps.

Coverage = $100,000

Loss   = $90,000

Value   = $165,000

$$\frac{\$100,000}{80\% \times \$165,000} \times \$90,000 = \text{Recovery}$$

STEPS

1. Multiply 80% by $165,000 (value of property).

    0.80 × $165,000 = $132,000 (amount of insurance required)

2. Divide the amount of coverage Lucia carried ($100,000) by the amount of insurance required. Carry your answers to 3 decimal places and round to 2 places then round to the nearest whole percent.

    $100,000 ÷ $132,000 = 0.76 or 76% recovery

3. Multiply 0.76 by the amount of damage ($90,000).

    0.76 × $90,000 = $68,400 recovered

─ PRACTICE PROBLEMS ─

Compute the amount recovered from the insurance company for fire damage given the following coinsurance policy coverages. Write your answers in the blanks provided.

Coverage	Value	Loss	Insurance Paid
1. $140,000	$250,000	$180,000	_____
2. $440,000	$900,000	$400,000	_____
3. $190,000	$366,900	$100,000	_____

Solutions: **1.** $126,000; **2.** $244,000; **3.** $65,000

MATH ALERT

```
                                            DECLARATIONS PAGE 1

FIRST STATE TEXAS INSURANCE
POLICY NUMBER   2 16 2908   11/26
NAMED INSURED/MAILING ADDRESS       DANIEL M & WIFE MICHELLE
                                    PARKS
                                    2009 OLA LANE
                                    GRAND PRAIRIE TX 75050-2233

AGENT NAME/ADDRESS      GREGORY MOREHEAD
                        1142 CARRIER PKWY155
                        GRAND PRAIRIE TX 75051
            PHONE       214 555-9090
MORTGAGE NAME/ADDRESS       NOT APPLICABLE


COVERAGES                                   LIMITS OF   ANNUAL
                                            LIABILITY   PREMIUMS

SECTION I PROPERTY
 COVERAGE B. PERSONAL PROPERTY              $   30,000
            PERSONAL PROPERTY OFF PREMISES  $    3,000
SECTION II LIABILITY
 COVERAGE C. PERSONAL LIABILITY
                    (EACH OCCURRENCE)       $   25,000
 COVERAGE D. MEDICAL PAYMENTS TO OTHERS
                    (EACH PERSON)           $      500
 LOSS OF USE COVERAGE                       $    6,000
                                    BASIC PREMIUM    $ 261


DEDUCTIBLES (SECTION I ONLY)    AMOUNT OF DEDUCTIBLE
DEDUCTIBLE CLAUSE 3                     $ 250           $   11
                            TOTAL POLICY PREMIUM    $ 272
```

Interpret the above renter's insurance policy and answer the following questions. Write your answers in the blanks provided.

1. If Daniel and Michelle Parks had $36,000 damage done to their personal property, how much could they collect from the insurance company?

   Insurance paid: _____

2. During a Christmas party at their home, a guest slipped on wet tile in the kitchen and incurred $4,600 in medical bills. How much would the Parks' policy pay?

   Insurance paid: _____

3. If Daniel bought a two-year policy, what would his premium be for the two years?

   Premium cost: _____

4. If Michelle's car was broken into and her fur coat valued at $7,000 was stolen, how much would the insurance pay?

   Insurance paid: _____

Study Guide

I.   Terminology

page 256	insurance	Protection against loss.
page 256	face value	The amount of insurance purchased.
page 256	premium	Payment to the insurance company for the purchase of an insurance policy.
page 256	term	The time period for which the insurance policy is in effect.
page 256	insured	Individual or company receiving the insurance protection.
page 256	policy	A written contract between the insurance company and the insured that explains the benefits of the protection purchased.
page 256	claim	A form filed to request payment for losses covered in an insurance policy.
page 256	carrier	The insurance company that "underwrites" or "insures".
page 256	policyholder	The person or business that purchases insurance.
page 256	hospitalization insurance	Insurance protection that covers all or part of the cost of a person's hospital stay.
page 256	beneficiary	The person who is designated to receive the face value of the policy upon the death of the insured.
page 258	automobile insurance	Insurance protection that provides coverage of an automobile and/or its passengers if in an accident.
page 259	automobile liability insurance	Insurance coverage for damages for bodily injury or property damage for which covered person becomes legally responsible because of an auto accident.
page 260	whole life	A premium is paid annually until the insured's death, at which time the face value of the policy is paid to the designated beneficiary.
page 260	term insurance	Life insurance coverage for a specified period of time. It is the least expensive type of life insurance. Upon the expiration of the time period, the coverage ends. If the death of the insured occurs during the term of the policy, the beneficiary receives the face value of the policy.
page 260	limited payment life insurance	Coverage for life, but allows payment of premiums for a specified period of time usually during the wage-earning years. The face value of the policy is payable to the beneficiary upon the death of the insured.
page 260	endowment insurance	Coverage during a specified period of time. If the insured dies during the specified period of time, the beneficiary receives the face value of the policy. If the insured does not die at the end of the specified time, he or she receives a predetermined cash value.

page 260	universal life insurance	Provides low-cost term insurance with the balance of each premium going into an account that would earn a minimum amount of interest–usually 4 to 5 percent. The face value of the policy and the premium payment can vary. As the account earns money, premiums can be paid from the account.
page 261	variable life insurance	Offers more investment options than universal life and this creates the possibility of higher rates of return on your money. The policy value and interest earned can fluctuate based on investment choices. Usually the insured pays a set premium.
page 261	variable universal life insurance	Guarantees a certain death benefit yet still allows the selection of investment choices. This policy may also be known as flexible-premium variable life insurance.
page 261	property insurance	Provides coverage of property against damage or loss.
page 262	coinsurance	Coverage equal to or less than 80% of property value. Provides a way to obtain cheaper insurance premiums when 100% coverage would be too expensive.

II. Health Insurance

page 257   *Calculating Health Insurance Premiums:* Divide individual annual premium by 12 months; for family, add amount to individual monthly premium; subtract amount of premium company pays.

Example: Annual premium: 100% coverage
Individual: $1,950
Family: Additional $75

$1,950 ÷ 12 = $162.50 per month

Family: $162.50 + $75 (for family) = $237.50 per month
$237.50 – $75 (company pays for individual) = $162.50

Individual: $162.50 – $75 (company pays) = $87.50

If there is a deductible, the employee must pay the deductible before the insurance begins paying.

III. Automobile Insurance

page 259   *Liability Insurance:* If there is an accident and the person has 25/50/10 liability coverage, the insurance company will pay $25,000 maximum payment for bodily injury to any one person in the accident, $50,000 total maximum payment for bodily injuries to all persons in the accident, and $10,000 maximum payment for property damage caused by the accident.

Example: Bodily injury:

	Landers	$28,000
	Eileen	$22,000
	Total	$50,000
Property damage:		$8,000

The insurance company will pay $25,000 of Landers $28,000 because the maximum to one person was $25,000. It will pay $22,000 to Eileen because her medical was less than the $25,000; therefore, bodily injury payment by the insurance company will be $25,000 + $22,000 = $47,000. The holder of the policy pays the $3,000 for Landers that the insurance company didn't pay.

The insurance company pays all of the $8,000 property damage because it was less than the $10,000 coverage.

IV. Life Insurance

page 261   *Calculating the annual premium:* Use the formula:

$$\frac{Amount\ of\ Coverage}{\$1,000} \times Cost\ per\ \$1,000 = Annual\ Premium$$

Example: Coverage: $200,000
Cost per $1,000: $3.75

$200,000 ÷ 1,000 × $3.75 = $750 cost of 1 year's premium

V. Property Insurance

page 262   *Calculating the Annual Premium:* Use the formula:

$$\frac{Amount\ of\ Coverage}{\$1,000} \times Cost\ per\ \$1,000 = Annual\ Premium$$

Example: Coverage: $50,000
Cost per $1,000: $5.10

$50,000 ÷ 1,000 × $5.10 = $255

VI. Coinsurance Coverage

page 263   *Calculating the Recovery Amount when Coverage is Equal to 80% Coinsurance:* Use the formula:

$$\frac{Amount\ of\ Coverage}{Amount\ of\ Insurance\ Required} \times Loss = Recovery$$

Example: Coverage: $76,000
Loss: $65,000
Value: $95,000

$95,000 × 0.80 = $76,000 coverage required

$$\frac{\$76,000}{\$76,000} \times \$65,000 = \$65,000\ recovery$$

Calculating the Recovery when Coinsurance is not equal to 80%: Use the formula:

$$\frac{Amount\ of\ Coverage}{80\% \times Value} \times Damage = Recovery$$

Example: Coverage: $100,000
Loss: $90,000
Value: $165,000

0.80 × $165,000 = $132,000 amount of insurance required
$100,000 ÷ $132,000 = 0.76 or 76% recovery (rounded)
0.76 × $90,000 = $68,400 recovered

Assignment 1

Name_____ Date _____

Complete the following problems. Write your answers in the blanks provided.

Determine the monthly premium for individual and family hospitalization coverage. The employer will pay $100 of the monthly premium for each employee. An additional $80 is added to the individual monthly premium for family coverage. Round all answers to the nearest penny.

A. Individual annual premium: $1,500 =

    1. _____ monthly premium

    Family premium =

    2. _____ monthly premium

B. Individual annual premium: $1,250 =

    3. _____ monthly premium

    Family premium =

    4. _____ monthly premium

C. Individual annual premium: $990 =

    5. _____ monthly premium

    Family premium =

    6. _____ monthly premium

D. Individual annual premium: $1,900 =

    7. _____ monthly premium

    Family premium =

    8. _____ monthly premium

E. Individual annual premium: $2,400 =

    9. _____ monthly premium

    Family premium =

    10. _____ monthly premium

F. Individual annual premium: $1,875 =

    11. _____ monthly premium

    Family premium =

    12. _____ monthly premium

G. Individual annual premium: $1,210 =

    13. _____ monthly premium

    Family premium =

    14. _____ monthly premium

H. Individual annual premium: $1,050 =

    15. _____ monthly premium

    Family premium =

    16. _____ monthly premium

Assignment 2   Name_____ Date _____

Complete the following problems. Write your answers in the blanks provided.

A. Determine the amount of money the insurance company and the individuals will pay on liability insurance coverage in the following automobile accidents. All insured motorists have 25/50/10.

Accident #1: Personal bodily injury: Corey Kramer = $32,000

Jane Kramer = $31,000

Property damage: Automobile = $15,000

1. Insurance pays for personal bodily injury: _____
2. Insurance pays for property damage: _____
3. Total paid by insurance company: _____
4. Total paid by individual causing accident: _____

Accident #2: Personal bodily injury: Dianne Greene = $22,000

Julian Greene = $40,000

Property damage: Automobile = $12,000

5. Insurance pays for personal bodily injury: _____
6. Insurance pays for property damage: _____
7. Total paid by insurance company: _____
8. Total paid by individual causing accident: _____

B. Determine the amount of annual premium due for life insurance.

9. $\dfrac{\$275,000}{\$1,000} \times \$3.25 =$ _____

10. $\dfrac{\$300,000}{\$1,000} \times \$2.95 =$ _____

11. $\dfrac{\$1,000,000}{\$1,000} \times \$2.50 =$ _____

12. $\dfrac{\$90,000}{\$1,000} \times \$3.00 =$ _____

C. Determine the amount of annual premium due for property insurance.

13. $110,000 coverage quoted at $6 per $1,000 = _____
14. $150,000 coverage quoted at $5.50 per $1,000 = _____
15. $45,000 coverage quoted at $1.10 per $1,000 = _____
16. $80,000 coverage quoted at $3.50 per $1,000 = _____
17. $100,000 coverage quoted at $2.15 per $1,000 = _____

Assignment 3

Name_____ Date _____

Complete the following problems. Write your answers in the blanks provided.

A. Determine the amount of coverage needed to be equal to 80% of property value.

Property Value	Needed Coverage		Property Value	Needed Coverage
1. $75,000	_____		6. $495,000	_____
2. $100,000	_____		7. $110,000	_____
3. $49,000	_____		8. $157,800	_____
4. $2,000,000	_____		9. $25,000	_____
5. $350,000	_____		10. $62,000	_____

B. Determine the amount paid by the insurance company on an 80% co-insurance policy for each of the following properties. Coverage is equal to 80% of property value.

	Coverage	Value	Loss	Insurance Paid
11.	$120,000	$150,000	$90,000	_____
12.	$288,000	$360,000	$200,000	_____
13.	$108,540	$135,675	$20,000	_____
14.	$184,000	$230,000	$140,000	_____
15.	$640,000	$800,000	$500,000	_____
16.	$48,000	$60,000	$15,000	_____
17.	$252,000	$315,000	$210,000	_____
18.	$19,200	$24,000	$19,200	_____
19.	$48,000	$60,000	$40,000	_____
20.	$88,000	$110,000	$60,000	_____

Assignment 4

Name_____ Date _____

Complete the following problems. Write your answers in the blanks provided.

Danielle Fox, a new employee at Hargrove, Estes, and Yates law firm, decided to enroll in the firm's health insurance program. The annual premium for Danielle will be $1,725. The firm will pay $150 of the premium per month. What will Danielle pay?

1.   Danielle will pay: _____

After Danielle worked at Hargrove, Estes, and Yates, two years, she married. To add her husband to her health insurance policy, she will have to pay an additional $110 per month. How much annual premium will Danielle now pay?

2.   Danielle will pay: _____

Hargrove, Estes, and Yates built and moved into a new office building valued at $295,000. Their insurance company quoted them $3.90 per $1,000. What would a two-year premium cost the firm?

3.   A two-year premium would be: _____

Each of the partners is provided an automobile owned by the firm. Estes was involved in an accident that was his fault while on the way to see a client. The firm has 50/100/25 automobile liability coverage. In the accident Hilliary Martin incurred $5,000 personal bodily injury and her three children riding with her were also hurt. Jane had $12,000, Sally had $5,400, and Bobby had $8,300 bodily injuries. The Martin automobile's damage was $11,500. Answer the following questions.

How much will the insurance pay for:

4.   Hilliary?   _____

5.   Jane?   _____

6.   Sally?   _____

7.   Bobby?   _____

8.   Total bodily injury?   _____

9.   Property damage?   _____

10. Total paid by insurance
     company?   _____

11. Total paid by Estes over what
     the insurance paid?   _____

Hargrove, Estes, and Yates decided to purchase an 80% coinsurance policy on their building. How much insurance do they need to purchase to cover it at 80%?

12. 80% coinsurance coverage amount: _____

Hargrove, Estes, and Yates decided to purchase life insurance for themselves. They purchased $200,000 for each partner, quoted at $1.78 per $1,000 for one year.

13. How much will each partner pay for the coverage?   _____

14. How much will the premium be for all three partners?   _____

15. What would a two-year premium be for all three partners?   _____

Assignment 5

Name_____ Date_____

**Complete the following word problems. Write your answers in the blanks provided.
Round answers to the nearest cent.**

1. Yates Manufacturing has contracted with Health Care Insurance Company for a group plan. Yates has agreed to pay $\frac{1}{2}$ of each employee's premium per month. The cost of this plan to individual employees is as follows: Annual premium – Individual – $2,460; Family – Additional $35 monthly. Martha West is married and has 2 children. Compute her monthly premium for health insurance.

   _____

2. Ozzie Adams runs into Evan Deat's automobile, injuring Evan and Mary Smith, a passenger in Evan's car. Ozzie has 25/50/10 liability coverage. Evan is awarded $32,000 for personal bodily injuries. Mary Smith is awarded $20,000. It will cost $4,309 to repair Evan's automobile. How much will Ozzie Adams' insurance company pay?

   Personal bodily injuries _____

   Property damage _____

   How much will Ozzie Adams pay? _____

3. Ned Yearling wants to purchase $100,000 worth of life insurance at a rate of $3.25 per $1,000. What is his annual premium?

   _____

4. The Southside Clothing Store is valued at $89,900. To satisfy the 80% coinsurance clause of the policy, for how much should the owner insure the building?

   _____

5. Kris Kretchen owns a building with a market value of $135,345. He has insured the building for $90,000. A fire has caused $61,352 in damages. How much will the insurance company pay as compensation if his policy contains an 80% coinsurance clause?

6. Hope Wilson has an auto liability insurance policy with 20/40/10 coverage. She is responsible for an accident in which Rhoda Ewing, who was riding in another car, is injured. Rhoda's medical expenses totaled $45,098 and damages to her car totaled $12,098. What is the total amount Hope's insurance will pay?

   _____

7. Tom Potts decides to buy a no-fault insurance policy worth $30,000 and an uninsured motorist policy worth $35,000. The cost is $3.50 per $1,000 for the no-fault insurance and $2.50 per $1,000 for the uninsured motorist coverage. What is the cost of Tom's total annual premium?

   _____

8. Find the face value of a fire-protection policy on a building worth $97,230 if it is insured for 80% of its market value.

   _____

9. If you have a collision insurance policy with a $500 deductible clause, how much of $1,206 damage to your car will the policy cover?

   _____

10. If a premium of $798 can be paid semi-annually with a 3% annual charge added, what is the amount to be paid every 6 months?

    _____

11. Julie Kelley wants to purchase $250,000 worth of life insurance at a rate of $2.10 per $1,000. What is her annual premium?

    _____

Assignment 6

Name_____ Date _____

Complete the following word problems. Write your answers in the blanks provided. Round answers to the nearest cent.

1. Kathy Chase has three young children and wants to purchase an additional $300,000 of 5-year term insurance. The rate is $1.87 per $1,000. What will her annual premium be?

   _____

2. IAD Inc. has contracted with Health Care Specialists for a group plan. IAD has agreed to pay the first $50 of each employee's premium per month. The cost of the plan will be $1,246 for an individual and an additional $45 per month for a family. Paul Hampton is married with 3 children. Compute his monthly premium for health care insurance.

   _____

3. Yorktown Incorporated owns a building that is valued at $125,900. To satisfy the 80% coinsurance clause of the policy, for how much should Yorktown insure the building?

   _____

4. Irene Mattingly has an auto liability insurance policy with 30/50/15 coverage. She is responsible for an accident in which Robert Hammonds, riding in another car is injured. His medical expenses totaled $27,896 and damages to his car totaled $12,309. What is the total amount Irene's insurance will pay?

   _____

5. Xavier Rogers owns a building with a market value of $300,000. He has insured the building for $250,000. A fire has caused $125,000 in damages. How much will the insurance company pay as compensation if his policy contains an 80% coinsurance clause?

   _____

6. If you have a collision insurance policy with a $250 deductible clause, how much of $2,367 damage to your car will the policy cover?

   _____

7. Find the face value of a fire-protection policy on a building worth $225,000 if it is insured for 80% of its market value.

   _____

8. Jack Unser wants to purchase $250,000 worth of life insurance. The rate is $2.25 per $1,000. What is his annual premium?

   _____

9. David Lee would like to purchase a no-fault insurance policy worth $45,000 and an uninsured motorist policy worth $25,000. The cost is $2.50 per $1,000 for the no-fault insurance policy and $1.25 per $1,000 for the uninsured motorist coverage. What would be the cost of David's total annual premium?

   _____

10. Shelly Gavins runs into Tricia Winston's automobile, injuring Tricia. Shelly has 25/50/10 liability coverage. Tricia is awarded $26,540 for personal bodily injuries. It will cost $12,323 to repair Tricia's automobile. How much will Shelly have to pay after the insurance company has paid its part?

    _____

11. BBR Marketers owns a building valued at $115,000. How much insurance do they need to satisfy their 80% coinsurance clause?

    _____

Proficiency Quiz
R E V I E W

Name_____ Date_____

$$\frac{\text{Student's Score}}{\text{Maximum Score}} = \frac{}{27} = \text{Grade}_____$$

A. Compute the amount of property tax due. Round your answers to nearest penny.

Market Value	Assessed Value	Tax Rate	Tax Due
1. $65,000	85%	$1.65 per $1,000	= _____
2. $145,000	65%	$2.35 per $1,000	= _____
3. $90,000	80%	$3.35 per $1,000	= _____

B. Compute the property tax rate. (Use 4 decimal places; round answers to nearest whole percent.)

Money Needed	Assessed Value	Tax Rate
4. $165,000	$4,000,000	= _____
5. $240,000	$1,500,000	= _____
6. $85,000	$2,000,000	= _____

C. Compute state unemployment tax due. Use 3% on the first $9,000.

7. $10,356 gross wages = _____ tax due

8. $6,555 gross wages = _____ tax due

9. $2,399 gross wages = _____ tax due

D. Compute federal unemployment tax due. Use 0.8% on the first $7,000.

10. $6,123 gross wages = _____ tax due

11. $2,986 gross wages = _____ tax due

12. $9,452 gross wages = _____ tax due

E. Compute federal income tax—Form 1040EZ. Use the form on page 247 as a guide and the tax table on pages 241-242.

13. Shelby Sutton is single. Her total wages for the year were $18,200 and $320 is interest income. She claims no dependents other than herself. How much federal tax will Shelby be required to pay?

Federal tax = _____

14. Kristen is single. Her total wages were $21,000 in wages and $350 in interest from her savings account. Kristen has no exemptions other than herself. How much federal tax will Kristen be required to pay?

Federal tax = _____

F. Compute federal income tax—Form 1040. Use the tax table on pages 241-242.

15. Susie South's salary is $23,560; her husband, Jon, makes 24,000. They have 5 children who are dependents. Their itemized deductions total $12,050. If they file a joint return, what is their federal tax?

Federal tax = _____

16. Carl Muldoon's salary is $13,500. His wife, Marcy, makes $14,000. They have 2 dependent children. Their itemized deductions total $5,400. If they file a joint return, what is their federal tax?

Federal tax = _____

G. Determine the monthly premium for individual and family hospitalization insurance coverage. The employer will pay $85 of the monthly premium for each employee. An additional $60 is added to the individual monthly premium for family coverage. Round all answers to the nearest penny.

17. Individual annual premium $1,728　　　= _____ monthly premium

18. Family premium　　　= _____ monthly premium

Hodges, Inc. will pay $45 of the monthly premium for each employee. An additional $40 is added to the individual monthly premium for family coverage.

19. Individual annual premium $986　　　= _____ monthly premium

20. Family premium　　　= _____ monthly premium

H. Determine the amount of money the insurance company and the individual will pay for damages in the following automobile accident. The insured motorist has 20/50/10 coverage.

Accident : Personal bodily injury　　- Sara Crown　= $23,000
　　　　　　Personal bodily injury　　- Ted Crown　= $16,000
　　　　　　Property damage　　　　- Automobile　= $11,050

21. Insurance payment for personal bodily injury =　　　　　_____

22. Insurance payment for property damage =　　　　　_____

23. Total paid by insurance company =　　　　　_____

24. Total paid by individual causing accident =　　　　　_____

I. Determine the amount of annual premium due for life insurance. Round all answers to nearest penny.

25. $\frac{160,000}{1,000} \times \$2.65 =$ _____　　　26. $\frac{40,000}{1,000} \times \$4.45 =$ _____

J. Determine the amount paid by the insurance company on an 80% coinsurance policy for the following property.

	Amt. Coverage	Value	Loss	Insurance Paid
27.	$70,000	$125,000	$45,000	_____

CHAPTER

Simple Interest and Compound Interest

14

NO BANK PAYS YOU MORE.

· Federal regulations prohibit compounding of interest on this certificate.
· $10,000 minimum deposit. · Substantial interest penalty for early withdrawal. · FDIC insured.

14

Simple Interest and Compound Interest

Individuals as well as businesses borrow money to buy a variety of goods and services. Borrowing money allows people to make a purchase now rather than later. Loans are made to businesses for purchases of equipment, buildings, land, and merchandise for resale. Loans also are made to individuals for vacations and educational expenses as well as for purchases of homes, cars and major appliances.

Of course, you must pay for the privilege of borrowing someone else's money. When you borrow money, an additional amount of money is added to the original amount to borrow; this additional amount is called *interest*.

Interest is also paid on invested money. If you deposit money into a savings account, the bank will pay you interest for the use of your money. For instance, a bank pays 5% interest on money you deposit in your account. If you deposit a sum of money in your account and leave it there for one year, you will gain 5% on your invested money. People should look closely at the interest rates they are earning on their invested money.

The two basic types of interest are simple and compound. In this chapter, we will discuss both types.

Simple interest is used when a loan or investment is repaid in a lump sum. The borrower has use of the full amount of money for the entire period of the loan or investment. Simple interest is commonly figured on loans of one year or less.

Compound interest most often applies to savings accounts, loans, installments, and credit cards.

Compound interest differs from simple interest because compound interest is calculated on the original principal plus interest earned to date. This means that the original principal and the interest, which already have been figured, are increased over a period of time. Additional interest is calculated and added to the beginning principal plus interest; therefore, compound interest results in a higher yield than simple interest.

14.1 Terms Used for Calculating Simple Interest

The amount borrowed or invested for a certain period of time, sometimes called *face value*, is the *principal* (P). *Interest* (I) is the amount paid only on the principal for the privilege of using someone's money or earned on investment money. The *rate* (R) of interest, charged or earned in one year, is expressed as a percent of the amount borrowed or earned. The *time* (T) is the length of time you have to repay the loan or earn interest on your invested money. Time is expressed in days, months or years and is usually written as a number of years, or a fraction of a year.

14.2 Calculating Simple Interest

Interest is based on the principal, interest rate, and amount of time. The amount of interest is determined by multiplying the principal by the interest rate by the amount of time. The following formula is used:

$$I = P \times R \times T$$

I = Interest
P = Principal (amount of loan)
R = Rate (percent of interest charged or earned per year)
T = Time

The actual amount of time for which money is being used is important in calculating interest.

EXAMPLE _____

Suppose Nadine wants to borrow $3.500 for one year at 12% interest.

$$I = P \times R \times T$$
$$I = \$3,500 \times 0.12 \times 1$$
$$I = \$420.00$$

Suppose Nadine was borrowing $3,500 for 8 months at 12% interest. What is the amount of interest?

Eight months is $\frac{8}{12}$ of a year.

$$I = \$3,500 \times 0.12 \times \frac{8}{12}$$
$$I = \$280.00$$

PRACTICE PROBLEMS

Determine the simple interest for these loans. Write your answers in the blanks provided.

	Interest		Interest
1. $1,450 at 15% for 8 months.	_____	**2.** $1,680 at 12% for 6 months.	_____
3. $800 at 13% for 3 months.	_____	**4.** $600 at 16% for 5 months.	_____

Solutions: **1.** $145.00; **2.** $100.80; **3.** $26.00; **4.** $40.00

14.3 Finding the Maturity Value of a Loan

Maturity value (*M*) is the full amount of money that must be repaid when a loan is due; that is, the principal plus the interest. To determine the maturity value, figure the simple interest first, then figure its maturity value.

EXAMPLE _____

Suppose you are borrowing $950 at 14% interest for 10 months. What is the maturity value?

STEPS

1. Determine the interest first.

$$I = P \times R \times T$$
$$I = \$950 \times 0.14 \times \frac{10}{12}$$
$$I = \$110.83$$

2. Determine the maturity value by adding the interest to the principal.

$$M = P + I$$
$$M = \$950 + \$110.83$$
$$M = \$1,060.83$$

— PRACTICE PROBLEMS —

Determine the simple interest and maturity value for these problems. Write your answers in the blanks provided.

Principal	Rate	Time	Interest	Maturity Value
1. $1,750	16%	3 months	_____	_____
2. $3,000	$14\frac{1}{2}$%	1 year	_____	_____
3. $1,800	15%	9 months	_____	_____
4. $950	13%	1 year	_____	_____

Solutions: **1.** $70.00; $1,820.00; **2.** $435; $3,435.00; **3.** $202.50; $2,002.50; **4.** $123.50; $1,073.50

14.4 Finding Ordinary and Exact Interest

Thus far the amount of time for which money is borrowed has been expressed as several months or one year. Now you will learn how to calculate the amount of interest when time is expressed in days. The two basic methods of computing time and interest are discussed below. They are:

1. 360-day, referred to as *ordinary* interest (12 months of 30 days each)
2. 365-day, referred to as *exact* interest (or 366 days in a leap year)

Loan of $4,000 at 9% for 60 days. What is the amount of interest?

Method 1: 360-day year (Ordinary)

Most businesses and banks in the United States use the 360-day formula because it yields a slightly higher amount. Time is expressed as a fraction, such as $\frac{120}{360}$, to calculate interest. Once the time is determined, you can figure interest using the basic formula.

$$I = P \times R \times T$$
$$I = \$4,000 \times 0.09 \times \tfrac{60}{360}$$
$$I = \$60.00$$

Method 2: 365-day year (Exact)

$$I = P \times R \times T$$
$$I = \$4,000 \times 0.09 \times \tfrac{60}{365}$$
$$I = \$59.18$$

Notice that the ordinary interest (360-day) method yields a slightly higher interest than the exact interest (365-day) method.

PRACTICE PROBLEMS

Calculate the ordinary and exact simple interest on these loans. Carry your answer to 4 decimal places and round to 2 decimal places.

	360-Day Year	**365-Day Year**
1. $2,100 at 12% for 60 days	_____	_____
2. $3,200 at 14% for 90 days	_____	_____
3. $10,500 at 13% for 30 days	_____	_____
4. $1,250 at 8% for 45 days	_____	_____

Solutions: **1.** $42.00; $41.42; **2.** $112.00; $110.47; **3.** $113.75; $112.19; **4.** $12.50; $12.33

In many cases, the amount borrowed and the interest rate on a loan are given, but an actual due date is not indicated. It is common to find terms of a loan expressed as "due in 90 days." When this occurs, you will need to count the actual number of days to determine the date of maturity. Once you know the date the loan began, you can use the following method to determine the maturity date. The actual number of days is determined by totaling the number of days in each month.

Work through the following example to learn how to calculate due dates. The table on page 282 shows the number of days in each month.

EXAMPLE _____

Loan date is April 15; loan is due in 120 days. What is the due date?

STEPS

1. Determine number of days in the specified period of time.

   a. April (30 days: 30 − 15 = 15 days remaining)
   b. May (31 days)
   c. June (30 days)
   d. July (31 days)

2. Add all the days counted.

$$15 + 31 + 30 + 31 = 107$$

3. Subtract those 107 from the 120 days due.

   a. August has 13 days to be counted for this time period.
   b. The maturity date would be August 13.

$$120 - 107 = 13$$

4. Add all the days counted to be sure they total 120.

$$15 + 31 + 30 + 31 + 13 = 120$$

You can also use these steps to determine the number of days between two dates, such as March 20 to July 28. The number of days is computed as follows.

March has 31 days (31 − 20 = 11 remaining days)
April = 30 days
May = 31 days
June = 30 days
July = 28 days (end of period)

$$11 + 30 + 31 + 30 + 28 = 130 \text{ days}$$

Number of Days in Each Month

28 Days	30 Days	31 Days
February (29 leap year)	April June September November	January March May July August October December

FIGURE 14-1

Number of Days in Each Month

PRACTICE PROBLEMS

Complete the following problems. Write your answers in the blanks provided. Assume no leap years.

1. Find the maturity date.

Date

(a) December 15 (90-day)   _____

(b) May 16 (120-day)   _____

2. Find the number of days from:

Days

(a) January 10 to April 10   _____

(b) May 8 to September 15   _____

Solutions: **1a.** March 15; **1b.** September 13 **2a.** 90; **2b.** 130

14.5 Finding Principal, Rate, or Time When Interest Is Known

In many cases, the interest is known; but the principal, rate, or time may not be given. You can determine the missing part if any two of the three remaining parts are given. The illustration at right will help you learn the formula needed to find the missing part.

Finding the Principal

To find the principal, cover the *P* in the box to determine what parts remain.

> *EXAMPLE* _____
>
> Suppose you want to determine the amount of principal it will take to earn $210 interest in 90 days at 8%. (The 360-day method is used in these calculations.)
>
> The formula is written like this:
>
> $$P = \frac{I}{R \times T}$$
>
> $$P = \frac{\$210}{0.08 \times \frac{90}{360}} = \$10,500$$

> ■ **TIP**   Use the basic simple interest formula to check your answer. $I = P \times R \times T$

Finding the Rate

To find the rate, cover the *R* in the box to determine what parts remain.

> *EXAMPLE* _____
>
> Suppose you borrow $7,700 for 90 days and the interest is $180. What is the interest rate?
>
> The formula is written like this:
>
> $$R = \frac{I}{P \times T}$$
>
> $$R = \frac{\$180}{\$7,700 \times \frac{90}{360}} = 9.4\%$$

Finding the Time

To find the amount of time, cover the *T* in the box to see what parts remain.

> *EXAMPLE* _____
>
> Barbara Jordan's loan is $4,000 at 11.5% and the interest is $230. What is the amount of time for this loan?
>
> The formula is written like this:
>
> $$T \text{ (in years)} = \frac{I}{P \times R}$$
>
> $$T \text{ (in years)} = \frac{\$230}{\$4,000 \times 0.115} = \frac{\$230}{\$460} = \frac{1}{2} \text{ year or 180 days}$$

Solve the following problems. Write your answers in the blanks provided. Use the 360-day method.

Interest	Rate	Time	Principal
1. $22.50	9%	60 days	_____
2. $220	_____	120 days	$5,000
3. $75.60	13%	_____	$2,650

Solutions: **1.** $1,500.00; **2.** 13.2%; **3.** 79 days

14.6 Promissory Notes

When you borrow money from a business or bank, you sign a document, usually called a promissory note. The signer agrees: (1) to pay a specific amount of money (2) by a specific date (3) to a specific individual or business.

If a promissory note states an interest rate, it is called an **_interest-bearing note_**. In some cases, however, a note may not specify an interest rate, which would mean it is a **_non-interest-bearing note_**. In other cases, the interest is collected in advance and is called a **_bank discount_** to distinguish it from interest paid at maturity. If a person signs a promissory note for a $2,000 bank loan for 60 days, the bank may discount it. The person would receive $1,960, which is the face value less the discount. The maturity value would be $2,000 in 60 days, which is what the borrower would repay. The interest ($40) is the same as would be paid for $2,000 at 12% for 60 days, but the borrower has use of only $1,960 for the 60 days, not $2,000. The amount the borrower receives (maturity value less the bank discount) is called the **_proceeds_**.

A typical promissory note is illustrated in the figure below. Study the major parts of this promissory note.

1. The _maker_ is the person or business borrowing the money.
2. The _payee_ is the person or business who lent the money and who will receive the repayment.
3. _Time_ or _term of the note_ may be expressed as a specific date, in days, or in months.
4. _Face value_ is the dollar amount, which may be written in word or number form or both.
5. _Interest rate_ per year. This particular note is interest bearing.
6. _Maturity date_ or _due date_ is the date the loan is to be repaid.

FIGURE 14-2
Promissory Note

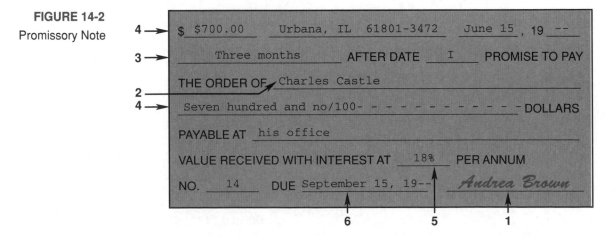

You already have figured due dates earlier in this chapter. Due dates on promissory notes may be calculated in the same manner. Here are some general procedures to follow.

1. When the note is expressed in terms of days, such as a 60-day note, determine the due date by using the exact number of days in the month.
2. When the time is expressed in months, such as a 3-month note dated July 15, determine the due date by counting months. The 3-month note dated July 15 would be due on October 15. If, however, a 3-month note is dated January 31, the due date is April 30 because April has only 30 days.

PRACTICE PROBLEMS

Find the due date, interest amount, and maturity value for each of these interest-bearing notes. Write your answers in the blanks provided. Use the 360-day method.

	Face Value	Rate	Date	Time	Due Date	Interest Amount	Maturity Value
1.	$1,550	14%	July 7	90 days	_____	_____	_____
2.	$2,275	12%	Sept. 8	6 months	_____	_____	_____

Solutions: **1.** Oct. 5; $54.25; $1,604.25; **2.** Mar. 8; $136.50; $2,411.50

A promissory note is a negotiable instrument. Suppose you own a business, and, rather than pay for goods received, you sign a note to George Marsch that you agree to pay for the goods at a later time. Mr. Marsch finds that he is running short of cash before you are due to pay him. Rather than request that you pay the note off early, he decides to take the note to his bank to cash the note so that he can obtain some money. The bank approves Mr. Marsch's request and agrees to give him cash. This procedure is called **discounting commercial paper**. For this privilege, the bank discounts the note, which means that the bank charges a fee based on the maturity value of the note.

To understand bank discounting, you will need to learn the following terms. **Maturity date** is when the note is due. **Discount date** is the date the bank discounts the note. **Term** is the period of time from the discount date to the maturity date. **Discount rate** is the percent rate the bank charges against the maturity value. The **discount** is the dollar amount of the bank discount. The actual amount of money received after the discount has been subtracted from the maturity value is called the **proceeds**, just as for a non-interest-bearing note.

The same concept for calculating simple interest ($I = P \times R \times T$) is applied to finding the maturity value and bank discount interest. The formula for figuring a bank discount is written as $B = M \times DR \times T$.

$$B = \text{Bank discount}$$
$$M = \text{Maturity value}$$
$$DR = \text{Discount rate}$$
$$T = \text{Term}$$

In order to determine the proceeds, first calculate the interest and maturity value. Next, determine the bank discount. Then find the proceeds by subtracting the bank discount from the maturity value. Figure the maturity value, bank discount, and proceeds by following this next example.

You have signed a promissory note for $6,000 at 9%, due in 4 months, payable to Mr. Marsch. The bank discounts the note for Mr. Marsch 16%, two *months* before maturity of the note.

STEPS

1. Calculate the interest.

$$I = P \times R \times T$$
$$I = \$6{,}000 \times 9\% \times \tfrac{4}{12}$$
$$I = \$180.00$$

2. Calculate the maturity value.

$$M = P + I$$
$$M = \$6{,}000 + \$180.00$$
$$M = \$6{,}180.00$$

3. Calculate the bank discount.

$$B = M \times DR \times T$$
$$B = \$6{,}180.00 \times 0.16 \times \tfrac{60}{360}$$
$$B = \$164.80$$

4. Calculate the proceeds.

$$P = M - B$$
$$P = \$6{,}180.00 - \$164.80$$
$$P = \$6{,}015.20$$

⌐ PRACTICE PROBLEMS

Find the bank discount and proceeds for these notes. The bank discounted these notes at 13% three months at maturity of the note. Write your answers in the blanks provided.

	Face Value	Interest Rate	Due	Bank Discount	Proceeds
1.	$2,500	12%	4 months	_____	_____
2.	$12,300	13%	3 months	_____	_____
3.	$15,750	12%	6 months	_____	_____
4.	$15,750	10%	7 months	_____	_____

Solutions: **1.** $84.50; $2,515.50; **2.** $412.74; $12,287.01;
3. $542.59; $16,152.41; **4.** $541.73; $16,127.02

14.7 Compound Interest

You have learned that for most short-term loans interest is computed by using the simple interest formula. If interest on a loan or investment is calculated more than once during the time period, this interest is added to the principal. This amount then becomes the principal for the next calculation of interest. This process is known as *compounding interest*

14.8 Terms Used for Compounding Interest

Compound interest is interest calculated on reinvested interest as well as on the original principal. The *compound amount* is the sum of the original principal and its compound interest. The *interest period* is the time (daily, monthly, quarterly, semiannually, or annually) for which interest has been computed. The *present value* of a future compound amount is the principal invested at a given rate today that will grow to the compound amount at a later date.

Financial institutions consider primarily four variables when calculating interest to be paid on deposits. The four variables are: 1) how much money is on deposit (the balance), 2) the interest rate applied (sometimes called *stated* rate of interest, 3) the method of determining the balance, 4) the frequency of compounding (annually semiannually, quarterly, weekly, daily). For example, with annual compounding, the amount of interest income earned on $1,000 deposited in a savings account paying 5.25 percent annual interest applied is calculated as follows:

$$\$1,000 \times 0.0525 \times 1 = \$1,052.50$$

If the amount is left on deposit for the second year, then the savings would be worth $1,107.76 ($1,052.50 × 0.0525 × 1 = $1,107.76).

If the amount were compounded quarterly (every 3 months) for the two years, you would receive a higher return on the amount. The table below provides a comparison of quarterly versus annual compounded interest.

TABLE 14-3

Comparison of Quarterly Versus Annual Compound Interest on Deposit at 6 percent Rate

Quarterly: (first year)		Annually
$1,000.00	6 ÷ 4 = 1.5 percent per quarter; (1,000 × 1.5 percent or 0.015 = 15.00)	
+ 15.00	1st quarter	
$1,015.00	(1,015.00 × 0.015 = 15.23)	
+ 15.23	2nd quarter	
$1,030.23		
+ 15.45	3rd quarter	
$1,045.68		
+ 15.69	4th quarter	
$1,061.37	$1,000 earns $61.37 (at the end of the first year)	$1,000 earns $60.00 (= $1,060.00)

Quarterly: (second year)		Annually
$1,061.37		
+ 15.92	5th quarter	
$1,077.29		
+ 16.16	6th quarter	
$1,093.45		
+ 16.40	7th quarter	
$1,109.85		
+ 16.64	8th quarter	
$1,126.49	$1,060.00 earns $65.12 (at the end of the second year)	$1,060.00 earns $63.60 (= $1,123.60)

In comparing the two methods of compounding, note the following:

1. You earn more interest on your deposited funds when compounding is used.
2. The more frequent the compounding, the greater is the return.
3. The greatest benefit or value of compounding is realized over longer time periods since greater differences will occur.

Because personal computers manipulate numbers quickly and reliably, they have had a tremendous impact on the financial field. Either by choice or from a competitive edge, most of the financial analysis and calculations have been turned over to personal computers.

In fact, calculations completed in this chapter could be done using a personal computer. However, the purpose of this chapter is to introduce you to the basic concepts underlying compound interest. The following summarizes the steps to determine compound interest. In this case, they are applied to find compound interest on original principal of $5,000 at $7\frac{1}{2}$% compounded annually for 3 years.

STEPS

1. Calculate simple interest.

$$I = P \times R \times T$$
$$I = \$5,000 \times 7\tfrac{1}{2}\% \text{ (first year)} \times 1$$
$$I = \$375 \text{ interest at end of first year}$$

2. Add the principal and interest to obtain the new amount on which to base the next period's (year's) interest.

$$\$5,000 + \$375 = \$5,375$$

3. Compound the interest for the second period (year).

$$I = \$5,375 \times 7\tfrac{1}{2}\% \text{ (second year)}$$
$$I = \$403.13$$

4. Add the compound interest to the principal plus the accrued interest to obtain the amount of investment at the end of the second period (year).

$$\$5,375 + \$403.13 = \$5,778.13$$

5. Continue this process for the remaining interest periods.

$$\$5,778.13 \times 7\tfrac{1}{2}\% = \$433.36 \ (\$5,778.13 + \$433.36 = \$6,211.49) \text{ (third year)}$$

6. Subtract the original amount principal from the compound amount to find the compound interest.

$$Compound\ Interest = Compound\ Amount - Original\ Principal$$
$$I = \$6,211.49 - \$5,000$$
$$I = \$1,211.49$$

PRACTICE PROBLEMS

Calculate the compound amount and interest at the end of a 2-year period. The original principal is $3,000 at $7\frac{1}{4}$% compounded annually.

1. Interest for first year   _____

2. Interest for second year   _____

3. Compound amount   _____

4. Compound interest   _____

Solutions: **1.** $217.50; **2.** $233.27; **3.** $3,450.77; **4.** $450.77

Note: If you did not obtain the same answers as given in the solutions, determine if an error occurred in the multiplication, addition, or subtraction process.

14.9 Compounding Interest Semiannually or Quarterly

As already mentioned, interest may be compounded more often than just once a year. For instance, financial institutions may compound interest daily, monthly, quarterly, or semiannually.

It is helpful to be able to convert the annual interest rate to the interest per compound period and determine the number of compound interest periods. You will divide the interest rate by the number of compounding periods per year. For instance, study these examples to see how the conversion is made.

Annual Rate	Time	Interest Compounded	Interest Rate	Number of Periods
10%	2 years	semiannually	5% (10 ÷ 2)	4 (2 years × twice a year)
12%	4 years	quarterly	3% (12 ÷ 4)	16 (4 years × 4 quarters)

Compounding Interest Semiannually

Calculate the compound amount and interest assuming the same information given earlier except that it is to be compounded semiannually.

Principal: $5,000
Interest Rate: 10%
Time: 2 years

Study the following illustration which shows the steps involved. With interest compounded semiannually, the interest rate is 5% for four compounding periods.

$5,000.00	original principal
+ 250.00	interest first compounding period ($5,000 × 0.05)
$5,250.00	
+ 262.50	interest second compounding period ($5,250 × 0.05)
$5,512.50	
+ 275.63	interest third compounding period ($5,512.50 × 0.05)
$5,788.13	
+ 289.41	interest fourth compounding period ($5,788.13 × 0.05)
6,077.54	compound amount
− 5,000.00	original principal
$1,077.54	compound interest

When comparing the examples showing interest compounded annually and semiannually, you see that the more frequently interest is compounded, the greater the amount of interest earned on the same original amount invested.

Calculate the number of periods, compound amount, and interest in the following problems. Interest is compounded semiannually.

Principal	Annual Rate	Time	No. of Periods	Compound Amount	Compound Interest
1. $6,200	$6\frac{1}{2}\%$	2 years	_____	_____	_____
2. $1,500	8%	2 years	_____	_____	_____
3. $18,400	$6\frac{1}{2}\%$	1 year	_____	_____	_____

Solutions: **1.** 4, $7,046.15, 846.15; **2.** 4, 1,754.79, $254.79; **3.** 2, $19,615.44, $1,215.44

Note: If you did not obtain the same answers given in the solutions, determine if the error occurred in the multiplication, addition, or subtraction process.

Compounding Interest Quarterly

To compound interest quarterly, you must know the number of compound periods for each year, and the interest rate for each period. There are four compound periods for each year, and the interest rate for each period is one-fourth of the yearly rate. Study this example illustrating how interest is compounded quarterly on $500 at 8% for 1 year.

STEPS

1. Divide the interest rate by 4 compound periods to obtain the quarterly interest rate.

    8% ÷ 4 = 2% or 0.02

2.
$500.00
+ 10.00   ($500 × 0.02)
$510.00
+ 10.20   ($510 × 0.02)
$520.20
+ 10.40   ($520.20 × 0.02)
$530.60
+ 10.61   (530.60 × 0.02)
$541.21   compound amount
−500.00   principal
$41.21   compound interest

Calculate the compound amount and interest in the following problems. Interest is compounded quarterly for one year. Write your answers in the blanks provided.

Principal	Annual Rate	Compound Amount	Compound Interest
1. $1,800	8%	_____	_____
2. $965	10%	_____	_____
3. $6,470	6%	_____	_____

Solutions: **1.** $1,948.37, $148.37; **2.** $1,065.19, $100.19; **3.** $6,867.02, $397.02

14.10 Using Compound Interest Tables

Even with a calculator, computing compound interest is very time consuming, particularly when a number of compound periods are involved. For instance, there are eight periods if interest is compounded quarterly for two years. Tables have been developed to help calculate compound interest, and today compounding of interest is figured using computers. Table 14-4 shows the amount of $1 compounded at the rates of 0% through 12% for interest periods for 1 to 50.

Let's use the table on the next page to calculate the compound amount and interest of this example.

Original Principal: $2,500
Interest Rate: 6% compounded annually
Time: 2 years

STEPS

1. Find the number of compound periods in the left column labeled *n* (meaning number). Move down (vertically) to 2.

2. Read across the table, horizontally to the proper interest rate column. Note the column headed 6%.

3. Where the two columns meet, the amount shown, 1.1236, is the compound value of $1 at 6% for 2 years. This means that $1.00 in 2 years compounded at 6% equals $1.12.

4. Multiply the compound value of $1 by the original principal to determine the compound amount of the principal.

   1.1236 × $2,500 = $2,809 compound amount

5. Subtract the original principal from the compound amount to determine the amount of compound interest.

   $2,809.00
   −2,500.00
   $309.00   compound interest

── PRACTICE PROBLEMS ──

Using the table, calculate the compound amount and compound interest on these problems. The interest is compounded annually. Write your answers in the blanks provided.

Principal	Rate	Time	Compound Amount	Compound Interest
1. $850	9%	2 years	_____	_____
2. $912	7%	2 years	_____	_____
3. $600	5%	1 year	_____	_____
4. $1,370	6%	2 years	_____	_____

Solutions: **1.** $1,009.89, $159.89; **2.** $1,044.15, $132.15; **3.** $630.00, $30.00; **4.** $1,539.33, $169.33

TABLE 14-4 Compound Interest

x	0.5%	1%	1.5%	2%	3%	4%	5%	6%	7%	8%	9%	10%	11%	12%
1	1.00500000	1.01000000	1.01500000	1.02000000	1.03000000	1.04000000	1.05000000	1.06000000	1.07000000	1.08000000	1.09000000	1.10000000	1.11000000	1.12000000
2	1.01002500	1.01010000	1.03022500	1.04040000	1.06090000	1.08160000	1.10250000	1.12360000	1.14490000	1.16640000	1.18810000	1.21000000	1.23210000	1.25440000
3	1.01507513	1.03030100	1.04567837	1.06120800	1.09272700	1.12486400	1.15762500	1.19101600	1.22504300	1.25971200	1.29502900	1.33100000	1.36763100	1.40492800
4	1.02015050	1.04060401	1.06136355	1.08243216	1.12550881	1.16985856	1.21550625	1.26247696	1.31079601	1.36048896	1.41158161	1.46410000	1.51807041	1.57351936
5	1.02525125	1.05101005	1.07728400	1.10408080	1.15927407	1.21665290	1.27628156	1.33822558	1.40255173	1.46932808	1.53862395	1.61051000	1.68505816	1.76234168
6	1.03037751	1.06152015	1.09344326	1.12616242	1.19405230	1.26531902	1.34009564	1.41851911	1.50073035	1.58687432	1.67710011	1.77156100	1.87041455	1.97382269
7	1.03552940	1.07213535	1.10984491	1.14868567	1.22987387	1.31593178	1.40710042	1.50363026	1.60578148	1.71382427	1.82803912	1.94871710	2.07616015	2.21068141
8	1.04070704	1.08285671	1.12649259	1.17165938	1.26677008	1.36856905	1.47745544	1.59384807	1.71818618	1.85093021	1.99256264	2.14358881	2.30453777	2.47596318
9	1.04591058	1.09368527	1.14338998	1.19509257	1.30477318	1.42331181	1.55132822	1.68947896	1.83845921	1.99900463	2.17189328	2.35794769	2.55803692	2.77307876
10	1.05114013	1.10462213	1.16054083	1.21899442	1.34391638	1.48024428	1.62889463	1.79084770	1.96715136	2.15892500	2.36736367	2.59374246	2.83942099	3.10584821
11	1.05639583	1.11566835	1.17794894	1.24337431	1.38423387	1.53945406	1.71033936	1.89829856	2.10485195	2.33163900	2.58042641	2.85311671	3.15175729	3.47854999
12	1.06167781	1.12682503	1.19561817	1.26824179	1.42576089	1.60103222	1.79585633	2.01219647	2.25219159	2.51817012	2.81266478	3.13842838	3.49845060	3.89597599
13	1.06698620	1.13809328	1.21355244	1.29360663	1.46853371	1.66507351	1.88564914	2.13292826	2.40984500	2.71962523	3.06580461	3.45227121	3.88328016	4.36349311
14	1.07232113	1.14947421	1.23175573	1.31947876	1.51258972	1.73167645	1.97993160	2.26090396	2.57853415	2.93719362	3.34172703	3.79749834	4.31044098	4.88711229
15	1.07768274	1.16096896	1.25023207	1.34586834	1.55796742	1.80094351	2.07892818	2.39655819	2.75903154	3.17216911	3.64248246	4.17724817	4.78458949	5.47356576
16	1.08307115	1.17257864	1.26898555	1.37228571	1.60470644	1.87298125	2.18287459	2.54035168	2.95216375	3.42594264	3.97030588	4.59497299	5.31089433	6.13039365
17	1.08848651	1.18430443	1.28802033	1.40024142	1.65284763	1.94790050	2.29201832	2.69277279	3.15881521	3.70001805	4.32763341	5.05447028	5.89509271	6.86604089
18	1.09392894	1.19614748	1.30734064	1.42824625	1.70243306	2.02581652	2.40661923	2.85433915	3.37993228	3.99601950	4.71712042	5.55991731	6.54355291	7.68996580
19	1.09939858	1.20810895	1.32695075	1.45681117	1.75350605	2.10684918	2.52695020	3.02559936	3.61652754	4.31570106	5.14166125	6.11590904	7.26334373	8.61276169
20	1.10489558	1.22019004	1.34685501	1.48594740	1.80611123	2.19112314	2.65329771	3.20713547	3.86968446	4.66095714	5.60441077	6.72749995	8.06231154	9.64629309
21	1.11042006	1.23239194	1.36705783	1.51566634	1.86029457	2.27876807	2.78596259	3.39956360	4.14056237	5.03383372	6.10880774	7.40024994	8.94916581	10.80384826
22	1.11597216	1.24471586	1.38756370	1.54597967	1.91610341	2.36991879	2.92526072	3.60353742	4.43040174	5.43654041	6.65860043	8.14027494	9.93357404	12.10031006
23	1.12155202	1.25716302	1.40837715	1.57689926	1.97358651	2.46471554	3.07152376	3.81974966	4.74052986	5.87146365	7.25787447	8.95430243	11.02626719	13.55234726
24	1.12715978	1.26973465	1.42950281	1.60843725	2.03279411	2.56330416	3.22509994	4.04893464	5.07236695	6.34118074	7.91108317	9.84973268	12.23915658	15.17862893
25	1.13279558	1.28243200	1.45094535	1.64060599	2.09377793	2.66583633	3.38635494	4.29187072	5.42743264	6.84847570	8.62308066	10.83470594	13.58546380	17.00006441
26	1.13845955	1.29525631	1.47270953	1.67341811	2.15659127	2.77246978	3.55567269	4.54938296	5.80735292	7.39635321	9.39915792	11.91817654	15.07986482	19.04007214
27	1.14415185	1.30820888	1.49480018	1.70688648	2.22128901	2.88336858	3.73345632	4.82234594	6.21386763	7.98806147	10.24508213	13.10999419	16.73864995	21.32488079
28	1.14987261	1.32129097	1.51722218	1.74102421	2.28792768	2.99870332	3.92012914	5.11168670	6.64883836	8.62710639	11.16713952	14.42099361	18.57990145	23.88386649
29	1.15562197	1.33450388	1.53998051	1.77584469	2.35656551	3.11865145	4.11613560	5.41838790	7.11425705	9.31727490	12.17218208	15.86309297	20.62369061	26.74993047
30	1.16140008	1.34784892	1.56308022	1.81136158	2.42726247	3.24339751	4.32194238	5.74349117	7.61225504	10.06265689	13.26767847	17.44940227	22.89229657	29.95992212
31	1.16720708	1.36132740	1.58652642	1.84758882	2.50000835	3.37313341	4.53803949	6.08810064	8.14511290	10.86766944	14.46176953	19.19434250	25.41044919	33.55511278
32	1.17304312	1.37494068	1.61032432	1.88454059	2.57508276	3.50805875	4.76494147	6.45338368	8.71527080	11.73708279	15.76332879	21.11377675	28.20555663	37.58172631
33	1.17890833	1.38869009	1.63447918	1.92223140	2.65233524	3.64838110	5.00318854	6.84058988	9.32533975	12.67604964	17.18202838	23.22515442	31.30821445	42.09153347
34	1.18480288	1.40257699	1.65899637	1.96067603	2.73190530	3.79431634	5.25334797	7.25102528	9.97811354	13.69013361	18.72841093	25.54766986	34.75211804	47.14251748
35	1.19072689	1.41660276	1.68388132	1.99988955	2.81386245	3.94608899	5.51601537	7.68608679	10.67658148	14.78534429	20.41396792	28.10243685	38.57485103	52.79961958
36	1.19668052	1.43076878	1.70913954	2.03988734	2.89827833	4.10393255	5.79181614	8.14725200	11.42394219	15.96817184	22.25122503	30.91268053	42.81808464	59.13557393
37	1.20266393	1.44507647	1.73477663	2.08068509	2.98522668	4.26808986	6.08140694	8.63608712	12.22363814	17.24562558	24.25383528	34.00394859	47.52807395	66.23184280
38	1.20867725	1.45952724	1.76079828	2.12229879	3.07478348	4.43881065	6.38547729	9.15425253	13.07927141	18.62527563	26.43680046	37.40434344	52.75616209	74.17966394
39	1.21472063	1.47412251	1.78721025	2.16474477	3.16702698	4.61636599	6.70475115	9.70350749	13.99482041	20.11529768	28.81598170	41.14477779	58.55933991	83.08122361
40	1.22079424	1.48886373	1.81401841	2.20803966	3.26203779	4.80102063	7.03998871	10.28571794	14.97445784	21.72452150	31.40942005	45.25925557	65.00086731	93.05097044
41	1.22689821	1.50375237	1.84122868	2.25220046	3.35989893	4.99306145	7.39198815	10.90286101	16.02266989	23.46248322	34.23626786	49.78511112	72.15096271	104.21708689
42	1.23303270	1.51878989	1.86884712	2.29724447	3.46069589	5.19278391	7.76158756	11.55703267	17.14425678	25.33948187	37.31753197	54.76369924	80.08756861	116.72313732
43	1.23919786	1.53397779	1.89687982	2.34318936	3.56451677	5.40049527	8.14966693	12.25045463	18.34435475	27.36664042	40.67610984	60.24006916	88.89720115	130.72991380
44	1.24539385	1.54931757	1.92533302	2.39005314	3.67145227	5.61651508	8.55715028	12.98548991	19.62845959	29.55597166	44.33695973	66.26407608	98.67589328	146.41750346
45	1.25162082	1.56481075	1.95421301	2.43785421	3.78159584	5.84117568	8.98500779	13.76461083	21.00245176	31.92044939	48.32728610	72.89048369	109.53024154	163.98760387
46	1.25787892	1.58045885	1.98352621	2.48661129	3.89504372	6.07482271	9.43425818	14.59048748	22.47262338	34.47408534	52.67674185	80.17953205	121.57856811	183.66611634
47	1.26416832	1.59626344	2.01327910	2.53634352	4.01189503	6.31781562	9.90597109	15.46591673	24.04570702	37.23201217	57.41764862	88.19748526	134.95221060	205.70605030
48	1.27048916	1.61222608	2.04347829	2.58707039	4.13225188	6.57052824	10.40126965	16.39387173	25.72890651	40.21057314	62.58523700	97.01723378	149.79695377	230.39077633
49	1.27684161	1.62834834	2.07413046	2.63881179	4.25621944	6.83334937	10.92133313	17.37750403	27.52992997	43.42741899	68.21790833	106.91895716	166.27461868	258.03766949
50	1.28322581	1.64463182	2.10524242	2.69158803	4.38390602	7.10668335	11.46739979	18.42015427	29.45702506	46.90161251	74.35752008	117.39085288	184.56482674	289.00218983

Using Compound Interest Tables for Periods Other Than One a Year

You can also use the table to calculate interest compounded for periods other than once a year. For instance,
- to compound *daily*, interest is figured on a balance each day.
- to compound *quarterly*, interest is figured on a balance every 3 months or every $\frac{1}{4}$ year.
- to compound *semiannually*, interest is figured on a balance every 6 months or every $\frac{2}{4}$ year.
- to compound *annually*, interest is figured on a balance once a year

The *rate* is the annual rate divided by the number of times compounded per year. For example, 8% compounded quarterly is 2% $\left(\frac{8\%}{4 \text{ (quarterly)}} \right)$. The *pay* period is figured by multiplying the number of years by the number of times compounded per year.

EXAMPLE —————————————————————————

Suppose you had a fund that was to be compounded quarterly for four years. The number of periods is 12 (3 years × 4 quarters). Follow these steps, which illustrate compounding interest quarterly.

$$\begin{array}{rl} \text{Original Principal:} & \$5,000 \\ \text{Interest:} & 8\% \text{ compounded quarterly} \\ \text{Time:} & 3 \text{ years} \end{array}$$

STEPS

1. Multiply 3 years by 4 quarters; then find 12 periods in the left column (*n*) of the table.

2. Take one-fourth of the interest rate (8% ÷ 4), then move across the table to the 2% column. The amount, 1.26824179 is the value of $1 compounded quarterly for 3 years at 8%.

3. Multiply the original principal by the compound value found in the table to compute compound amount.

$$\$5,000 \times 1.26824179 = \$6,341.21$$

4. Subtract the original principal from the compound amount to determine the compound interest.

$$\begin{array}{r} \$6,341.21 \\ -5,000.00 \\ \hline \$1,341.21 \quad \text{compound interest} \end{array}$$

PRACTICE PROBLEMS

Using the compound interest table, calculate the compound amount and interest on these problems. The interest is compounded quarterly.

	Principal	Annual Rate	Time	Compound Amount	Compound Interest
1.	$4,000	8%	2 years	_____	_____
2.	$8,900	8%	5 years	_____	_____
3.	$2,550	6%	3 year	_____	_____

Solutions: **1.** $4,686.64, $686.64; **2.** $13,224.93, $4,324.93; **3.** $3,048.83, $498.83

14.11 Calculating Present Value Using a Present Value Table

The present value of a future compound amount is the principal invested at a given rate today that will grow to the compound amount at a later date. Understanding the present value concept will help you to answer the following question.

EXAMPLE _____

How much money should I invest now in the first year (present value) to yield $10,000 (future value) in three years at 8% compounded quarterly?

A table showing the present value of $1 at different interest rates is commonly used for speed and accuracy. The table on the next page shows the present value of $1 with compound interest rates from 0.5% to 12%. The amount shown in a present value table is used for multiplying the desired amount by the present value of $1. Let's determine the present value using the preceding example.

Future Amount: $10,000
Interest Rate: 8% compounded quarterly
Time: 3 years

STEPS

1. Find the number of periods.   3 years × 4 (quarterly) = 12 compound periods

2. Determine the rate.   8% ÷ 4 (quarters) = 2%

3. Determine the table amount. (Read down the left column to 12 (periods) then across to 2%.)

0.78849318

4. Multiply the desired amount by the table amount.

0.78849318 × $10,000 (future value) = $7,884.93 (present value)

Thus, if you would invest $7,884.93 today at 8% compounded quarterly for three years, you would have $10,000 at the end of three years.

■ **TIP** To prove, use the compound interest table. 12 compound periods,
2% = 1.26824179 × $7,884.93 = $10,000.

─ PRACTICE PROBLEMS ─

Using the present value table, calculate the present value for these problems.

Future Amount	Period of Time	Annual Interest	Compound Period	Present Value
1. $16,000	4 years	9%	annually	_____
2. $1,700	1 year	8%	quarterly	_____
3. $20,000	4 years	12%	quarterly	_____

Solutions: **1.** $11,334.80; **2.** $1,570.54; **3.** $12,463.34

TABLE 14-5   Present Value

n	0.5%	1%	1.5%	2%	3%	4%	5%	6%	7%	8%	9%	10%	11%	12%
1	0.99502488	0.99009901	0.98522167	0.98039216	0.97087379	0.96153846	0.95238095	0.94339623	0.93457944	0.92592593	0.91743119	0.90909091	0.90090090	0.89285714
2	0.99007450	0.98029605	0.97066175	0.96116878	0.94259591	0.92455621	0.90702948	0.88999644	0.87343873	0.85733882	0.84167999	0.82644628	0.81162243	0.79719388
3	0.98514876	0.97059015	0.95631699	0.94232233	0.91514166	0.88899636	0.86383760	0.83961928	0.81629788	0.79383224	0.77218348	0.75131480	0.73119138	0.71178025
4	0.98024752	0.96098034	0.94218423	0.92384543	0.88848705	0.85480419	0.82270247	0.79209366	0.76289521	0.73502985	0.70842521	0.68301346	0.65873097	0.63551808
5	0.97537067	0.95146569	0.92826033	0.90573081	0.86260878	0.82192711	0.78352617	0.74725817	0.71298618	0.68058320	0.64993139	0.62092132	0.59345133	0.56742686
6	0.97051808	0.94204524	0.91454219	0.88797138	0.83748426	0.79031453	0.74621540	0.70496054	0.66634222	0.63016963	0.59626733	0.56447393	0.53464084	0.50663112
7	0.96568963	0.93271805	0.90102679	0.87056018	0.81309151	0.75991781	0.71068133	0.66505711	0.62274974	0.58349040	0.54703424	0.51315812	0.48165841	0.45234922
8	0.96088520	0.92348322	0.88771112	0.85349037	0.78940923	0.73069021	0.67683936	0.62741237	0.58200910	0.54026888	0.50186628	0.46650738	0.43392650	0.40388323
9	0.95610468	0.91433982	0.87459224	0.83675527	0.76641673	0.70258674	0.64460892	0.59189846	0.54393374	0.50024897	0.46047778	0.42409762	0.39092477	0.36061002
10	0.95134794	0.90528695	0.86166723	0.82034830	0.74409391	0.67556417	0.61391325	0.55839478	0.50834929	0.46319349	0.42241081	0.38554329	0.35218448	0.32197324
11	0.94661487	0.89632372	0.84893323	0.80426304	0.72242128	0.64958093	0.58467929	0.52678753	0.47509280	0.42888286	0.38753285	0.35049390	0.31728331	0.28747610
12	0.94190534	0.88744923	0.83638742	0.78849318	0.70137988	0.62459705	0.55683742	0.49696936	0.44401196	0.39711376	0.35553473	0.31863082	0.28584082	0.25667509
13	0.93721924	0.87866260	0.82402702	0.77303253	0.68095134	0.60057409	0.53032135	0.46883902	0.41496445	0.36769792	0.32617865	0.28966438	0.25751426	0.22917419
14	0.93255646	0.86996297	0.81184928	0.75787502	0.66111781	0.57747508	0.50506795	0.44230096	0.38781724	0.34046104	0.29924647	0.26333125	0.23199482	0.20461981
15	0.92791688	0.86134947	0.79985150	0.74301473	0.64186195	0.55526450	0.48101710	0.41726506	0.36244602	0.31524170	0.27453804	0.23939205	0.20900435	0.18269626
16	0.92330037	0.85282126	0.78803104	0.72844581	0.62316694	0.53390818	0.45811152	0.39364628	0.33873460	0.29189047	0.25186976	0.21762914	0.18829220	0.16312166
17	0.91870684	0.84437749	0.77638526	0.71416256	0.60501645	0.51337325	0.43629669	0.37136442	0.31657439	0.27026895	0.23107318	0.19784467	0.16963262	0.14564434
18	0.91413616	0.83601731	0.76491159	0.70015937	0.58739461	0.49362812	0.41552065	0.35034379	0.29586392	0.25024903	0.21199374	0.17985879	0.15282218	0.13003959
19	0.90958822	0.82773992	0.75360747	0.68643076	0.57028603	0.47464242	0.39573396	0.33051300	0.27650833	0.23171206	0.19448967	0.16350799	0.13767764	0.11610678
20	0.90506290	0.81954447	0.74247042	0.67297133	0.55367675	0.45638695	0.37688948	0.31180473	0.25841900	0.21454821	0.17843089	0.14864363	0.12403391	0.10366677
21	0.90056010	0.81143017	0.73149795	0.65977582	0.53754928	0.43883360	0.35894236	0.29415540	0.24151309	0.19865575	0.16369806	0.13513057	0.11174226	0.09255961
22	0.89607971	0.80339621	0.72068763	0.64683904	0.52189250	0.42195539	0.34184987	0.27750510	0.22571317	0.18394051	0.15018171	0.12284597	0.10066870	0.08264251
23	0.89162160	0.79544179	0.71003708	0.63415592	0.50669175	0.40572633	0.32557131	0.26179726	0.21094688	0.17031528	0.13778139	0.11167816	0.09069252	0.07378796
24	0.88718567	0.78756613	0.69954392	0.62172149	0.49193374	0.39012147	0.31006791	0.24697855	0.19714662	0.15769934	0.12640494	0.10152560	0.08170498	0.06588210
25	0.88277181	0.77976844	0.68920583	0.60953087	0.47760557	0.37511680	0.29530277	0.23299863	0.18424918	0.14601790	0.11596784	0.09229600	0.07360809	0.05882331
26	0.87837991	0.77204796	0.67902052	0.59757928	0.46369473	0.36068923	0.28124073	0.21981003	0.17219549	0.13520176	0.10639251	0.08390545	0.06631359	0.05252081
27	0.87400986	0.76440392	0.66898574	0.58586204	0.45018906	0.34681657	0.26784832	0.20736793	0.16093037	0.12518682	0.09760781	0.07627768	0.05974197	0.04689358
28	0.86966155	0.75683557	0.65909925	0.57437455	0.43707675	0.33347747	0.25509364	0.19563014	0.15040221	0.11591372	0.08954845	0.06934335	0.05382160	0.04186927
29	0.86533488	0.74934215	0.64935887	0.56311231	0.42434636	0.32065141	0.24294632	0.18455674	0.14056282	0.10732752	0.08215454	0.06303941	0.04848793	0.03738327
30	0.86102973	0.74192292	0.63976243	0.55207089	0.41198676	0.30831867	0.23137745	0.17411013	0.13136712	0.09937733	0.07537114	0.05730855	0.04368282	0.03337792
31	0.85674600	0.73457715	0.63030781	0.54124597	0.39998715	0.29646026	0.22035947	0.16425484	0.12277301	0.09201605	0.06914783	0.05209868	0.03935389	0.02980172
32	0.85248358	0.72730411	0.62099292	0.53063330	0.38833703	0.28429685	0.20986624	0.15495718	0.11474143	0.08520006	0.06343838	0.04736244	0.03545395	0.02660868
33	0.84824237	0.72010307	0.61181568	0.52022083	0.37702625	0.27409417	0.19987254	0.14618622	0.10723470	0.07888893	0.05820035	0.04305676	0.03194050	0.02375775
34	0.84402226	0.71297334	0.60277407	0.51002817	0.36604490	0.26355209	0.19035480	0.13791153	0.10021934	0.07304531	0.05339481	0.03914251	0.02877522	0.02121227
35	0.83982314	0.70591420	0.59386608	0.50002761	0.35538340	0.25341547	0.18129029	0.13010522	0.09366294	0.06763454	0.04898607	0.03558410	0.02592363	0.01893953
36	0.83564492	0.69892495	0.58508974	0.49022315	0.34503243	0.24366872	0.17265741	0.12274077	0.08753546	0.06262458	0.04494135	0.03234918	0.02334918	0.01691029
37	0.83148748	0.69200490	0.57644309	0.48061093	0.33498294	0.23429685	0.16443563	0.11579318	0.08180884	0.05798572	0.04122059	0.02940835	0.02104020	0.01509848
38	0.82735073	0.68515337	0.56792423	0.47118719	0.32522615	0.22528543	0.15660536	0.10923885	0.07645686	0.05369048	0.03782623	0.02673486	0.01895513	0.01348078
39	0.82323455	0.67836967	0.55953126	0.46191680	0.31575355	0.21662061	0.14914797	0.10305552	0.07145501	0.04971341	0.03470296	0.02430442	0.01707670	0.01203641
40	0.81913886	0.67165314	0.55126232	0.45289042	0.30655684	0.20828904	0.14204568	0.09722219	0.06678038	0.04603093	0.03183758	0.02209493	0.01538441	0.01074680
41	0.81506354	0.66500311	0.54311559	0.44401021	0.29762800	0.20027793	0.13528160	0.09171905	0.06241157	0.04262123	0.02920879	0.02008630	0.01385983	0.00959536
42	0.81100850	0.65841892	0.53508925	0.43530413	0.28895922	0.19257493	0.12883962	0.08652740	0.05832857	0.03946411	0.02679706	0.01826027	0.01248633	0.00856728
43	0.80697363	0.65189992	0.52718153	0.42676875	0.28054294	0.18516820	0.12270440	0.08162962	0.05451268	0.03654084	0.02458446	0.01660025	0.01124869	0.00764936
44	0.80295884	0.64544546	0.51939067	0.41840074	0.27237178	0.17804635	0.11686133	0.07700908	0.05094643	0.03383341	0.02255455	0.01509113	0.01013419	0.00682978
45	0.79896402	0.63905492	0.51171494	0.41019680	0.26443862	0.17119841	0.11129651	0.07265007	0.04761349	0.03132788	0.02069224	0.01371921	0.00912990	0.00609802
46	0.79449907	0.63272764	0.50415265	0.40215373	0.25673653	0.16461386	0.10599668	0.06853781	0.04449859	0.02900730	0.01898371	0.01247201	0.00822513	0.00544466
47	0.79103390	0.62646301	0.49670212	0.39426836	0.24925876	0.15828256	0.10094921	0.06465831	0.04158747	0.02685861	0.01741625	0.01133819	0.00741003	0.00486131
48	0.78709841	0.62026041	0.48936170	0.38653761	0.24199880	0.15219476	0.09614211	0.06099840	0.03886679	0.02486908	0.01597821	0.01030745	0.00667570	0.00434045
49	0.78318250	0.61411921	0.48212975	0.37895844	0.23495029	0.14634112	0.09156391	0.05754566	0.03632410	0.02302693	0.01465891	0.00937041	0.00601415	0.00387540
50	0.77928607	0.60803882	0.47500468	0.37152788	0.22810708	0.14071262	0.08720373	0.05428836	0.03394776	0.02132123	0.01344854	0.00851855	0.00541815	0.00346018

MATH ALERT

1. Suppose your uncle lent you $1,800 for tuition and books for your first year of college at $2\frac{1}{2}\%$ interest for one year. Find the interest and maturity value.

2. What if you wanted your son to be able to go to college in 10 years. You have estimated he will need $40,000. How much money should you invest this year to yield $40,000 in 10 years at 6% interest compounded quarterly?

Future amount: $40,000

Interest Rate: 6%

Time: 10 years

1. Compound periods: _____

2. Rate: _____

3. Table amount: _____

4. Present value: _____

Study Guide

I.   Terminology

page 278	interest	The amount paid only on the principal for the privilege of using someone's money.
page 278	simple interest	Interest based on principal, interest rate, and time.
page 278	compound interest	Interest is calculated on reinvested interest as well as on the original principal.
page 279	principal	The amount borrowed or invested for a certain period of time. Sometimes called face value.
page 279	time	The length of time you have to repay a loan or earn interest on your money invested.
page 280	maturity value	The full amount of money that must be repaid when the loan is due, that is, principal + interest.
page 284	promissory note	A written promise to pay a specific amount of money by a specific date to an individual or business.
page 284	interest-bearing note	A promissory note that states an interest rate.
page 284	non-interest bearing note	A promissory note that does not state an interest rate.
page 284	bank discount	When a bank collects the interest on a note in advance.
page 284	proceeds	The amount the borrower receives (maturity value less the bank discount).
page 284	maker	The person or business borrowing the money.
page 284	payee	The person or business who loaned the money and who will receive the repayment.
page 284	interest rate	The percent of interest charged for borrowing money.
page 285	discount date	The date the bank collects the note.
page 285	discounting commercial paper	The process of discounting a note.
page 285	discount rate	The percent rate the bank charges against the maturity value of a note.
page 285	discount	The dollar amount of the bank discount.
page 286	compound amount	The sum of the original principal and its compound interest.
page 286	interest period	The time (daily, monthly, quarterly, semiannually, or annually) for which interest has been computed.
page 286	present value	The principal invested at a given rate today that will grow to the compound amount at a later date.
page 288	quarterly	Every three months.
page 288	monthly	Every month.
page 293	semiannually	Twice a year.
page 293	annually	Once a year.
page 294	present value	The principal invested at a given rate today that will grow to the compound amount at a later date.

II. Calculating Simple Interest

page 279 Use the formula $I = P \times R \times T$ where:
I = Interest
P = Principal (amount of the loan)
R = Rate (percent of interest charged or earned per year)
T = Time

page 279 *Calculating Interest for Less Than One Year:* Convert the interest rate to a decimal; multiply the principal by the rate; calculate the time by converting the months to a fraction of a year.

Example: Principal: $3,500
Rate: 12%
Time: 8 months

$3,500 × 0.12 = $420
8 months = $\frac{8}{12}$
$420 × $\frac{8}{12}$ = $280.00 interest

page 280 *Finding the Maturity Value:* Determine the interest first; determine the maturity value by adding the interest to the principal. Use the formula $I = P \times R \times T$.

Example: Principal: $950
Rate: 14%
Time: 10 months or $\frac{10}{12}$

$950 × 0.14 × $\frac{10}{12}$ = $110.83 interest
$950 + $110.83 = $1,060.83 maturity value

page 281 *Finding Interest when Time is Expressed in Days:* Use the formula $I = P \times R \times T$ divided by 360; time is expressed as a fraction of 360, such as $\frac{120}{360}$.

Example: Principal: $4,000
Rate: 9%
Time: 60 days (360 days)

$4,000 × 0.09 × 60 ÷ 360 = $60.00
Calculation for 365 days: $4,000 × 0.09 × 60 ÷ 365 = $59.18

page 281 *Finding the Due Dates:* Determine the number of days in the first partial month; add the remaining months actual days; subtract the total from 120 days to find the actual due date.

Example: Loan date: April 15
Number of days: 120

April (30 − 15): 15 days left in April
May: 31 days
June: 30 days
July: 31 days
Total days: 107 days

120 − 107 = 13; therefore, the due date is August 13.

As a check, add all the days in the full months and partial months. They should equal 120.

page 283   *Finding Principal when Interest is Known:* Reduce the fraction expressed as time; multiply the rate by the time; divide the interest by the answer. Use the formula:

$$P = \frac{I}{R \times T}$$

Example: Interest: $210
            Time: 90 days
            Rate: 8%

$\frac{90}{360} = \frac{1}{4}$
$0.08 \times \frac{1}{4} = 0.2$
$210 ÷ 0.02 = $10,500 principal

page 283   *Finding the Rate:* Use the same formula as above; simplify the denominator and change to a fraction; divide the interest by the answer.

Example: Principal: $7,700
            Interest: $180
            Time: 90 days

$\frac{90}{360} = \frac{1}{4}$
$7,700 \times \frac{1}{4} = $1,925
$180 ÷ $1,925 = 0.0935 or 9.4%

page 283   *Finding the Time:* Use the same formula as above; multiply the principal by the rate; divide the interest by the answer found; to change the answer to days, multiply the answer found by 360.

Example: Principal: $4,000
            Interest: $230
            Rate: 11.5%

$4,000 \times 0.115 = $460
$230 ÷ $460 = 0.5
$0.5 \times 360 = 180 days

page 285   *Calculating Due Dates on Promissory Notes:* When the due date is expressed in terms of a specific date, determine the number of days between the date of the note and the due date.

Example: A note signed on July 19 and due on November 14 is a 120-day note.

When the note is expressed in terms of days, such as a 60-day note, determine the due date by using the exact number of days in each month.

Example: A 60-day note signed on November 10 is due January 9.

When the time is expressed in months, such as a 3-month note, the due date is three months from that date on the same day.

Example: A 3-month note signed on July 15 is due October 15.

page 286   *Discounting a Note:* Calculate the interest ($I = P \times R \times T$); calculate the maturity value ($M = P + I$); calculate the bank discount ($B = M \times DR \times T$); calculate the proceeds ($P = M - B$).

Example: Principal: $6,000   Discount rate: 16%
            Rate: 9%   Discount date: 2 months before maturity
            Time: 4 months   or $\frac{2}{12}$

$6,000 \times 0.09 \times \frac{4}{12} = $180.00 interest
$6,000 + $180.00 = $6,180.00 maturity value
$6,180.00 \times 0.16 \times \frac{2}{12} = $164.80 bank discount
$6,180.00 - $164.80 = $6,015.20 proceeds

III. Compounding Interest

page 287 *Calculating Compound Interest:* Calculate the interest; add the interest and principal; compound the interest by calculating the interest on the original principal plus interest earned previously; continue the process for the designated periods for which the interest is figured; find the compound interest by subtracting the original principal from the compound amount.

Example: Principal invested: $5,000
Rate: $7\frac{1}{2}$%
Time: Compounded for 3 years

$5,000 × 0,075 = $375
$5,000 + $375 = $5,375 (first year)
$5,375 × 0.075 = $403.13
$5,375 + $403.13 = $5,778.13 (second year)
$5,778.13 × 0.075 = $433.36
$5,778.13 + $433.36 = $6,211.49 (third year)

page 289 *Compounding Interest Semiannually:* Semiannually means twice a year; divide the interest rate by 2; calculate the compound interest following the procedure just given except calculate it twice per year.

page 290 *Compounding Interest Quarterly:* Quarterly means four times a year; divide the interest rate by 4; calculate the compound amount and interest following the procedure given except calculate it four times per year.

page 291 *Using Compound Interest Tables:* Find the number of compound periods in the left column labeled *n* (meaning number); read across the table horizontally to the proper interest rate column; where the two columns meet, the amount shown is the compounded value of $1; multiply the compound value of $1 by the original principal; subtract the original principal from the compounded amount to determine the amount of compound interest.

Example: Compound value from table: 1.1236
Principal: $2,500
Time: 2 years

1.1236 × $2,500 = $2,809
$2,809 − $2,500 = $309 compound interest

IV. Present Value

page 294 *Calculating Present Value Using a Present Value Table:* Find the number of periods; determine the rate per period; determine the table amount; multiply the desired amount by the table amount.

Example: Future amount: $10,000
Rate: 8% compounded quarterly
Time: 3 years

3 × 4 = 12 compound periods
0.08 ÷ 4 quarters (periods in 1 year) = 2%
Look up table value; 12 periods at 2% = 0.78849318
0.78849318 × $10,000 = $7,884.93 present value to be invested

Assignment 1

Name_____ Date _____

Write your answers in the blanks provided. **Round money amounts to two decimal places.**

A. Find the simple interest.

	Principal	Rate	Time	Interest
1.	$5,700	11%	8 mos.	_____
2.	$9,460	15%	6 mos.	_____
3.	$4,500	11%	9 mos.	_____
4.	$2,900	16%	7 mos.	_____

B. Find the interest and maturity value.

	Principal	Rate	Time	Interest	Maturity Value
5.	$820	13%	3 mos.	_____	_____
6.	$860	14%	4 mos.	_____	_____
7.	$3,700	12%	6 mos.	_____	_____
8.	$5,600	14%	1 year	_____	_____
9.	$980	9%	3 mos.	_____	_____

C. Calculate the ordinary and exact simple interest on these loans.

	Principal	Rate	Time	360-Day Year	365-Day Year
10.	$2,400	13%	60 days	_____	_____
11.	$1,750	12%	90 days	_____	_____
12.	$980	13%	60 days	_____	_____
13.	$3,600	13%	90 days	_____	_____
14.	$4,850	14%	120 days	_____	_____

D. Find the maturity date.

		Date
15.	March 10 (90-day)	_____
16.	January 20 (120-day)	_____
17.	September 13 (90-day)	_____

E. Find the number of days from:

		Days			Days
18.	February 15 to April 4	_____	**19.**	June 13 to September 7	_____

Assignment 2

Name_____ Date _____

Write your answers in the blanks provided. Round money amounts to two decimal places.

A. Find the principal, rate, or time for the following loans. Assume 360-day method.

	Interest	Principal	Rate	Time
1.	$81.90	_____	13%	120 days
2.	$15.76	$675.50	14%	_____
3.	$296.10	$9,870	_____	60 days
4.	$125.00	_____	15%	60 days
5.	$136.00	$3,400	_____	90 days
6.	$29.70	$990.00	9%	_____
7.	$110.00	$6,000.00	11%	_____
8.	$39.60	_____	12%	60 days
9.	$123.67	$2,650.00	14%	_____
10.	$51.35	$1,580	_____	90 days

B. Find the bank discount and proceeds. The bank discounts rates at 12% 2 months before the maturity of the loan.

	Face Value	Interest Rate	Due	Bank Discount	Proceeds
11.	$5,410	9%	3 mos.	_____	_____
12.	$975	12%	4 mos.	_____	_____
13.	$2,186	13%	6 mos.	_____	_____
14.	$1,540	11%	4 mos.	_____	_____
15.	$1,200	12%	3 mos.	_____	_____
16.	$3,400	12%	6 mos.	_____	_____
17.	$7,000	13%	3 mos.	_____	_____
18.	$1,890	14%	4 mos.	_____	_____
19.	$3,400	15%	6 mos.	_____	_____
20.	$4,375	14%	8 mos.	_____	_____

Assignment 3

Name_____ Date _____

Complete the calculations for the following word problems in the space provided. Write your answers in the blanks provided.

1. Joyce Snider's savings account shows $3,282.50 on deposit. Her account earns $5\frac{1}{2}$% annual interest rate. After three months, the bank has computed her interest. How much interest has Joyce's account earned?

    _____

2. James Johnson has deposited a total of $980 in his savings account. At the end of six months, he has earned $5\frac{3}{4}$% annual interest rate on his account. How much interest has he earned?

    _____

3. Maria Morales borrowed $3,650 at 15% interest for 120 days. How much interest did she pay based on a 360-day year?

    _____

4. Hector Sanchez borrowed money at 14% interest for 60 days. His interest for the loan amounted to $39.20, based on a 360-day year. How much had he borrowed?

    _____

5. Tien Chu borrowed $2,000 for 60 days (based on a 360-day year). Her interest amounted to $43.33. What was the interest rate?

    _____

6. Lydia's loan date is March 20 and her loan is due in 120 days. What is the due date?

    _____

7. Chihuho Enami obtained a loan on September 5 and it is due December 20. How many days are there between the dates?

    _____

8. John Raymond's loan is $1,490 at 14% interest rate and the interest amount is $69.53. What is the amount of time for this loan?

    _____

9. **a.** A note has been signed for $3,800, the interest rate is 14%, and the time is 120 days (360-day year). How much is the interest?

    _____

    b. How much is the maturity value?

    _____

    c. The note was discounted at a rate of 18.5% two months before its maturity. How much is the discount?

    _____

    d. How much are the proceeds?

    _____

10. The face value of a note is $7,500, the interest rate is 12%, and the time is based on 90 days (360-day year). The note has been discounted at 15% four months before the maturity of the note. How much were the discount and the proceeds?

    _____ discount   _____ proceeds

Assignment 4

Name_____ Date _____

Complete the following problems. Write your answers in the blanks provided.

A. Compute the interest, compounded annually without using the compound interest table.

	Principal	Rate	Time (years)	Compounded Interest Amount
1.	$8,100	8%	2	_____
2.	$590	9%	3	_____
3.	$2,900	7%	5	_____
4.	$1,800	8.5%	2	_____

B. Calculate the following problems using the compound interest table.

	Principal	Rate	Time (years)	Compounded	Compounded Amount	Compounded Interest
5.	$5,000	8%	2	quarterly	_____	_____
6.	$9,200	10%	3	annually	_____	_____
7.	$1,000	8%	4	annually	_____	_____
8.	$2,750	8%	2	quarterly	_____	_____

C. Using the present value table, find the present value in each of the following problems.

	Future Amount	Period of Time (years)	Interest Rate	Compound Period	Present Value
9.	$8,000	3	6%	quarterly	_____
10.	$7,150	4	8%	annually	_____
11.	$11,200	6	8%	quarterly	_____
12.	$1,000	1	11%	annually	_____

D. Compute the interest, compounded quarterly without using the compound interest table.

	Principal	Interest Rate	Period of Time (years)	Compounded Interest
13.	$1,800	12%	1	_____
14.	$2,000	16%	1	_____
15.	$7,500	12%	2	_____
16.	$3,470	8%	1	_____

Assignment 5   Name_____ Date _____

Write your answers in the blanks provided.

A.   Complete the following using the compound interest table.

	Rate	Time (years)	Principal	Compounded	Compounded Amount	Compounded Interest
1.	10%	4	$1,600	annually	_____	_____
2.	12%	3	$16,000	quarterly	_____	_____
3.	6%	2	$3,000	semiannually	_____	_____
4.	8%	3	$2,450	quarterly	_____	_____
5.	8%	1	$500	quarterly	_____	_____

B.   Complete the following word problems. Write your answers in the blanks provided.

6.   First Savings and Loan pays 6% interest compounded quarterly on regular savings accounts. Suppose you deposited $350 and made no other deposits or withdrawals. How much interest did you earn at the end of two years?

_____

7.   Sylvia wants $3,600 after 5 years. Her fund yields a rate of 8% compounded quarterly. How much money will she need to invest today to yield the desired amount in the future?

_____

8.   Yang Chu has received $5,000 from her uncle. She has visited with two different banks. The first bank offers a 10% rate compounded semiannually while the second bank offers a 8% rate compounded quarterly. Yang expects to invest her money for 5 years. Which bank should Yang select?

_____

9.   James Mendoza wants to purchase a new computer system next year. He will need $5,000 for his purchase. His bank will offer him 8% interest compounded quarterly. How much must James invest now so that he will have $5,000 next year?

_____

10.   Margarita Gomez deposits $3,000 into a fund which pays 12% interest compounded quarterly. How much money will she have in her account at the end of 3 years? She does not plan to withdraw or make any further deposits or withdrawals.

_____

11.   Andrea Brown wants to enter college in three years. She will need $20,000. What will she need to invest today to meet her goal of $20,000 in three years if she finds an investment that pays 8% interest compounded quarterly?

_____

Assignment 6

Name_____ Date _____

**Complete the following word problems. Write your answers in the blanks provided.
Round answers to the nearest cent or tenth of percent where needed.**

1. Elaine Miller purchased a $5,000 certificate of deposit paying 12% interest compounded monthly for 2 years. What is the value at maturity and the amount of interest earned on this investment for two years? (Use compound interest table on page 292.)

   Maturity Value _____
   Interest Earned _____

2. Henry Rhoades wants to retire in 5 years. He wants to have $500,000 in the bank at that time. If his investment is paying 10% with interest being compounded semi-annually, how much does he need to invest now? (Use present value table on page 295.)

   _____

3. Chelsea Yates deposited $2,000 in a savings account which was paying interest at 12%, compounded quarterly. What will be the amount of interest earned at the end of one year?

   _____

4. Caleb Frye invested $7,000 in a bond paying 12% interest, compounded quarterly. What will be the maturity value of the bond in 4 years? (Use the compound interest table on page 292.)

   _____

5. Westside Bank is currently paying 18% on a certificate of deposit with a minimum deposit of $100,000. The interest is compounded monthly. What is the amount of interest earned on a $200,000 deposit at the end of one year? (Use the compound interest table on page 292.)

   _____

6. Velma Vasquez inherited $25,000 and invested it into an IRA savings account paying 16%, compounded quarterly. What total amount of money will she have at the end of 5 years? (Use the compound interest table on page 292.)

   _____

7. In four years, Lucy Mae Dillon's granddaughter will go to college. She will need $25,000 for tuition. The bank is paying 8% on a certificate of deposit with interest compounded semi-annually. How much will Lucy Mae need to deposit for a yield of $25,000 in four years? (Use the present value table on page 292.)

   _____

8. John Hall purchased a $6,000 certificate of deposit paying 12% interest, compounded quarterly for two years. What is the value at maturity and the amount of interest earned on this investment for two years?

   Value at Maturity _____
   Interest Earned _____

9. Bill Williams opened a savings account with an initial deposit of $100. The bank paid 6% interest, compounded quarterly. If the money was on deposit for 1 year, what was the amount of interest earned?

   _____

10. Yolanda Soto will need to have $1,000 in three years to buy a computer. If the interest rate is 6% and it is compounded semi-annually, what will she need to deposit in her savings account? (Use present value table on page 295.)

   _____

Proficiency Quiz
R E V I E W

Name_____ Date_____

$\dfrac{\text{Student's Score}}{\text{Maximum Score}} = \dfrac{}{35} = \text{Grade}_____$

A. Calculate simple interest.

	Principal	Rate	Time	Interest
1.	$2,350	8%	10 months	_____
2.	$4,500	10%	1 year	_____
3.	$1,500	10%	3 months	_____
4.	$6,000	9%	1 year	_____

B. Determine the simple interest and maturity value.

Principal	Rate	Time	Interest	Maturity Value
$1,500	9%	1 year	**5.** _____	**6.** _____
$2,450	10%	1 year	**7.** _____	**8.** _____
$5,800	13%	1 year	**9.** _____	**10.** _____

C. Find the simple interest on the following loans using the 360- and 365-day methods.

Principal	Rate	Time	360-Day	365-Day
$4,650	8%	90 days	**11.** _____	**12.** _____
$2,800	13%	60 days	**13.** _____	**14.** _____
$5,350	9%	90 days	**15.** _____	**16.** _____

D. Determine the principal, rate, or time. Use the 360-day method.

	Interest	Principal	Rate	Time
17.	$125	1,250	15%	_____
18.	$264	4,400	_____	180 days
19.	$97.50	_____	13%	90 days

E. Find the bank discount and proceeds for these notes. The bank discounted these notes at 13% two months before maturity of the note.

Face Value	Interest Rate	Due	Bank Discount	Proceeds
$3,500	10%	4 months	**20.** $78.38	**21.** $3,538.20
$12,000	13%	4 months	**22.** $271.32	**23.** $12,248.60

F. Compute the interest, compounded annually.

	Principal	Rate	Time	Compound Interest
24.	$2,600	10%	3 years	$860.60
25.	$4,600	12.5%	3 years	$1,949.62
26.	$1,230	11%	2 years	$285.48
27.	$6,490	15%	3 years	$3,380.48
28.	$9,960	8%	2 years	$1,657.34

G. Compute the interest, compounded as noted.

	Principal	Rate	Time	Compound Period	Compound Interest
29.	$2,900	13%	1 year	quarterly	$395.78
30.	$1,750	14%	1 year	semiannually	$253.58
31.	$5,100	13%	2 years	semiannually	$1,460.99
32.	$1,460	12%	1 year	quarterly	$183.24
33.	$2,550	14%	2 years	quarterly	$807.86

H. Manson Inc. borrowed $3,500 from the bank for 90 days paying an interest rate of 13%. What was the interest amount paid for this loan? Use the 360-day method for calculating the interest amount.

34. $113.75

I. Sync Inc. borrowed $14,900 for 180 days paying interest of $894. At what rate were they charged? Use the 360-day method for calculating the interest rate.

35. 12%

CHAPTER

Charge Accounts and Consumer Loans

15

15

Charge Accounts and Consumer Loans

We live in a fast-paced era, and many consumer products are designed to help us keep up the pace and enjoy our leisure time. To afford these products, we have taken advantage of various credit sources, one in particular—the *charge account.* Charge accounts are popular because they allow the use of a product or service and provide an extra 30, 60 or more days between the time of purchase and the time of payment.

Businesses, such as retail stores and oil companies, provide a convenient way to shop by actually allowing customers to possess goods or services that are to be paid for at a future time. When a retail store allows its customers to receive goods with payment to be made in the future, both the store and the customer are establishing a form of trust. The store believes that the customer has both the ability and the willingness to make payment for those goods. The customer must live up to that trust and make payment. This chapter discusses terms used with charge accounts, types of charge accounts, and methods of calculating finance charges.

In the previous chapter, you learned to calculate the interest to be paid on loans that are to be paid in full on the date of maturity. However, there are loans that allow the borrower to pay a given amount in regular payments. Loans with regular payments are called installment loans. In this chapter, you will learn how to compute the interest on installment loans.

15.1 Terms Used with Charge Accounts

As the use of charge accounts became popular, businesses calculated finance charges using different methods. To assist customers in obtaining information about finance charges, the Federal Consumer Credit Protection Act of 1968 (also known as the *Truth in Lending Act*) was passed so customers would be told the specifics of the total cost of the finance charges. This act requires that certain conditions of a credit agreement be stated. This act applies to *any* credit extended for personal, family, household, or agricultural use whenever the credit is under $25,000—or over $25,000 in the case of real estate transactions. This act requires that all creditors and sellers use the same methods for stating financial costs so that consumers can compare credit terms more easily.

In order to meet this requirement, the creditor must give each customer: (a) the exact amount of the finance charge, such as interest, insurance, and special fees, and (b) the interest rate in terms of annual percentage rate. Both amounts must appear on the credit agreement before it is signed. The *finance charge* is the dollar amount paid on an account by the purchaser for the use of credit. The *annual percentage rate (APR)* is the total finance charge over a full year and is expressed as a percentage. For example, an APR of 18% means that the interest expense of using $1 for a year is $0.18.

The purpose of the Consumer Credit Protection Act is to specify the annual percentage rate to be charged and define the method of calculating interest. Within a state, businesses are regulated by state law and cannot exceed the maximum annual percentage rate set by their state government.

15.2 Types of Charge Accounts

Charge accounts that allow customers to use goods and services repeatedly without having to reapply each time for credit are referred to as *open-ended credit*. There are two characteristics of open-ended credit: a maximum limit on debt, known as a *credit limit*, and a flexible repayment plan. Customers who use charge accounts generally use a credit card which identifies the holder as the participant in the charge account plan of a lender, such as a department store or an oil company. There are several types of charge accounts.

A *30-day charge account* is an agreement allowing the consumer to make purchases during a 30-day period and pay the full amount within 30 days. With a 30-day charge account, a maximum amount that a consumer may charge may or may not be set, but if the consumer does not pay the specified amount due on the bill within the time limit, the lender can assess a late fee on the amount past due or cancel the account.

A *revolving charge account* allows the customer to either pay a portion of the bill or pay the bill in full. More than two-thirds of all charge cardholders maintain a balance. This account is called a revolving account because the customer may continue to charge on the account as long as the balance does not exceed the credit limit. Types of revolving charge accounts include the following: option accounts and bank credit card accounts.

An *option account* allows the customer to either pay the bill in full with no additional cost or pay a portion of the bill over several months at an interest rate per month, for example 1.5 percent each month or an 18 percent APR. Examples of option accounts include credit cards issued by Sears Roebuck and J.C. Penney.

Bank credit card accounts, offered through affiliated banks, savings and loan associations, and credit unions, permit consumers to make purchases or to obtain a cash advance. Bank credit card accounts, such as VISA, MasterCard, and Discover, can be used worldwide. Again, the consumer has the option of paying the bill in full when it arrives or paying a portion of the bill over several months. Cash loans, known as cash advances, are obtained from banks issuing the bank credit cards and from 24-hour automated teller machines (ATMs).

Examples of travel and entertainment charge accounts include American Express and Diners Club. To obtain one of these credit cards, the consumer must pay an annual membership fee. This type of charge account is normally used by business people for traveling and entertainment expenditures and requires the consumer to pay the entire account balance within 30 days.

As already mentioned, those who provide credit must comply with the Consumer Credit Protection Act. They must provide the following information: when the finance charge is calculated, what method is used to calculate the finance charge, the monthly and annual percentage rates, the minimum amount the customer must pay each month, and any late charges to be assessed.

15.3 Annual Percentage Rate (APR) Computed Monthly

Commonly used annual percentage rates may be 21%, 18%, and 15%. For example, for an annual percentage rate of 18%, the monthly interest rate is $1\frac{1}{2}\%$ or 1.5% (18% ÷ 12 months = 1.5% per month). Generally, interest charges are calculated using the monthly rate; therefore, the monthly rate can be calculated using the percentage formula, $R = P \div B$. In some cases, the annual percentage rate may not be known. In this case, you can apply the percentage formula again, $P = R \times B$. When the monthly interest rate is known, you can determine the annual percentage rate by multiplying the monthly rate (R) by 12 months (B).

EXAMPLE

If the monthly interest rate is 2%, what is the APR?

$$0.02 \times 12 = 0.24 = 24\%$$

15.4 Methods of Calculating Finance Charges

Two basic methods of calculating finance charges are: (1) previous balance, and (2) average daily balance. These methods are described in the following sections.

Previous Balance Method

The *previous balance* is the balance in a charge account on the final billing date of the previous month. Monthly rates can range from 1.5% to 2%. The following example shows how the finance charges were computed as a percent of the previous balance.

EXAMPLE

Corene Tunnell, who is a charge customer at Lady Castor Fashions, has an unpaid beginning balance on her account of $532.60. Lady Castor Fashions computes 1.5% a month on the previous balance.

STEPS

1. Compute the finance charge by multiplying the previous balance by the monthly rate.

    $532.60 × 0.015 = $7.99 finance charge

2. Add the finance charge to the unpaid balance for the new account balance.

    $7.99 + $532.60 = $540.59 new account balance

PRACTICE PROBLEMS

Using the previous balance method, compute the monthly rate, finance charge, and account balance for these problems. Write your answers in the blanks provided.

	Previous Balance	APR	Monthly Rate	Finance Charge	Account Balance
1.	$430.50	18%	_____	_____	_____
2.	$370.40	21%	_____	_____	_____
3.	$235.60	19.2%	_____	_____	_____

Solutions: **1.** 1.5%; $6.46; $436.96; **2.** 1.75%; $6.48; $376.88; **3.** 1.6%; $3.77; $239.37

Average Daily Balance

This commonly used method computes the finance charge by applying the monthly rate to the average daily balance. The *average daily balance* (**ADB**) is calculated by adding the daily balances and dividing by the number of days in the billing cycle.

To get the average daily balance, study the following example.

EXAMPLE

A customer's credit card statement shows an unpaid balance of $350.00 on the billing date of November 2. A finance charge of 1.5% on the daily balance is entered. A payment was received on November 17 for $100, a charge of $175 is shown on November 7, and another charge of $100 is shown for November 23.

STEPS

1. The beginning balance was $350. This balance was in effect for 5 days: November 2 to November 6 — $350

2. A charge was made on November 7; add charge to beginning balance. — + $175

3. This balance was in effect for 10 days: November 7 to November 16. — $525

4. A payment was made on November 17; subtract payment from balance. — − $100

5. This balance was in effect for 6 days: November 17 to November 22. — $425

6. A purchase was made on November 23; Add this purchase to the balance in effect. — + $100

7. This balance was in effect through December 1, the last day of the billing period. — $525

8. Find the daily balance by multiplying balances by days in effect as shown and then add the daily balances and the days.

	Days		Daily Balance
$350 ×	5	=	$1,750
$525 ×	10	=	$5,250
$425 ×	6	=	$2,550
$525 ×	9	=	$4,725
	30		$14,275

9. Divide the total of the daily balances by the total number of days in the billing cycle to obtain the average daily balance.

$14,275 ÷ 30 = $475.83 average daily balance

10. Compute the finance charge.

$475.83 × 0.015 = $7.14 finance charge

11. Add the last account balance and finance charge to get the new account balance on the billing date.

$525.00 + $7.14 = $532.14 account balance

PRACTICE PROBLEM

Compute the average daily balance, finance charge, and the new account balance for this problem. Write your answers in the blanks provided. As of March 1, the credit card balance shows a previous balance of $875.90; a payment of $125 was made on March 10; a purchase of $175 was made on March 15. Finance charges are computed at 1.5% per month. This balance was in effect through the billing date, April 1.

1. average daily balance ______ **2.** finance charge ______ **3.** new account balance ______

Solutions: **1.** $883.40; **2.** $13.25; **3.** $939.15

15.5 Calculating the New Balance on an Account

Customers receive a monthly statement at the end of each billing period. This statement itemizes purchases along with previous balance, finance charges, payments, credits, new balance, and minimum payment information. Although this information is given on all monthly statements, the forms themselves may vary. In the figure below, the finance charge, $1.50, is computed on the previous balance, and added to it before payments and credits are deducted.

FIGURE 15-1
Monthly Statement

Previous Balance	Finance Charge	Payments	Credits	Purchases and Cycle Closing Date	New Balance	Minimum Payment
100.00	1.50	25.00	10.00	80.75 7/24/--	147.25	30.00

If we receive payment of the full amount of the new balances before the next cycle closing date, shown above, you will avoid a finance charge next month. The finance charge, if any, is calculated on the previous balance before deducting any payments or credits shown above. The periodic rates are $1\frac{1}{2}$% of the balance on amounts under $1,000 and 1% of amounts in excess of $1,000, which are annual percentage rates of 18% and 12% respectively.

A retail store may state its credit terms as follows: The finance charge, if any, is computed on the previous balance before payments or credits are deducted. The monthly rate is 2% on the first $1,000 (24% APR) and 1.75% on amounts over $1,000 (21% APR). There is no finance charge if the full amount of the new balance is paid on or before the next month's bill's closing date.

Assuming you did not pay the full amount on time, compute the finance charge and the new balance in the following example.

EXAMPLE

Your account shows a previous balance of $1,325.17; a payment of $290.00 was received during the month; charges amounted to $375.30; a credit was given in the amount of $37.90.

Follow these steps to make your calculations.

STEPS

1. Multiply the first $1,000 of the previous amount by the monthly rate of 2%.

$$\$1{,}000.00 \times 0.02 = \$20.00$$

2. Multiply the difference between the first $1,000 and the amount over $1,000 by the monthly rate of 1.75%.

$$\$325.17 \times 0.0175 = \$5.69$$

3. Add the individual finance charges to obtain the total finance charge.

$$\$20.00 + \$5.69 = \$25.69$$

4. Determine the new balance as shown.

$$
\begin{array}{rl}
\$1,325.17 & \text{previous balance} \\
+ \quad \$25.69 & \text{finance charge} \\
- \quad \$290.00 & \text{payment} \\
+ \quad \$375.30 & \text{purchases} \\
- \quad \$37.90 & \text{credit} \\
\hline
\$1,398.26 & \text{new balance}
\end{array}
$$

PRACTICE PROBLEM

For these problems, the finance charge is based on the previous balance before payments or credits are deducted. The monthly rate is 1.5% on amounts up to $1,000 and 1% on amounts over $1,000. Compute the finance charge and new balance.

	Previous Balance	Finance Charge	Payment	Credits	Purchases	New Balance
1.	$968.50	_____	$230.50	– 0 –	$89.50	_____
2.	$1,645.19	_____	$425.00	$122.70	$290.74	_____
3.	$1,656.20	_____	$589.40	– 0 –	$187.60	_____
4.	$860.30	_____	$150.50	– 0 –	$92.75	_____
5.	$1,127.90	_____	$325.60	$11.23	– 0 –	_____

Solutions: **1.** $14.53; $842.03; **2.** $21.45; $1,409.68; **3.** $21.56; $1,275.96; **4.** $12.90; $815.45; **5.** $16.28; $807.35

15.6 Consumer Loans

In the previous sections, you studied the concept of open-end credit or the charge account. Open-end credit is sometimes referred to as a *credit card loan*— using such credit cards as Visa, MasterCard, or Discover. This section discusses a second type of credit, the consumer or closed-end account, which is used most often for buying items such as large appliances, furniture, stereo equipment, or cars. Many people choose this type of credit for these more expensive items because they are unable to pay cash or do not want to use a credit card or charge account.

With a consumer or installment loan, you sign an agreement called a *credit agreement.* A type of credit agreement that includes a *retail installment contract* is used to finance items such as television sets or appliances. You are protected by the Consumer Credit Protection Act when you sign a credit agreement, the same as you are when using an open-end agreement. These facts include the following: the purchase price, the down payment (or trade-in) amount, the total amount being financed, the number of payments, the payment due dates, the finance charge shown in dollars, any penalties for early repayment or late payments, and the annual percentage rate in numbers.

In order to understand the calculations involved in consumer loans, study the following terms.

An *installment loan* is one in which the consumer agrees to pay back a certain amount of money in a series of payments over a period of time, usually monthly. These payments are known as *installments*. A *down payment* is an initial payment that the consumer may be required to make toward a purchase. It is usually a percent of the purchase or cash price, such as 20% of the price. The *remaining balance* or the *amount to be financed* is the remainder of the purchase price after any down payment has been deducted. The *amount to be repaid* is the amount to be financed plus any installment charges, also known as finance charges or interest. The *finance charge* is the money paid to use the installment plan and includes costs such as interest or special charges on the installment loan. The special charges may include the insurance to cover any payments not made by the consumer and bookkeeping costs for handling the installment plan. In this section, interest will be the only finance charge computed.

A *stated interest rate* is the rate of interest quoted by a financial institution, whereas an *annual percentage rate* is the true rate of interest charged over a year. Additional terms are explained as they are presented.

15.7 Calculations Used When Buying on an Installment Plan

As a consumer, it is important that you are able to make certain calculations when you buy on an installment plan. Study the following examples to learn how to make these calculations.

Finding the Monthly Payment

Suppose a company sells appliances on the installment plan. They have advertised a microwave oven for $400. You want to purchase the microwave oven on an installment plan. The appliance company covers the following terms for your installment loan.

1. A down playment of 10% of the cash price (sometimes a flat sum is stated as the down payment).
2. Nine monthly payments.
3. Stated interest rate of 20%.

The steps presented next will help you learn how to compute the following amounts:

1. Amount to be financed (unpaid balance) on an installment loan.
2. Amount of interest charged for the installment loan.
3. Amount to be repaid.
4. Amount of each monthly installment.

STEPS

1. Find the amount to be financed (the unpaid balance) by subtracting the down payment from the cash price.

$$\$400 - \$40 = \$360 \text{ amount to be financed}$$

2. Find the amount of interest on the amount financed at 20% by applying the interest formula.

$$I = \$360 \times 0.20 = \$72$$

3. To find the amount to be repaid, add the total interest to the unpaid balance. For example:

$$\$360 + \$72 = \$432 \text{ amount to be repaid}$$

4. To find the amount of each monthly payment, divide the amount to be repaid by the number of monthly payments.

$$\frac{\$432}{9} = \$48 \text{ monthly installment}$$

— PRACTICE PROBLEM —

Find the amount to be financed, the total interest, the amount to be repaid, and the monthly installment on these problems. The terms are 10% down, 24 equal monthly installments, and an interest rate of 20%. Write your answers in the blanks provided.

	Cash Price	Amount to Be Financed	Total Interest	Amount to Be Repaid	Monthly Installment
1.	$540	_____	_____	_____	_____
2.	$2,000	_____	_____	_____	_____
3.	$1,200	_____	_____	_____	_____
4.	$650	_____	_____	_____	_____

Solutions: **1.** $486; $97.20; $583.20; $24.30; **2.** $1,800; $360; $2,160; $90; **3.** $1,080; $216; $1,296; $54; **4.** $585; $117; $702; $29.25

15.8   Determining the Total Installment Price and Finance Charge Amount

The *total installment price* is the sum of the down payment and all of the installment payments. The *finance charge* or the additional cost of installment buying is the difference between the cash price and the installment price.

EXAMPLE

Carmen is buying stereo equipment on installment. Her down payment is $127; there are 24 monthly payments of $61.91. The cash price is $1,270. How much was the additional cost of installment buying?

STEPS

1. Determine the total installment price by multiplying the monthly payment by the number of payments and adding any down payment. In the example given, a down payment of $127 was required.

$$(\$61.91 \times 24) + \$127.00 = \$1,612.84 \text{ installment price}$$

2. Determine the cost or finance charge by subtracting the cash price from the total installment price.

$$\$1,612.84 - \$1,270 = \$342.84 \text{ finance charge}$$

15.9 Rebate

Occasionally, consumers like to pay off an installment loan early. If you pay off a loan early, you may be entitled to a rebate. *Rebate* means to give money back or, in the following example, give back some of the interest prepaid on the loan. According to federal law, the lender must compute the amount of your interest rebate by using a method called the Rule of 78ths, which is also called the Sum-of-Digits method.

The *Sum-of-Digits* method is based on the number 78 which is the sum of the digits for the months of the loan. For example, if you were to add the digits in the numbers 1 through 12 for a 12-month loan, it would equal 78 which represents a fractional part—the denominator or the sum of the digits of the total payments. For instance,

$$1 + 2 + 3 + 4 + 5 + 6 + 7 + 8 + 9 + 10 + 11 + 12 = 78$$

The numerator is the sum of the digits of the remaining payments.

EXAMPLE _____

Let's suppose you have a one-year loan with a finance charge of $138 which you pay off four months early. The lender could determine your interest rebate in the following manner.

STEPS
_____

1. Find the sum of the digits of the remaining payments (numerator).
$$1 + 2 + 3 + 4 = 10$$

2. Find the sum of the digits of the total payments (denominator).
$$1 + 2 + 3 + 4 + 5 + 6 + 7 + 8 + 9 + 10 + 11 + 12 = 78$$

3. Multiply the finance charge by the fraction.
$$\frac{10}{78}\left(\text{or}\,\frac{5}{39}\right) \times \$138 = \$17.69 \text{ interest rebate}$$
_____

You can see that steps 1 and 2 can be cumbersome, especially if you are finding the sum of payments for a loan that is longer than one year. Rather than using these steps, you can use the following formula to save time.

$$\text{Sum} = \frac{N(N+1)}{2}, \text{ where } N = \text{the number of payments.}$$

Finding the Rebate Fraction

Let's apply this formula to find the rebate fraction using the previous example.

STEPS

1. Determine the sum of the digits of the remaining payments (numerator).

$$N = 4 \text{ remaining payments}$$

$$\frac{4(4+1)}{2} = \frac{20}{2} = 10$$

2. Find the sum of the digits of the total payments (denominator).

$$N = 12 \text{ total payments}$$

$$\frac{12(12+1)}{2} = \frac{156}{2} = 78$$

3. Multiply the rebate fraction by the finance charge.

$$\frac{10}{78}\left(\text{or } \frac{5}{39}\right) \times \$138 = \$17.69 \text{ interest rebate}$$

PRACTICE PROBLEMS

Find the interest rebate for the following problems. Write your answers in the blanks provided.

Finance Charge	Length of Loan	Time Left in Loan	Interest Rebate
1. $190	12 months	6 months	_____
2. $525	2 years	10 months	_____

Solutions: **1.** $51.15; **2.** $96.25

Finding the Final Payment

Let's suppose you want to pay off a 12-month installment loan at the end of 8 months with payments of $86.50 and an interest charge of $72.30. By paying off the loan early, you want to know the amount of the final payment. Study the following steps to determine the amount of the final payment.

STEPS

1. Find the rebate fraction using the formula.

$$\frac{4(4+1)}{2} = \frac{20}{2} = 10 \qquad \frac{12(12+1)}{2} = \frac{156}{2} = 78$$

2. Find the interest rebate by multiplying the rebate fraction by the interest charged.

$$\frac{10}{78}\left(\text{or } \frac{5}{39}\right) \times \$72.30 = \$9.27 \text{ interest rebate}$$

3. Determine the amount still owed on the loan by multiplying the amount of the payment by the number of payments still owed.

$$\$86.50 \times 4 = \$346.00$$

4. Find the final payment by subtracting the interest rebate from the amount still owed.

$$\$346.00 - \$9.27 = \$336.73 \text{ final payment}$$

Length of Loan	Amount of Payments	Interest Charged	Time Left in Loan	Final Payment
1. 12 months	$67.30	$110.70	6 months	_____
2. 12 months	$112.20	$220.00	9 months	_____
3. 18 months	$167.40	$450.00	12 months	_____

Solutions: 1. $374;   2. $882.88;   3. $1,803.54

15.10 Comparing Cash and Installment Purchases

As mentioned earlier, buying on an installment plan allows the customer the privilege of using an item now and paying for it later. On occasion, as a buyer you must decide whether it is better to buy an item on an installment plan or wait and purchase it with cash when the price may be higher. The following example illustrates the difference between installment buying and purchasing an item by paying cash.

EXAMPLE

You are considering purchasing a word processor. One store sells the model you want for $475 cash, while another store has the same one for $45 a month for 12 months.

Find the dollar amount difference in buying on an installment plan compared to paying cash for the same item by following these steps.

STEPS

1. Multiply the amount of the monthly payment by the number of monthly payments to find the installment price charged by the second store.

$$\$45 \times 12 = \$540 \text{ installment price}$$

2. Subtract the cash price from the installment price to find the difference.

$$\$540 - \$475 = \$65 \text{ savings by paying cash}$$

In other words, the cost of using the installment plan is $65 in actual cash outlay. However, other factors should certainly be considered, such as need, available cash, and other opportunities for which the cash can be used.

When you want to purchase an item on credit, you may obtain an installment loan or a personal loan from a lending institution. In seeking the best finance costs, you should consider these basic elements:

1. The dollar amount of the finance charge.
2. The stated rate of interest.
3. The true annual percentage rate.
4. The monthly payment.

If you consider these four basic elements, you will become more knowledgeable in making comparisons of the finance costs charged by various retail businesses and financial institutions.

15.11 Mortgage Loans

Because buying a home involves an investment of thousands of dollars, most people buy a home by making a down payment and borrowing the remainder of the purchase price to be paid in monthly installments. A *mortgage loan* is the amount of money lent to a borrower (mortgagor) by a lender (mortgagee) to purchase real estate, with the real estate itself serving as collateral for the loan. *Collateral* means that the lender has the legal right to obtain the property in the event the borrower defaults on the loan; that is, the borrower cannot repay the mortgage loan. *Foreclosure* is the process of the lender suing the borrower to prove that the borrower cannot repay the loan and asking the court to order the sale of the property in order to pay the debt. You may obtain a mortgage loan from savings and loan associations, credit unions, mutual savings banks, and commercial banks.

The process by which the amount of money and principal that are paid off in each monthly payment is called *amortization*. Each payment includes the principal (the original amount borrowed) and the interest charged by the lender. In addition to the principal and interest, taxes and insurance are included in monthly mortgage payments. When potential buyers are considering the purchase of a home, for example, one of the first questions they usually ask is "How much is the monthly payment?"

Computing Monthly Mortgage Payments

It is not practical to calculate monthly payments manually for a home loan that has a term of 15, 20, 25, or 30 years because of the lengthy calculations. Specific computer software can provide this type of information to a potential buyer in a matter of minutes rather than taking hours to calculate by hand. You can calculate the monthly payment for mortgage loans by using the amortization table below. This table gives the amount of monthly payment required for each $1,000 of a mortgage loan at different interest rates for the term of a loan.

FIGURE 15-2
Amortization Table
Estimate your mortgage payment
(monthly payment based on each $1,000 debt)

Interest Rate	Payment Period (years)			
	15	20	25	30
8	$ 9.5565	$ 8.3644	$ 7.7182	$ 7.3376
8.5	9.8474	8.6782	8.0528	7.6891
9	10.1427	8.9973	8.3920	8.0462
9.5	10.4422	9.3213	8.7370	8.4085
10	10.7461	9.6502	9.0870	8.7757
10.5	11.0539	9.9838	9.4418	9.1474
11	11.3660	10.3219	9.8011	9.5232
11.5	11.6819	10.6643	10.1647	9.9030
12	12.0017	11.0109	10.5322	10.2861
12.5	12.3252	11.3614	10.9035	10.6726
13	12.6524	11.7158	11.2784	11.0620
13.5	12.9832	12.0737	11.6564	11.4541
14	13.3174	12.4352	12.0376	11.8487
14.5	13.6550	12.8000	12.4216	12.2456
15	13.9959	13.1679	12.8083	12.6444
15.5	14.3399	13.5388	13.1975	13.0452
16	14.6870	13.9126	13.5889	13.4476

Use the following example to become familiar with finding the monthly payment using the amortization table above.

EXAMPLE

A $70,000 mortgage loan at 12% over a period of 30 years.

Follow these steps to find the monthly payment necessary to amortize this loan. Notice that this monthly payment does not include taxes or insurance.

STEPS

1. Using the table, locate the term of the loan, 30 years, and read down to find the column labeled 12%.

2. Where the two columns meet, find $10.2861. Multiply this amount by the loan amount divided by 1,000. (70,000 ÷ 1,000 = 70)

$$10.2861 \times 70 = \$720.03 \text{ monthly payment on loan}$$

PRACTICE PROBLEMS

Using the table, find the monthly payment for the following problems. The interest rate is 9.5%. Round your answers to the nearest penny and write your answers in the blanks provided.

Loan Amount	Term	Monthly Payment
1. $80,000	20 years	_____
2. $65,000	25 years	_____

Solutions: **1.** $745.70; **2.** $567.91

15.12 Preparing a Loan Payment Schedule

When the monthly payment is found, a loan payment schedule (or loan amortization schedule) is prepared by the lending institution. This schedule shows the monthly payments necessary to pay off a mortgage loan. The schedule shown below shows the loan amortization calculated for 2 months on this loan: $70,000, 12% interest, 30 year term.

FIGURE 15-3
Amortization Schedule

Payment Schedule				
Payment Number	Monthly Payment	Interest Payment	Principal Payment	Balance of Principal
1	$720.03	$700.00	$20.03	$69,979.97
2	$720.03	$699.80	$20.23	$69,959.74

As a review, the monthly payment of $720.03 as shown in this schedule was computed using the amount in the table: $10.2861 × 70 = $720.03.

Study these steps to learn how the interest and principal payments and balance of the principal in the payment schedule are computed:

STEPS

1. For the loan payment, compute the interest for 1 month by multiplying the amount of the loan by the interest rate and then dividing by 12 (months). For instance:

$$\$70,000 \times 0.12 \div 12 = \$700.00$$

2. Compute the monthly payment on the principal by subtracting the interest payment from the monthly payment.

$$\$720.03 - \$700 = \$20.03$$

3. Compute the balance of the principal after the first payment by subtracting the principal payment from the amount of the loan.

$$\$70,000.00 - \$20.03 = \$69,979.97 \text{ balance after first payment}$$

4. For all payments after the first one, compute the interest on the new principal balance. For the second payment in the example:

$$\$69,979.97 \times 0.12 \div 12 = \$699.80$$

5. Continue with steps 2, 3, and 4. (In step 3, subtract the principal payment from the new principal balance.)

These calculations show that the payments in the early years of a mortgage loan consist mostly of interest; at this point, only a small portion of each payment is applied to paying off the principal. As each payment is made, the amount of interest decreases so that a larger amount of payment applies to the principal. Only during the last few years of the loan will most of the monthly payment apply toward the principal.

Although these calculations are made very quickly using computers, the following problem gives you practice in setting up a payment schedule.

PRACTICE PROBLEMS

Prepare a payment schedule for the first and second months of a loan of $80,000 at 12% interest for 30 years. The monthly payment on this loan is $822.89.

Monthly Payment	Interest Payment	Principal Payment	Balance of Principal
1. $822.89	_____	_____	_____
2. $822.89	_____	_____	_____

Solutions: **1.** $800.00; $22.89; $79,977.11; **2.** $799.77; $23.12; $79,953.99

MATH ALERT

1. A department store advertised a home satellite system for a cash price of $1,994 or an installment plan with 95 monthly payments of $55 and no money down. Find the finance charge and prove the monthly payment.

2. Here are the rates paid on credit cards at the 25 largest-deposit financial institutions in the Dallas-Fort Worth area as of February 7, 19--. Credit card rates are for conventional cards and for purchases only. Cash advances frequently are charged interest from the date of transaction. Additional card fees may also be charged.

Institution	Rate	Fee	Grace Period
American Airlines Emp FCU	NA	NA	NA
American Federal Bank	17.90	$ 0	25 days
Bank One Texas	18.00	$20	30 days
Bluebonnet Savings Bank	17.90	$ 0	25 days
Central Bank & Trust	17.90	$ 0	0 days
Colonial S&L	NA	NA	NA
Comerica Bank	18.00	$20	25 days
Dallas Teachers CU	14.90	$ 0	30 days
First City Texas	19.80	$18	25 days
First Gibralter Bank	17.90	$ 0	25 days
First Interstate Bank of TX	18.00	$20	30 days
First State Bank-Denton	18.90	$25	30 days
Guaranty Fed Savings Bank	18.90	$20	25 days
Guardian Savings	NA	NA	NA
Hibernia Nat'l Bank-Texas	NA	NA	NA
Nations Bank of Texas	17.90	$18	25 days
North Dallas Bank & Trust	17.90	$ 0	25 days
NorthPark Nat'l Bank	17.90	$ 0	25 days
Overton Park Nat'l Bank	NA	NA	NA
Saving of America	NA	NA	NA
Sunbelt Federal Savings	NA	NA	NA
Team Bank	17.90	$ 0	25 days
Texas Commerce Bank	17.80	$20	30 days
Texins Credit Union	NA	NA	NA
United Saving of Texas	17.90	$ 0	25 days

Grace period refers to a specified period of time when no interest or fees are charged on bills.

Determine the monthly rate for the following institutions:

Monthly Rate

a. First City Texas _____

b. First Gibralter Bank _____

c. Guaranty Fed Savings Bank _____

d. Dallas Teachers CU _____

Study Guide

I.   Terminology

page 310	charge account	An account offered by a business that allows you to charge purchases and pay all or a portion of the amount at a later time, usually monthly.
page 311	finance charge	The dollar amount paid on an account by the purchaser for the use of credit.
page 311	annual percentage rate (APR)	The total finance charge over a full year and is expressed as a percentage.
page 311	Consumer Credit Protection Act	An act whose purpose is to specify the annual percentage rate to be charged and the method of calculation of interest.
page 311	open-ended credit	Charge accounts that allow customers to use goods and services repeatedly without having to reapply each time for credit.
page 311	30-day charge account	Where the consumer agrees to make purchases within 30 days and to pay the full amount within 30 days.
page 311	revolving charge account	Allows the customer to either pay a portion of the bill or pay the full bill.
page 311	option account	Allows the customer to either pay the bill in full with no additional cost or pay a portion of the bill over several months at an interest rate per month.
page 311	bank credit card accounts	Offered through affiliated banks, savings and loan associations, and credit unions; permit the customer to make purchases or to obtain cash advances.
page 312	previous balance	The balance in a charge account on the final billing date of the previous month.
page 313	average daily balance (ADB)	Computes finance charge by applying the monthly rate to the average daily balance.
page 315	credit agreement	An agreement between the business and the customer that outlines what each agree about purchasing and payment.
page 316	installments	Monthly payments.
page 316	total installment price	The sum of the down payment and all of the installment payments.
page 316	finance charge	The dollar amount paid on an account by the purchaser for the use of credit.
page 318	rebate	The amount given back or returned for early payoff of a loan.
page 318	sum-of-digits method	Also called the Rule of 78ths; according to federal law, the lender must compute the amount of interest rebate using this method.
page 321	mortgage loan	The amount of money loaned to a borrower by a lender to purchase real estate, with the real estate itself serving as collateral for the loan.
page 321	mortgagor	The borrower.
page 321	mortgagee	The lender.

page 321	collateral	The lender has the legal right to obtain the property in the event the borrower defaults on the loan.
page 321	foreclosure	The process of the lender suing the borrower to prove that the borrower cannot repay the loan and asking the court to order the sale of the property in order to pay the debt.
page 321	amortization	The process by which the amount of money and principal that is paid off in each monthly payment is calculated.

II. Computing Percentage Rate

page 312   *Monthly:* Use the formula $R \times 12 = APR$

Example: 1.67% monthly $\times$ 12 = 20% APR

page 312   *Annual:* Use the formula $APR \div 12 =$ monthly percentage rate

Example: 21.6% APR $\div$ 12 = 1.8% monthly

III. Methods of Calculating Finance Charges

page 312   *Previous Balance Method:* Compute the finance charge by multiplying the previous balance by the monthly rate; add the finance charge to the unpaid balance for the account balance.

Example: Balance: $532.60   Monthly rate: 1.5%

$532.60 $\times$ 0.015 = $7.99 finance charge
$7.99 + $532.60 = $540.59 new account balance

page 313   *Adjusted Balance Method:* Subtract the payment from the unpaid balance to find the adjusted balance; multiply the monthly interest rate by the adjusted balance to find the finance charge; add the finance charge to the adjusted balance to get the new account balance.

Example: Balance: $540.59   Payment: $125.00   Monthly rate: 1.5%

$540.59 − $125.00 = $415.59 adjusted balance
$415.59 $\times$ 0.015 = $6.23 finance charge
$6.23 + $415.59 = $421.82 new balance

page 313   *Average Daily Balance:* Calculate the average daily balance (ADB) by adding the daily balances and dividing by the number of days in the billing cycle. (See page 313 for the example and steps.)

IV. Calculating the Finance Charge and the New Balance When the Amount Due is not Paid in Full

page 314   Assume a retail store has stated the finance charge is computed on the previous balance before payments or credits are deducted. The monthly rate is 2% on the first $1,000 (24% APR) and 1.75% on amounts over $1,000 (21% APR).

Example: Previous balance: $1,325.17
Payment: $290.00   Charges: $375.30   Credit: $37.50

Multiply the first $1,000 of the previous balance by the monthly rate of 2%; multiply the difference between the first $1,000 and the amount over $1,000 by the monthly rate of 1.75%; add the individual finance charges to obtain a total finance charge; take the previous balance, add the finance charge, subtract the payment, add the purchases, and subtract the credit to get the new balance.

$1,000 $\times$ 0.02 = $20.00
$325.17 $\times$ 0.0175 = $5.69
$20.00 + $5.69 = $25.69 total finance charge
$1,325.17 + $25.69 − $290.00 + $375.30 − $37.90 = $1,398.26

V. Calculations Used When Buying on an Installment Plan

page 316   *Finding the Monthly Payment:* Find the amount to be financed (the unpaid balance) by subtracting the down payment from the cash price; find the amount of interest on the amount to be financed at whatever the APR is by applying the interest formula; find the amount to be repaid, add the total interest to the unpaid balance; find the amount of each monthly payment by dividing the amount to be repaid by the number of monthly payments.

Example: Cash price: $400
         Down payment: 10% of amount borrowed ($40)
         Number of payments: 9
         Interest rate: 20%

$400 − $40 = $360 amount to be financed
$360 × 0.20 = $72 finance charge
$360 + $72 = $432 amount to be financed
$432 ÷ 9 = $48 monthly installment

page 317   *Calculating the Total Installment Price Amount:* Multiply the monthly payments by the number of payments and add any down payment.

Example: Use the example above.

$48 × 9 = $432 installment price

page 317   *Calculating the Total Price:* Add the amount financed and the down payment.

Example: Use the example on the previous page.

$432 (amount financed) + $40 (down payment) = $472 (total price)

VI. Rebate

page 318   A rebate means money is given back for early payoff of a loan. Use the formula:

$Sum = \dfrac{N(N+1)}{2}$, where N is the number of payments. Determine the sum of the digits of the remaining payments (numerator of the rebate fraction); find the sum of the total number of payments (denominator of the rebate fraction); multiply the finance charge by the rebate fraction.

Example: 1-year loan
         $138 finance charge
         Paid off 4 months early

$N = 4$ remaining payments
$\dfrac{4(4+1)}{2} = \dfrac{4(5)}{2} = \dfrac{20}{2} = 10$ (numerator)

$N = 12$ total payments
$\dfrac{12(12+1)}{2} = \dfrac{12(13)}{2} = \dfrac{156}{2} = 78$ (denominator)

$138 \times \dfrac{10}{78} \left(or \dfrac{5}{39} \right) = \$17.69$ interest rebate

page 319   *Finding the Final Payment:* Find the rebate fraction by using the formula; find the rebate interest by multiplying the rebate fraction by the interest charges; multiply the amount of the payment by the number of payments still owed; subtract the interest rebate from the amount still owed.

Example: Use the example above for the rebate fraction of $\frac{10}{78}$.
         Payment per month: $86.50   Interest charged: $72.30

$72.30 \times \frac{10}{78} = \$9.27$ interest rebate
86.50×4 payments = $346 amount still owed
$346 − \$9.27 = \$336.73$ final payment

VII. Mortgage Loans

page 322

Preparing a Loan Payment Schedule: The schedule shows the monthly payments necessary to pay off a mortgage loan. For the first payment, compute the interest for 1 month by multiplying the amount of the loan by the interest rate and then dividing by 12(months); compute the monthly payment on the principal by subtracting the interest payment on the principal from the monthly payment; compute the balance of the principal after the first payment by subtracting the principal payment from the amount of the loan; for all payments after the first one, compute the interest on the new principal balance; continue repeating the procedure for succeeding payments.

Example:
Principal:	$70,000
Rate:	12%
Time:	30 years
Monthly payment:	$720.03

$70,000.00 × 0.12 ÷ 12 = $700.00
$720.03 − $700.00 = $20.03
$70,000.00 − $20.03 = $69,979.97

Assignment 1

Name _____ Date _____

Complete the following problems. Write your answers in the blanks provided.

A. Figure the finance charge using the previous balance method.

	Previous Balance	Monthly Rate	Finance Charge
1.	$365.30	1.8%	_____
2.	$190.40	1.5%	_____
3.	$295.19	1.75%	_____
4.	$690.35	1.5%	_____

B. Determine the finance charge using the average daily balance method for this situation. John Richard's charge account shows an unpaid balance of $500 on July 2. A payment of $180 was made on July 10. Purchases of $175 were made on July 15. This amount is in effect until August 1, the last day of the billing cycle. An annual rate of 18% is applied to this account.

  5. Average daily balance: _____

  6. Finance charge: _____

  7. New balance: _____

C. Use the following information to compute the new balance.

Previous balance:	$313.40
Finance charge:	4.70
Payments:	113.40
Credits:	12.50
Purchases:	37.58

  8. New balance: _____

D. Determine the finance charge and the new balance for these problems. The finance charge is based on the previous balance before payments or credits are subtracted. The monthly rates are 2% on amounts up to $1,000 and 1.5% on amounts over $1,000. Assume the balance was not paid within the specified period.

	Previous Balance	Finance Charge	Payments	Credits	Purchases	New Balance
9.	$1,215.60	_____	$378.50	– 0 –	$161.30	_____
10.	$1,015.40	_____	$487.65	$39.70	$62.00	_____
11.	$1,265.30	_____	$315.00	– 0 –	$146.00	_____
12.	$1,172.40	_____	$198.50	– 0 –	$97.60	_____
13.	$1,348.20	_____	$392.80	$42.90	$136.28	_____
14.	$1,476.15	_____	$295.00	– 0 –	– 0 –	_____

Assignment 2

Name_____ Date _____

Complete the following word problems. Write your answers in the blanks provided. Round answers to the nearest cent or tenth of percent where needed.

1. Flair Fashions charges 1.83% per month on the previous balances of its accounts. Account #181-909 shows a beginning balance of $574.65. Show the finance charge and the balance due.

    a) Finance charge _____

    b) Beginning balance due _____

2. A charge account statement shows a beginning balance of $278.00 and a billing date of September 5th. On September 10th a payment of $60.00 was made. Later two purchases were made: $23.50 on September 20th and $17.38 on September 25th. Find the average daily balance on the next billing date of October 5th.
    Average daily balance _____

3. A credit card account shows a previous balance of $1,740.35. The finance charge is 1.7% on the previous balance of $1,000. A rate of 1.5% is applied to balances that exceed $1,000. Show the total finance charges to be appplied to this account.
    Finance charge _____

4. An account with a billing date of June 8th shows a beginning balance of $486.32. A charge of $39.64 was made on June 10th; a credit of $14.23 was recorded on June 12th; and a payment of $150.00 was made on June 23rd. Show the average daily balance as of the next billing date.
    Average daily balance _____

5. A finance charge of 1.75% is applied to monthly account balances that are $500.00 or less. The rate drops to 1.5% on previous balances that exceed $500.00. Mr Caster's account has a beginning balance of $368.23. Calculate the finance charge.
    Finance charge _____

Practical Math Applications

Assignment 3

Name_____ Date _____

Complete the following word problems. Write your answers in the blanks provided. Round answers to the nearest cent or tenth of percent where needed.

1. Find the finance charge of Wanda Lowe's credit card if the unpaid balance is $567 and the rate is 1.5% _____

2. Find the monthly rate of interest on John Clevinger's credit card account if the unpaid balance is $159 and the finance charge is $2.86. There were no charges or payments this month. _____

3. On April 1, the unpaid balance on a credit card was $236. During the month, purchases of $24, $55, and $12.75 are made. Using the previous balance method, find the unpaid balance on May 1 if the finance charge is 1.6% of the unpaid balance and a payment of $45 is made on April 15. _____

4. Horatio Jeves had a previous balance of $129.88 on his department store credit card. He made purchases of $26.77 and a payment of $55. The interest charge was 1.5%. Find the new unpaid balance of Horatio's credit card using the previous balance method.

   _____

5. On January 1, the previous balance for Donna Small's charge account was $342.23. On the following days, she made the following purchases: January 12, $23.90; January 20, $99.56. On January 15, Donna made a $60 payment. Using the average daily balance method, find the finance charge and unpaid balance of February 1 if the interest is 1.5% per month.

   Finance charge _____
   Unpaid Balance _____

6. Gloria has a charge account at Taylor's Department Store with a current balance of $458.76. The store requires a minimum monthly payment of 10% or $20 whichever is greater, and a finance charge of 1.5% on the unpaid balance. If Gloria makes a minimum monthly payment and additional purchases amounting to $35, what will be the amount of the ending unpaid balance?

   _____

7. Calculate the average daily balance if on the 1st of the month the beginning balance is $298.50, there was a payment of $20 made on the 15th of the month, it is a 30 day billing cycle and a charge of $45 was made on the 25th of the month.

   _____

8. Nina Jessup has an unpaid balance of $265 on her credit card with an interest rate of 1.8%. What is the interest charged on the account for the month?

   _____

9. On May 1, the unpaid balance of Tawny's credit card was $305. During the month, she made purchases of $100 and a payment of $50. Using the previous balance method and a finance charge of 1.5%, find Tawny's unpaid balance as of June 1.

   _____

Assignment 4

Name_____ Date _____

Complete the following problems. Write your answers in the blanks provided

A. Find the installment price and the finance charge. The down payment is 10% of the cash price.

	Cash Price	Monthly Payment	Length of Loan	Installment Price	Finance Charge
1.	$485	$50.12	12 months	_____	_____
2.	$780	$90.60	9 months	_____	_____
3.	$299	$62.30	6 months	_____	_____
4.	$500	$36.40	24 months	_____	_____

B. Find the rebate fraction in these problems.

	Length of Loan	Time Left in Loan	Rebate Fraction
5.	18 months	12 months	_____
6.	12 months	3 months	_____
7.	24 months	15 months	_____
8.	14 months	3 months	_____

C. Find the interest rebate for these problems. Hint: Remember, to calculate the interest rebate, first determine the time left in loan.

	Finance Charge	Length of Loan	Paid in Full	Interest Rebate
9.	$157.14	24 months	14 months	_____
10.	$125.39	12 months	8 months	_____
11.	$224	12 months	6 months	_____
12.	$387	14 months	10 months	_____

D. Find the final payment for these loans.

	Length of Loan	Amount of Payments	Interest Charge	Paid in Full	Final Payment
13.	16 month	$110.00	$232.40	12 months	_____
14.	12 month	$97.80	$197.30	8 months	_____
15.	14 month	$119.00	$187.20	10 months	_____
16.	6 month	$320.00	$110.50	3 months	_____

Assignment 5

Name_____ Date _____

Solve the following problems. Write your answers in the blanks provided.

1. John and Rachel purchased new furniture for their sun room on an installment plan of 18 monthly payments of $176.67 each, with a down payment of $300. The actual cash price was $2,995. How much more did they pay for buying on the installment plan? _____

2. Ping is going to purchase a new truck priced at $12,990, minus the trade-in value of his car at $1,300. The interest is stated at 18% for four years. Calculate his monthly payment. _____

3. What would be the final payment for Ping's truck should he pay it off after making 36 payments? _____

4. A loan is being paid in 14 equal payments. At the end of the ninth month, the loan is paid off. What is the rebate fraction? _____

5. A computer system was purchased for $1,450 on an installment plan. A down payment of $145 was made and the unpaid balance was financed for 6 months at 16% interest. What is the monthly payment? _____

6. If the cash price of an item is $499, the down payment is $50, and there are 18 monthly payments of $29.43, how much was the additional cost of using the installment plan? _____

7. Using the table provided, determine the monthly payment, excluding taxes and insurance, of a $68,000 mortgage loan over a period of 25 years at 10% interest. _____

8. Find the interest payment for one month based on an $85,000 mortgage loan at 10% interest for a period of 30 years. _____

9. Determine the interest payment, principal payment, and balance of principal on a monthly payment of $539.84. The mortgage loan was $60,000 at 9% interest for 20 years. _____

_____

_____

Assignment 6

Name_____ Date_____

Complete the following word problems. Write your answers in the blanks provided.

1. Rudy Herrera bought a new boat for $8,500. Rudy put down $1,200 and financed the balance at 10% for 60 months. What is his monthly payment?

    _____

2. Keith Smith borrowed $4,200 from Easy Loan Company. The loan is to be repaid in 48 monthly installments of $127.50. At the end of 46 months, Keith decided to pay off the loan. What is Keith's rebate and payoff amount?

    Rebate _____
    Payoff Amount _____

3. A video camcorder with a cash price of $999 can be purchased on the installment plan in 18 monthly payments of $75. Find the amount of finance charge.

    _____

4. What is the rebate fraction on a four year loan if it is paid off in 28 months?

    _____

5. Shirley Lindale borrowed money from her brother-in-law and is repaying it at $55 a month for 12 months. What is the finance charge rebate after making 8 payments if the finance charge is $435?

    _____

6. Sara Benson purchased an electric stove on the installment plan at a cost of $976. She paid 10% of the total cost as a down payment and promised to pay the unpaid balance in 24 equal monthly installments. What will be the amount of each monthly installment?

    _____

7. Hal Jacobsen purchased a lawnmower that was priced to sell for $340 on an installment plan. The installment agreement provided for a 10% down payment of the selling price, and the unpaid balance to be paid in 12 monthly payments of $36 each. What amount of money did Hal pay as finance charges?

    _____

8. Winnie Polk can buy a car for $10,250.00. The dealer will allow her a trade-in of $1,500 on her old car. Winnie can finance the balance with the dealer by making 36 equal monthly payments of $275 or she can borrow the money from a bank at a total cost of $576.50. What amount of money will Winnie save by borrowing the money from a bank instead of financing the car with the dealer?

    _____

9. Tim purchased a new camera that had a cash price of $299.99. He made a down payment of 10% of the cash price. There was a carrying charge of $76.50 on the unpaid balance. The unpaid balance, plus the carrying charges were to be paid in 12 equal monthly installments. What was the amount of each monthly payment? _____

10. Barbara purchased a computer for her home based business. The finance charges were $398.55 and it was financed in 24 monthly installments. If she pays it off after 20 payments, what is her rebate due on the finance charge? _____

11. If you purchase a ski boat for 36 monthly payments of $135 and an interest charge of $349, how much is the rebate after 12 payments? What is the payoff amount?

    Rebate _____
    Payoff Amount _____

Proficiency Quiz
R E V I E W

Name_____ Date_____

$\dfrac{\text{Student's Score}}{\text{Maximum Score}} = \dfrac{}{37} = \text{Grade}_____$

A. Determine the annual percentage rate or monthly rate for these problems. Write your answers in the blanks provided.

	Monthly Rate	Annual Percentage Rate		Monthly Rate	Annual Percentage Rete
1.	_____	18.5%	**2.**	1.5%	_____
3.	1.67%	_____	**4.**	_____	24%

B. Using the previous balance method, calculate the finance charge and account balance. Round answers to nearest penny.

Previous Balance	APR	Finance Charge		Account Balance	
$287.44	12%	**5.**	_____	**6.**	_____
$542.18	18.5%	**7.**	_____	**8.**	_____

C. Using the average daily balance method, determine the average daily balance and finance charge. A customer's charge account shows a balance of $255.50 on June 1. A payment of $155.50 was made on June 10. Purchases of $340 were made on June 15. An annual rate of 18.5% is applied to your account. The month of June has 30 days. The next billing date is July 1. Round answers to nearest penny.

   9. Average daily balance _____   **10.** Finance charge _____

D. Solve the following problem using the adjusted balance method. If necessary, round answers to nearest penny.

   11. Wynonna has a charge account at Barne's Ladies Wear with a current balance of $55.94. The store charges a 1.5% interest rate on the unpaid balance. If Wynonna makes a $15 payment and an additional purchase of $23, what will be the amount of the ending unpaid balance?

   _____

E. Determine the finance charge and the new balance for these problems. The finance charge is based on the previous balance before payments or credits are subtracted. The monthly rates are 2% on amounts up to $1,000 and 1.5% on amounts over $1,000. Assume the balance was not paid within the specified period.

Previous Balance	Finance Charge		Payments	Credits	Purchases	New Balance	
$885.50	**12.**	_____	455.50	– 0 –	$160	**13.**	_____
$1,015.00	**14.**	_____	500.00	$40.00	$75	**15.**	_____

F. Calculate the following information on these installment purchases. A 10% cash down payment has already been calculated. The amount in the first column is the amount to be financed. Terms of installment buying are:

   a. The balance paid in 12 monthly installments.

   b. Interest is 18%. Round answers to nearest penny.

Amount to be Financed	Total Interest	Amount to Be Repaid	Monthly Installment
$2,660	16. _____	17. _____	18. _____
$999	19. _____	20. _____	21. _____
$1,350	22. _____	23. _____	24. _____
$675	25. _____	26. _____	27. _____
$3,620	28. _____	29. _____	30. _____

G. Find the installment price and expense of using the installment plan for this purchase. The cash price is $4,509; the down payment is $451; the number of payments is 36 and the monthly payment is $129.63.

31. Installment price = _____

32. Expense of installment buying = _____

H. Find the rebate fraction in these problems.

Length of Loan	Paid in Full	Rebate Fraction
16 months	10 months	33. _____
18 months	9 months	34. _____

I. Find the interest rebate for these problems. Round answers to nearest penny.

Finance Charge	Length of Loan	Paid in Full	Interest Rebate
$150.00	12 months	10 months	35. _____
$224.00	24 months	12 months	36. _____

J. Find the monthly payment necessary to amortize a 10% mortgage loan of $80,000 over a 30 year period. Compute the monthly basic payment using this information: $8.78 is the monthly payment required to repay a $1,000 loan.

37. Monthly payment = _____

CHAPTER
Metrics and Currency
16

16

Metrics and Currency

OBJECTIVES

After completing this chapter, you will be able to:

1. Use the standard metric units for length, area, weight, volume, and temperature.
2. Use the standard metric prefixes and abbreviations.
3. Calculate the amount of foreign currency based upon current exchange rates.

The United States began adopting the metric system of measurement when President Gerald Ford signed the Metric Conversion Act of 1975. Since that time, you have been purchasing soft drinks by the liter, hearing weather forecasters report temperatures by Celsius as well as Fahrenheit, and reading highway signs showing distances in both miles and kilometers. The Metric Conversion Act of 1975 did not identify a specific date when the United States would switch completely to the metric system, nor did it make it mandatory for companies to do so. Congress did recognize, however, that the United States must move in that direction to be able to compete in world trade.

On August 23, 1988, President Ronald Reagan signed the Omnibus Trade and Competitiveness Act of 1988. This act included a call for Federal agencies to strive to complete their transition to metric by the end of fiscal year 1992. In passing this act, Congress officially designated metric as our preferred measurement system for trade and commerce. However, individual groups and industries are still free to decide whether or not to convert to the metric system.

Over 98 percent of the world's population already uses the metric system. The main reason that the United States is adopting the metric system is to facilitate trade with other nations. This means that we will be able to compete more easily in a competitive world market. The absence of metric measurements in U.S. goods and services could directly reduce our ability to compete successfully in the world market.

The metric system is easy to learn. Here are some interesting facts about the metric system:

1. It is a decimal system that makes using fractions practically obsolete.
2. Like our monetary system, each metric unit is related to larger or smaller units by powers of 10.
3. The most commonly used units of measure are the meter to measure length, the gram for mass (or weight), the liter for volume, and degrees Celsius to measure temperature.

16.1 Terms Used in Metrics

The terminology used in metric measure is shown in the following table. Note that periods are dropped from all abbreviated terms except inches (in.).

Metric Terms

Measure	Name	Symbol	Approximate English Size	Symbol
length	meter	m	39.37 inches	in.
	kilometer	km	0.6217 mile	mi
	centimeter	cm	0.3937 inch	in.
	millimeter	mm	0.0394 inch	in.
area	are	a	119.60 square yards	yd^2
	hectare	ha	2.4711 acres	A
weight (mass)[1]	gram	g	0.353 ounce	oz
	kilogram	kg	2.2046 pounds	lb
	metric ton	t	1.1023 short tons	T
volume[2]	liter	L	1.0567 quart	qt
	milliliter	mL	0.2 teaspoon	tsp
temperature	generally expressed in degrees Celsius	°C	0°C is the freezing point as opposed to 32°F	

[1] *Mass* defines the quantity of material in an object and is commonly referred to as *weight*.
[2] The term *volume* is interchangeable with *capacity*. For our purposes in this chapter, the term *volume* will refer to liquid capacity.

16.2 English System versus Metric System

In the English system measurements are expressed in inches, feet, yards, ounces, pounds, quarts, and so on. For example, you must memorize how many inches are in a foot (12 inches), then how many feet are in a yard (3 feet). You need to learn many numbers to determine length, area, weight, volume, and temperature correctly.

The metric system is a decimal system based on one number–10. This is a simple system once you understand its use.

16.3 Metric Base Units

The following table shows the units used to measure length, weight, volume, and temperature.

Common Units of Measure

Quality	Unit of Measure	Symbol
length	meter	m
area	are	a
weight	gram	g
volume	liter	L
temperature	degrees Celsius	°C

16.4 Metric Prefixes

You should become familiar with the more common prefixes used in the metric system. When a prefix is added to the base unit of measure, larger or smaller units are created. For example, study this table of common prefixes.

Common Prefixes

Prefix	Symbol	Base of 10	Factor
mega	M	10^{6}*	1,000,000
kilo	k	10^{3}	1,000
hecto	h	10^{2}	100
deka	da	10^{1}	10
deci	d	10^{-1}	0.1
centi	c	10^{-2}	0.01
milli	m	10^{-3}	0.001
micro	μ	10^{-6}	0.000001

*10^{6} means 10 to the 6 power or expressed mathematically: $10 \times 10 \times 10 \times 10 \times 10 \times 10 = 1,000,000$. Negative integers (−1) are integers less than 1.

16.5 Measuring Length Using Metrics

The *meter* is the base unit for measuring length metrically. A meter is a little longer than a yard. Study the following table that shows increasing and decreasing length in meters.

Length in Meters

Unit of Measure	Symbol	Meters
kilometer	km	1,000
hectometer	hm	100
dekameter	dam	10
meter	m	1
decimeter	dm	0.1
centimeter	cm	0.01
millimeter	mm	0.001

The kilometer, centimeter, and millimeter are the most commonly used measures. For example, long distances are measured in kilometers; centimeters and millimeters are used in science.

Converting from Smaller to Larger Lengths

To convert from a smaller unit (meter) to the next larger unit (dekameter), simply divide by 10. Study this example:

$$2,000 \text{ m} \div 10 = 200.000^* \text{ dam}$$

*For all problems in this chapter, round your answers to 3 decimal places unless otherwise directed.

■ **TIP**   Notice when dividing by 10, all you have to do is move the decimal point 1 place to the left in the dividend to get the correct answers.

     Practical Math Applications

PRACTICE PROBLEMS

Complete the following problems and write your answers in the blanks provided.

1. 4,392 mm = _____ cm   **2.** 2,196 dam = _____ hm   **3.** 240 mm = _____ cm

Solutions: **1.** 439.200; **2.** 219.600; **3.** 24.000

Converting from Larger to Smaller Lengths

To convert from a larger unit (meter) to the next smaller unit (decimeter), simply multiply by 10. Study this example:

$$5,000 \text{ m} \times 10 = 50,000 \text{ dm}$$

■ **TIP**   Notice when multiplying by 10, all you have to do is move the decimal point 1 place to the right in the multiplicand to get the correct answer.

PRACTICE PROBLEMS

Complete the following problems and write your answers in the blanks provided.

1. 295 m = _____ dm   **2.** 1,070 dam = _____ m   **3.** 395 hm = _____ dam

Solutions: **1.** 2,950.000; **2.** 10,700.000; **3.** 3,950.000

Converting English Measures to Metric Measures: Length

The following table shows the metric equivalent for 1 inch, 1 foot, 1 yard, and 1 mile. For simplicity, 3 decimal places have been used; therefore, these are approximate equivalents.

English to Metric: Length

TABLE 16-5
English to Metric

Unit of Measure	Metric Equivalent
1 in.	2.540 cm
1 ft	0.305 m
1 yd	0.914 m
1 mi	1.609 km

Using the preceding list of metric equivalents for length, you can convert English measurements for length to metric measures simply by multiplying the English measure (inch, foot, yard, mile) by its equivalent. Study this example:

$$24 \text{ in.} \times 2.540 \text{ cm per in.} = 60.960 \text{ cm}$$

PRACTICE PROBLEMS

Complete the following problems and write your answers in the blanks provided.

1. 81 in. = _____ cm   **2.** 512 ft = _____ m

3. 6 yd = _____ m   **4.** 12 mi = _____ km

Solutions: **1.** 205.740; **2.** 156.160; **3.** 5.484; **4.** 19.308

16.6 Measuring Area Using Metrics

When measuring area or surface, the *are* is the base unit. One *are* equals 100 square meters. Land is an example of area that might need to be measured. Usually metric land area is sold in *hectares* One *hectare* is equivalent to approximately 2.4711 acres (A). One acre is equal to 0.4047 (0.405 rounded) hectares. One hectare (ha) equals 10,000 square meters. Study the following example.

FIGURE 16-1
One Hectare

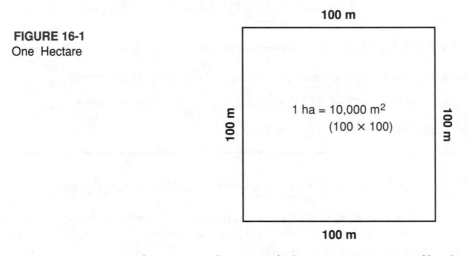

100 m

1 ha = 10,000 m² (100 × 100)

To determine what part of a hectare an amount of land measured in meters is, divide each side by 100, then multiply the products. In other words, if a plot of land is 56 m × 400 m, its hectare equivalent would be computed as follows:

$$56 \div 100 = 0.560$$
$$400 \div 100 = 4.000$$
$$0.560 \times 4.000 = 2.240 \text{ ha}$$

■ **TIP**   When dividing by 100, move the decimal point 2 places to the left in the dividends; then multiply the two answers.

PRACTICE PROBLEMS

Complete the following problems and write your answers in the blanks provided.

1. 660 m by 590 m = _____ ha   **2.** 240 m by 3,800 m = _____ ha

Solutions: **1.** 38.940; **2.** 91.200

Converting English Measures to Metric Measures: Area

The following table shows the metric equivalents for 1 square inch, 1 square foot, 1 square yard, 1 square mile, and 1 acre. Three decimal places have been used for simplicity; therefore, these are approximate equivalents.

English to Metric: Area

TABLE 16-6

English to Metric: Area

Unit of Measure	Symbol	Metric Equivalent	Symbol
1 square inch	in.^{2}	6.452 square cm	cm^2
1 square foot	ft^2	0.093 square m	m^2
1 square yard	yd^2	0.836 square m	m^2
1 square mile	mi^2	2.590 square km	km^2
1 acre	A	0.405 hectare	ha

By using the preceding list of metric equivalents for area, you can convert English measurements for area to metric measures simply by multiplying the English measure (square inch, foot, yard, etc.) by the metric equivalent. Study this example:

$$246 \text{ in.}^2 \times 6.452 = 1{,}587.192 \text{ cm}^2$$

PRACTICE PROBLEMS

Complete the following problems and write your answers in the blanks provided.

1. 631 ft^2 = _____ m^2 **2.** 45 mi^2 = _____ km^2

Solutions: **1.** 58.683;   **2.** 116.550

16.7   Measuring Weight Using Metrics

The unit for measuring weight using metrics is the *gram* The gram is a very small measurement (0.0353 or about $\frac{1}{28}$ of a U.S. ounce); therefore, the base measure is the *kilogram*(1,000 g or 2.2046 lb). Prefixes are added to *gram* to show multiples and submultiples, just as in length. The most often used units are the kilogram, the gram, and the milligram. Study the following table.

Weight in Grams

TABLE 16-7

Weight in Grams

Unit of Measure	Symbol	Grams Equivalent
kilogram	kg	1.000
hectogram	hg	100
dekagram	dag	10
gram	g	1
decigram	dg	0.1
centigram	cg	0.01
milligram	mg	0.001

To convert grams to the next larger amount, divide by 10 as shown in this example:

$$590.000 \text{ g} \div 10 = 59.000 \text{ dag}$$

To convert grams to the next smaller amount, multiply by 10 as shown in this example:

$$590.000 \text{ g} \times 10 = 5,900.000 \text{ dg}$$

■ **TIP**   When converting to larger metric measures, you divide (think: divide up); when converting to smaller metric measures, you multiply.

Converting English Measures to Metric Measures: Weight

The following table show the equivalent in grams for 1 ounce, 1 pound, and 1 short ton. Study this table. These equivalent amounts are rounded to 3 decimal places and are therefore approximate.

English to Metric: Weight

TABLE 16-8
English to Metric: Weight

Unit of Measure	Metric Equivalent
1 ounce (oz)	28.350 g
1 pound (lb)	0.454 kg
1 short ton (2,000 pounds) (T)	0.907 t

Using the preceding list of metric equivalents for weight, you can convert English measurements for weight to metric measures simply by multiplying the English measure (ounce, pound, ton) by its metric equivalent. Study this example:

$$24 \text{ oz} \times 28.350 = 680.400 \text{ g}$$

PRACTICE PROBLEMS

Complete the following problems and write your answers in the blanks provided.

1. 75 lb = _____ kg   **2.** 60 T = _____ t

Solutions: **1.** 34.050; **2.** 54.420

16.8   Measuring Volume Using Metrics

For measuring volume the English system of measurement has one set of measures for dry and one set for liquid. The metric system requires only three units of measure: the liter (a little more than a quart), the milliliter $\left(\dfrac{1}{1,000} \text{ of a liter} \right)$, and a cubic meter (the volume of a box measuring 1 m on each side). The liter and the milliliter are commonly used in measuring smaller amounts of volume, while the cubic meter is commonly used to measure larger capacities, such as the volume of a railroad car. Study the following table.

Volume in Liters

TABLE 16-9
Volume in Liters

Unit of Measure	Symbol	Liters Equivalent
kiloliter	kL	1,000
hectoliter	hL	100
dekaliter	daL	10
liter	L	1
cubic decimeter	dm^3	1
deciliter	dL	0.1
centiliter	cL	0.01
milliliter	mL	0.001

Converting liters from smaller units to larger units or from larger units to smaller units is completed by dividing or multiplying by 10, respectively.

PRACTICE PROBLEMS

Complete the following problems and write your answers in the blanks provided.

1. 3,820 daL = _____ hL   **2.** 6,000 L = _____ dL

Solutions: **1.** 382.000; **2.** 60,000

Converting English Measures to Metric Measures: Volume

The following table shows the metric equivalents for 1 teaspoon, 1 tablespoon, 1 cubic inch, 1 fluid ounce, 1 cup, 1 pint, 1 quart, 1 gallon, 1 cubic foot, and 1 cubic yard. For simplicity, the equivalents have been rounded to 3 decimal places and are therefore approximate. Study this table.

English to Metric: Volume

TABLE 16-10
English to Metric: Volume

Unit of Measure	Metric Equivalent
1 teaspoon	4.929 mL
1 tablespoon	14.787 mL
1 cubic inch (in.^{3})	16.387 cm^3
1 fluid ounce (fl oz)	29.574 mL
1 cup	0.237 L
1 pint (pt)	0.473 L
1 quart (qt)	0.946 L
1 gallon (gal)	3.785 L
1 cubic foot (ft^3)	0.028 m^3
1 cubic yard (yd^3)	0.765 m^3

Using the preceding list of metric equivalents for volume, you can convert English measurements for volume to metric measures simply by multiplying the English measure (teaspoon, tablespoon, cup, etc.) by its metric equivalent. Study this example:

$$28 \text{ cups} \times 0.237 = 6.636 \text{ L}$$

PRACTICE PROBLEMS

Complete the following problems and write your answers in the blanks provided.

1. 22 gal = _____ L   **2.** 10 tablespoons = _____ mL

Solutions: **1.** 83.270; **2.** 147.870

16.9 Measuring Temperature Using Metrics

In the United States, temperature is commonly measured in degrees *Fahrenheit*. However, more and more, degrees **Celsius** are being quoted as well. *Celsius* is the commonly used metric measure for temperature. Study this comparison of Fahrenheit and Celsius:

	Degrees Fahrenheit	Degrees Celsius
Water freezes	32	0
Water boils	212	100
Normal body temperature	98.6	37

Converting English Measures to Metric Measures: Temperature

To convert Fahrenheit to Celsius, subtract 32 and multiply by $\frac{5}{9}$ or 0.556; then round to the nearest degree. Study this example:

72° Fahrenheit = ? Celsius
72° − 32° = 40°
40° × 0.556 = 22.240 rounded to 22°C

If you multiplied by $\frac{5}{9}$, your answer would be 22.22 rounded to 22°C. For problems in this book, use the decimal equivalent to convert Fahrenheit to Celsius.

PRACTICE PROBLEMS

Complete the following problems and write your answers in the blanks provided.

1. 110°F = _____ °C   **2.** 75°F = _____ °C

3. 212°F = _____ °C   **4.** 32°F = _____ °C

Solutions: **1.** 43.368 or 43°C; **2.** 23.908 or 24°C; **3.** 100.080 or 100°C°; **4.** 0°C

16.10 Foreign Currency Exchange

In today's world, everyone should become familiar with how to handle foreign currency. The reasons one should know about foreign currency are because world travel has become so common, more and more companies have established offices overseas, and, increasingly, companies are conducting business internationally. Should you enter the work force and become employed by one of the larger companies, the chances are greater now than ever before that that company will conduct business internationally.

U.S. dollars may be exchanged for foreign currency at most of the larger banks in the U.S. Rates are quoted on a daily basis based on the value of the dollar on the world market. Also, most of the larger airports have a Foreign Currency Dealer to exchange currency. Notice the word "currency" is used. Only currency is exchanged, not coin, because of the lack of demand for it. Most major credit cards advertise a better rate of exchange for charges rather than cash. You may have noticed a table showing foreign exchange rates in the business section of the local paper. These rates are based on exchanges of $1 million or more.

When traveling, you might want to obtain from a local bank some of the foreign currency for the country in which you will be traveling. However, some countries, such as India or Latvia (one of the former Soviet Baltic states), will not allow their currency to leave their country. In these cases, you would have to wait until you are in the country to make your currency exchange. If you have some foreign currency before you leave the U.S., you will be able to get through the airport, pay taxi fare, and get to your hotel before you must change money. Also, remember if you are traveling on weekends, most banks are closed; you should carry enough foreign currency to pay weekend expenses.

Here are some exchange rates based on one U.S. dollar. The figures quoted here are as of May 8, 1995. The following exchange rates for each U.S. dollar will apply for our example and should be used in all practice problems and assignments, unless otherwise specified.

Country	Per $1 Exchange Rate	Country	Per $1 Exchange Rate
Japan (Yen)	84.2200	Canada ($)	1.3542
Germany (Mark)	1.3748	Italy (lira)	1,657.0000
England (Pound)	0.6263	Netherlands (Guilder)	1.5360
Mexico (Peso)	5.8450	Switzerland (Franc)	1.1340

What matters most is the equivalent cost of goods and services in U.S. dollars. Notice in most countries the rates give us more than our dollar's worth, but in England we get a little over half a pound for every dollar. Therefore, the cost of travel is much cheaper in Japan and Italy than in England. Of course, Mexico would, by far, be the cheapest country in which to travel, based on these rates.

To determine the amount of currency you would receive follow this procedure:

Exchange Rate Per 1 dollar = Amount of Foreign Currency

$1 = 84.2200 Yen; therefore, $5 × 84.2200 = 421.1000 Yen

$1 = 1.3748 Mark; therefore, $5 × 1.3748 = 6.8740 Marks

— PRACTICE PROBLEMS

Complete the following problems and write your answers in the blanks provided.
Carry your answers to 4 decimal places.

1. $125 U.S. @ 84.2200 per dollar = _____ Japanese Yen

2. $300 U.S. @ 1.3748 per dollar = _____ German Marks

3. $100 U.S. @ 5.8450 per dollar = _____ Mexican Pesos

4. $75 U.S. @ 1.1340 per dollar = _____ Swiss Francs

Solutions: **1.** 10,527.5000; **2.** 412.4400; **3.** 584.5000; **4.** 85.0500

MATH ALERT

1. A biology class is studying the temperature changes in its area, measuring the temperature in degrees Celsius rather than in degrees Fahrenheit. Compute the Celsius equivalents for the first week in January.

Temperature: January 1 to January 7

Day	Fahrenheit	Celsius
1	20°	_____
2	18°	_____
3	22°	_____
4	32°	_____
5	38°	_____
6	39°	_____
7	45°	_____

2. When Chris Kelly crossed the U.S.–Canadian border into British Columbia, he found all the speed limits posted in kilometers. Convert the following distances for Chris from kilometers to miles. Carry your answers to three decimal places; then round your answers to the nearest mile.

Town	Kilometers	Miles
Coutts to Lithbridge	117.477	_____
Lithbridge to Fort Macleod	56.324	_____
Fort Macleod to Claresholm	37.013	_____
Claresholm to High River	65.980	_____
High River to Calgary	61.152	_____
Total miles		_____

3. Chris wanted to exchange $200 at a bank in Calgary, Canada. Using the following figures from his newspaper, determine the amount of Canadian dollars he will get on February 6. Round your answer to the nearest cent.

FOREIGN EXCHANGE

	Feb. 3	Feb. 4	Feb. 5	Feb. 6	Feb. 7
BRITISH POUND*	1.794	1.804	1.815	1.819	1.836
CANADIAN DOLLAR*	.8524	.8452	.8485	.8485	.8490
FRENCH FRANC**	5.495	5.475	5.421	5.385	5.377
GERMAN MARK**	1.602	1.588	1.580	1.578	1.559
ITALIAN LIRA**	1206	1200	1194	1187	1188
JAPANESE YEN**	126.3	126.0	125.4	125.7	125.2
MEXICAN PESO**	3067	3062	3062	3062	3062
SWISS FRANC**	1.428	1.417	1.410	1.407	1.390

*Foreign currency in U.S. dollars.
**U.S. dollar in foreign currency.
All rates for trades of $1 million minimum, New York prices.

SOURCE: Associated Press

Chapter 16   Metrics and Currency
Study Guide

I.   Terminology

page 338	metric system	A decimal system based on one number–10
page 339	meter	A measurement of length; abbreviated m
page 339	liter	A measurement of volume; abbreviated L
page 340	mega	Meaning 1,000,000; abbreviated M
page 340	kilo	Meaning 1,000; abbreviated k
page 340	hecto	Meaning 100; abbreviated h
page 340	deka	Meaning 10; abbreviated da
page 340	deci	Meaning 0.1; abbreviated d
page 340	centi	Meaning 0.01; abbreviated c
page 340	milli	Meaning 0.001; abbreviated m
page 340	micro	Meaning 0.000001; abbreviated μ
page 340	negative integers	Integers less than 1 (–1)
page 340	kilometer	1,000 meters; abbreviated km
page 340	hectometer	100 meters; abbreviated hm
page 340	dekameter	10 meters; abbreviated dam
page 340	meter	1 meter; abbreviated m
page 340	decimeter	0.1 meter; abbreviated dm
page 340	centimeter	0.01 meter; abbreviated cm
page 340	millimeter	0.001 meter; abbreviated mm
page 342	are	A measurement of area; 100 square meters; abbreviated a
page 342	hectares	Approximately 2.4711 acres; abbreviated ha
page 343	gram	A measurement of weight; abbreviated g
page 343	kilogram	1,000 grams or 2.2046 pounds; abbreviated kg
page 343	hectogram	100 grams; abbreviated hg
page 343	dekagram	10 grams; abbreviated dag
page 343	decigram	0.1 grams; abbreviated dg
page 343	centigram	0.01 grams; abbreviated cg
page 343	milligram	0.001 grams; abbreviated mg
page 345	kiloliter	1,000 liters; abbreviated kL
page 345	hectoliter	100 liters; abbreviated hL
page 345	dekaliter	10 liters; abbreviated daL
page 345	deciliter	0.1 liters; abbreviated dL
page 345	centiliter	0.01 liters; abbreviated cL
page 345	milliliter	0.001 liters; abbreviated mL
page 346	degrees Celsius	A measurement of temperature; abbreviated C
page 346	currency	Paper money
page 346	coin	Silver money
page 347	exchange rate	The rate at which one country's money is exchanged for that of another country's

II. Calculating Using Metric Measures

page 340   *Converting from Smaller to Larger Lengths:* divide by 10.

Example: 2,000 m ÷ 10 = 200 dam

page 341   *Converting from Larger to Smaller Lengths:* Multiply by 10.

Example: 5,000 m × 10 = 50,000 dm

page 341   *Converting English Measures to Metric Measures–Length:* 1 in. = 2.54 cm;
1 ft = 0.305 m; 1 yd = 0.914 m; 1 mi = 1.609 km

Example: 24 in. × 2.54 cm per in. = 60.960 cm

page 342   *Measuring Area Using Metrics:* To determine what part of a hectare an amount of land measured in meters is, divide each side by 100, then multiply the products.

Example: A plot of land 56 m × 400 m^2

56 ÷ 100 = 0.56

400 ÷ 100 = 4

0.56 × 4 = 2.240 ha

page 343   *Converting English Measures to Metric Measures–Area:* 1 in.^{2} = 6.452 cm^2;
1 ft^2 = 0.093 m^2; 1 yd^2 = 0.836 m^2; 1 mi^2 = 2.590 km^2; 1 A = 0.405 ha

Example: 246 in.^{2} × 6.542 = 1,587.192 cm^2

page 343   *Measuring Weight Using Metrics:* 1 kg = 1.000 g; 1 hg = 100 g; 1 dag = 10 g;
1 dg = 0.1 g; 1 cg = 0.01 g; 1 mg = 0.001 g

Example: 590 g ÷ 10 = 59 dag (smaller to larger)

590 g × 10 = 5,900 dg (larger to smaller)

page 344   *Converting English Measures to Metric Measures–Weight:* 1 oz = 28.350 g;
1 lb = 0.454 kg; 1 T (short ton) = 0.907 t (metric ton)

Example: 24 oz × 28.350 = 680.400 g

page 344   *Measuring Volume Using Metrics:* 1 kL = 1,000 L; 1 hL = 100 L; 1 daL = 10 L;
1 dL = 0.1 L; 1 cL = 0.01 L; 1 mL = 0.001 L

Example: 3,820 daL = 382 hL

page 346   *Measuring Temperature Using Metrics:* Subtract 32 degrees from Fahrenheit and multiply by 0.556.

Example: 72 degrees – 32 degrees = 40 degrees
40 degrees × 0.556 = 22.24 rounded to 22 degrees Celsius

III. Foreign Currency Exchange

page 346   To convert U.S. dollars to foreign currency, multiply the current exchange rate times each U.S. dollar.

Example: Mexican peso exchange rate: 5.8450 per U.S. dollar
Number of U.S. dollars to be exchanged: $150
$150 × 5.8450 = 876.7500 Mexican pesos

Assignment 1

Name _____ Date _____

Complete the following problems. Write your answers in the blanks provided. Be sure to identify thousands and millions with commas. Carry your answers to one decimal place unless otherwise directed.

A. Convert from smaller to larger lengths.

1.	396 cm	= _____ dm		**2.**	986 dm	= _____ m
3.	4,241 m	= _____ dam		**4.**	325 hm	= _____ km
5.	41,000 m	= _____ cm		**6.**	251 m	= _____ dam
7.	80,140 mm	= _____ cm		**8.**	605 dm	= _____ m
9.	602 hm	= _____ km		**10.**	57 cm	= _____ dm

B. Convert from larger to smaller lengths.

11.	577 m	= _____ dm		**12.**	21,903 dm	= _____ cm
13.	600 cm	= _____ mm		**14.**	983 hm	= _____ dam
15.	3,111 km	= _____ hm		**16.**	413 cm	= _____ mm
17.	15,209 km	= _____ hm		**18.**	641 dm	= _____ cm
19.	411 m	= _____ dm		**20.**	792	= _____ dam

C. Convert English measures to metric measures: length. Carry answers to 3 decimal places.

21.	901 ft	= _____ m		**22.**	63 yd	= _____ m
23.	85 mi	= _____ km		**24.**	214 ft	= _____ m
25.	29 in.	= _____ cm				

D. Determine what part of a hectare an amount of land measured in meters is. Carry answers to 3 decimal places.

26.	12,000 m by 600 m = _____ ha		**27.**	380 m by 807 m = _____ ha
28.	37 m by 365 m = _____ ha		**29.**	255 m by 299 m = _____ ha

E. Convert English measures to metric measures: area. Carry answers to 3 decimal places.

30.	149 in.^{2} = _____ cm^2		**31.**	300 yd^2 = _____ m^2
32.	232 in.^{2} = _____ cm^2		**33.**	9,746 mi^2 = _____ km^2
34.	803 ft^2 = _____ m^2		**35.**	169 in.^{2} = _____ cm^2

Assignment 2

Name_____ Date _____

Complete the following problems. Write your answers in the blanks provided. Be sure to identify thousands and millions with commas. Carry your answers to one decimal place unless otherwise directed.

A. Convert from smaller to larger weights.

1. 473 cg   = _____ dm   **2.** 171 hg   = _____ kg

3. 1,100 dag   = _____ hg   **4.** 200 dg   = _____ g

5. 213 mg   = _____ cg

B. Convert from larger to smaller weights.

6. 605 g   = _____ dg   **7.** 61,216 dg   = _____ cg

8. 141 cg   = _____ mg   **9.** 741,000 dag   = _____ g

10. 803 kg   = _____ hg

C. Convert English measures to metric measures: weight. Carry answers to three decimal places.

11. 757 lb   = _____ kg   **12.** 88 oz   = _____ g

13. 2,000 T   = _____ t   **14.** 629 lb   = _____ kg

15. 69 oz   = _____ g

D. Convert from smaller to larger volumes.

16. 504 mL   = _____ cL   **17.** 5,120 dL   = _____ L

18. 490 daL   = _____ hL   **19.** 212 cL   = _____ dL

20. 69,411 L   = _____ daL

E. Convert English measures to metric measures: volume. Carry answers to three decimal places.

21. 12 yd^3 = _____ m^3 **22.** 18 cups   = _____ L

23. 509 gal   = _____ L

F. Convert Fahrenheit to Celsius: temperature. Round your answers to the nearest degree.

24. 98°F   = _____ °C   **25.** 47°F   = _____ °C

Assignment 3

Name_____ Date_____

Complete the following problems. Write your answers in the blanks provided. Be sure to identify thousands and millions with commas. Carry your answers to one decimal place.

A. Convert the following lengths to smaller or larger measurements.

   1. 14,000 mm   = _____ cm   **2.** 933 km   = _____ hm

   3. 628 hm   = _____ dam   **4.** 255 hm   = _____ km

   5. 400 cm   = _____ mm   **6.** 6,984 m   = _____ dam

   7. 6,879 dm   = _____ cm   **8.** 747 m   = _____ dm

   9. 193 dm   = _____ m   **10.** 347 cm   = _____ dm

B. Convert the following weights to smaller or larger measurements.

   11. 472 mg   = _____ cg   **12.** 100 kg   = _____ hg

   13. 941 dag   = _____ g   **14.** 23 dg   = _____ g

   15. 29 dag   = _____ hg   **16.** 301 hg   = _____ kg

   17. 999 g   = _____ dg   **18.** 867 cg   = _____ dg

   19. 386 cg   = _____ mg   **20.** 12,905 dg   = _____ cg

C. Convert the following foreign currency amounts based on the exchange rates shown. Carry your answers to one decimal place.

21. $60 U.S. @ 6.3 per dollar   = _____ Danish Kroner

22. $95 U.S. @ 4.5 per dollar   = _____ Finnish Marks

23. $27 U.S. @ 5.359 per dollar   = _____ French Francs

24. $129 U.S. @ 189 per dollar   = _____ Greek Drachmas

25. $150 U.S. @ 0.67 per dollar   = _____ Jordanian Dinars

26. $175 U.S. @ 2.57 per dollar   = _____ Malaysian Ringgits

27. $80 U.S. @ 24.25 per dollar   = _____ Pakistani Rupees

28. $50 U.S. @ 142 per dollar   = _____ Portuguese Escudes

29. $200 U.S. @ 2.85 per dollar   = _____ South African Rands

30. $500 U.S. @ 64 per dollar   = _____ Venezuelan Bolivars

Assignment 4

Name_____ Date _____

**Complete the following problems. Write your answers in the blanks provided.
Be sure to identify thousands and millions with commas.**

1. Darrian Lopez, a high school senior, decided to begin his own business of painting house numbers on curbs. Darrian quickly had several orders. He purchased 5 gallons of white paint for the background and 5 gallons of black paint for the numbers. He needed the equivalent of three gallons of florescent paint, but it came in liters rather than gallons. How many liters of florescent paint would equal three gallons?

    Liters purchased: _____

2. Joan has just moved to the U.S. from Canada. She passed her bank building on her way to work and noticed the bank had put up a new sign giving the temperature in Fahrenheit. Joan noticed the temperature was 83 degrees. Joan was used to degrees Celsius. What is the equivalent in Celsius for the temperature? Round your answer to the nearest degree Celsius.

    Celsius: _____

3. Peter's company has decided to send him to Paris, France for a month to work in their office there. Peter has learned the exchange rate per dollar for French francs is 5.60. If Peter exchanged $300 at his local bank, how many francs would he receive in return?

    French francs: _____

4. Courtney's company is sponsoring a marathon for charity on Saturday. Courtney runs 5 miles each day. The track for the marathon will be marked in kilometers. How many kilometers will be equivalent to her 5 miles that she runs each day?

    Kilometers: _____

5. The marathon Courtney will enter is 32.180 kilometers. How many miles is the marathon?

    Miles: _____

6. If Courtney ran the 32.180 kilometers in 3 hours, how many kilometers did she average per hour?

    Kilometers per hour: _____

7. Paul has decided to recarpet his office. His office requires 225 square yards of carpet. How many square meters of carpet would he have to purchase?

    Square meters: _____

8. Ruth's company is having a picnic on Saturday. She is to bring enough liters of cola for 50 people. Ruth estimated she would need an average of 16 ounces (1 quart) of cola per person for the day. Ruth decided to purchase the equivalent cola in liters. How many liters must Ruth purchase?

    Liters: _____

9. To conserve energy, Janet's office has agreed to keep the temperature at 68 degrees this winter. The thermostat was replaced with a new one that displayed degrees Celsius rather than Fahrenheit. What degree Celsius is equivalent to 68 degrees Fahrenheit? Round your answer to the nearest degree.

    Celsius: _____

10. When Peter returned from Paris, France, he brought back 452 French francs. He found the exchange rate had changed to 5.3599. How much U.S. money did he receive in exchange? Round to the nearest cent.

    U.S. dollars: _____

Assignment 5

Name_____ Date _____

**Complete the following problems. Write your answers in the blanks provided.
Be sure to identify thousands and millions with commas.**

1. Hector drove by the bank and noticed the temperature was given in degrees Celsius. The temperature was 36 degrees Celsius. What is the equivalent in Fahrenheit for the temperature? (Round your answer to the nearest tenth degree)

    _____

2. Kelsey and Anita bicycled their way across the state for their vacation. They estimated they traveled 532.4 miles. How far did they travel in kilometers?

    _____

3. Todd measured a piece of rope to make a tree swing. The rope is 3.2 meters long. How long is the rope in feet and inches.?

    _____

4. Louise purchased a tract of land in France. It measured 2.1 hectares. How many acres has Louise bought?

    _____

5. Nancy and Bob remodeled their home and added an extra room. It measured 422 square feet. How much carpet should Bob order in square meters?

    _____

6. Dr. Marvin Frankston is a pediatrician. He requires each child be measured in kilograms. The nurse must convert each weight into pounds for the parents' records. If Elizabeth weighs 23.1 kg, what does she weigh in pounds?

    _____

7. The new Ballpark in Arlington required 12,900 tons of steel supports. What does this equal in metric tons?

    _____

8. Pierre DuBois is a french pastry chef. His recipes are all in metric measurements. The restaurant only has English measuring spoons. Convert his recipe and give the English equivalents.

    Cinnamon Lemon Rolls

9 mL salt	= _____ tsp
0.5 L sugar	= _____ cups
4.929 L cinnamon	= _____ tsp
1 L flour	= _____ cups
$\frac{1}{2}$ kg butter	= _____ lb
23.1 mL lemon juice	= _____ oz

9. The weather channel gave these temperatures for the day across Europe. Convert these temperatures to degrees Celsius. Round your answer to the nearest tenth degree.

    France 89°F = _____

    Germany 76°F = _____

    Brussels 32°F = _____

    England 55°F = _____

    Italy 47°F = _____

    Iceland 22°F = _____

10. Rashonda Lewis will travel to Mexico this summer on vacation. The exchange rate for the peso is 3.12 per 1 American dollar. If she has $500 to spend, how many pesos will she have in Mexico?

    _____

Assignment 6

Name_____ Date _____

Complete the following problems. Write your answers in the blanks provided. Be sure to identify thousands and millions with commas.

1. Wayne Shipman, a high school junior, will compete this year in a Science Quiz. A part of the contest is metric conversion. Convert these measurements from English to metric so he can study for the quiz.

   3.2 mi   = _____ km

   55 T   = _____ t

   12.11 oz   = _____ g

   1.45 gal   = _____ L

2. Jenna Wade will travel to Germany for two weeks to see her cousins who are stationed there with the military. She has $1,450 saved for her vacation. How many German marks will she receive if the exchange rate is 1.586 per dollar?

   _____

3. Rebecca Maxwell purchased land in Canada for a bird sanctuary. The land measured 125.3 acres. What would this land measure in hectares?

   _____

4. Matthew Moore has vacationed in Japan for three weeks. He will exchange his remaining 1,000 yen into American dollars at the airport. The exchange rate is 107.25 yen per 1 American dollar. How much money does Matthew have left after vacation?

   _____

5. Yvette needs $4\frac{1}{2}$ yards of silk to make a sarong for the cultural festival at school. The silk is only sold in meters. How many meters of silk will Yvette need to purchase?

   _____

6. Terrence will travel to Greenland for an archaeological expedition. He watched the world weather program on television to find out how cold it is there. The weatherman stated Greenland's high temperature would be 3° C. Should Terrence pack warm clothes or cool clothes? _____

   What is the temperature in °F?   _____

7. Susan will compete in a 26 km race to raise money for the American Cancer Society. She normally runs 6 miles a day. How many miles must she add to her daily routine to equal the length of the race?

   _____

8. A 14k gold bracelet is sold by the inch. The cost is $1.55 per inch. Sheila wants 25 cm of gold chain for her wrist. How many inches will she buy and how much will it cost?

   25 cm   = _____ in.

   cost   = _____

9. Paul and Rhonda will recarpet their home. The carpet company estimated their home to be 3,250 sq ft. How many meters of carpet should they purchase?   _____

10. Pam found a Hummel figurine in a magazine priced at 30 pounds. The exchange rate is 0.640 pounds per American dollar. How much money in American dollars does the figurine cost?

    _____

Proficiency Quiz
R E V I E W

A. Convert from smaller to larger lengths. Show 1 decimal place in your answers except for whole numbers.

1. 50,000 mm = _____ cm

2. 459 dm = _____ m

3. 2,359 m = _____ dam

4. 211 hm = _____ km

5. 29 cm = _____ dm

6. 36,590 mm = _____ cm

7. 869 dm = _____ m

8. 400 hm = _____ km

9. 395 cm = _____ dm

10. 254 m = _____ dam

B. Convert from larger to smaller lengths. Show 1 decimal place in your answers except for whole numbers.

11. 8,611 km = _____ hm

12. 7,896 dm = _____ cm

13. 88 hm = _____ dam

14. 539 m = _____ dm

15. 99 cm = _____ mm

16. 4,339 dm = _____ cm

17. 123 m = _____ dm

18. 489 cm = _____ mm

C. Convert English measures to metric measures–length. Show 2 decimal places in your answers.

19. 258 ft = _____ m

20. 49 yd = _____ m

21. 45 mi = _____ km

22. 88 in. = _____ cm

23. 894 ft = _____ m

24. 99 mi = _____ km

D. Determine what part of a hectare an amount of land measured in meters is.
Show 4 decimal places in your answers except for whole numbers.

25. 13,000 m by 900 m = _____ ha

26. 47 m by 466 m = _____ ha

E. Convert English measures to metric measures–area. Show 1 decimal place in your answers.

27. 48 ft^2 = _____ m^2

28. 408 mi^2 = _____ km^2

29. 888 in.^{2} = _____ cm^2

30. 300 yd^2 = _____ m^2

F. Convert from smaller to larger weights. Show 3 decimal places in your answers, except for whole numbers. Round your answers when necessary.

31. 0.88 mg = _____ cg   **32.** 4,400 dag = _____ hg

33. 598 dg = _____ g   **34.** 887 hg = _____ kg

35. 333 mg = _____ cg   **36.** 300 dag = _____ hg

G. Convert from larger to smaller weights. Round your answers when necessary.

37. 596 kg = _____ hg   **38.** 243 cg = _____ mg

39. 45,869 dg = _____ cg   **40.** 857,465 dag = _____ g

H. Convert English measures to metric measures–weight.

41. 56 oz = _____ g   **42.** 555 lb = _____ kg

43. 2,000 T = _____ t

I. Convert from smaller to larger volumes. Carry answers to one decimal place.

44. 786 mL = _____ cL   **45.** 342 daL = _____ hL

46. 5,897 dL = _____ L   **47.** 28,546 L = _____ daL

J. Convert English measures to metric measures–volume. Show 3 decimal places in your answers.

48. 46 tsp = _____ mL   **49.** 99 cups = _____ L

50. 23 gal = _____ L   **51.** 13 T = _____ mL

K. Convert Fahrenheit to degrees Celsius–temperature. Round your answers to the nearest whole number.

52. 88°F = _____ °C   **53.** 43°F = _____ °C

54. 64°F = _____ °C   **55.** 35°F = _____ °C

L. Compute the following foreign currency amounts based on the exchange rates shown. Carry your answers to one decimal place.

56. $50 U.S. @ 6.3 per dollar = _____ Danish Kroner

57. $88 U.S. @ 4.5 per dollar = _____ Finnish Marks

58. $12 U.S. @ 5.6 per dollar = _____ French Francs

59. $495 U.S. @ 189 per dollar = _____ Greek Drachmas

60. $290 U.S. @ 0.67 per dollar = _____ Jordanian Dinars

61. $98 U.S. @ 24.25 per dollar = _____ Pakistani Rupees

62. $49 U.S. @ 2.85 per dollar = _____ South African Rands

63. $898 U.S. @ 64 per dollar = _____ Venezuelan Bolivars

Index

direct deposit, 159, 172
discount date, 285, 297
discount rate, 285, 297
discounting commercial paper, 285–86, 297, 299
discounts, 182, 183–87, 192
distribution, percentage, 138, 143–44, 146
dividend, 25, 26, 32, 34
division: averaging, 26, 33; checking of, 26, 34; with decimals, 28, 34; definition of, 25, 32; of fractions, 86, 91–94, 96; quotients, estimating, 26–27, 33, 34; with remainders, 27–28, 34; symbols, 25, 32; terms used in, 25–26; of whole numbers and decimals, 26–28, 34
divisor, 25, 26, 32, 34
double time, 204, 206, 219, 220
down payment, 316
due dates, 281–82, 298, 299

E

earned income credit, 239
employee's earnings record, 210, 214–16, 219
Employee"s Withholding Allowance Certificate (W-4 form), 210, 211–12, 219, 237–38, 248
endorsement of checks, 161, 172
endowment insurance, 260, 266
English system of measurement, 339
estimating quotients, 26–27, 33, 34
exact interest, 280, 281
exemptions, 239, 248
extending a sales slip, 30, 33

F

face value: of an insurance policy, 256, 266; of a loan, 279; of a promissory note, 284
factors, 22, 32
Fair Labor Standards Act, 204, 205, 219
Federal Consumer Credit Protection Act of 1968, 310, 311, 315, 325
Federal Deposit Insurance Corporation (FDIC), 159
federal income tax, 210, 211–14, 219, 221, 237, 239–45, 249
Federal Insurance Contributions Act (FICA), 210, 219
Federal Reserve Bank routing number, 161

federal unemployment insurance tax (FUTA), 236, 249
finance charges, 311, 312–14, 316–17, 325, 326
f.o.b. (destination), 182, 192
f.o.b. (shipping point), 182, 192
footing (cross-footing), 29, 33
foreclosure, 321, 326
Form 1040, 239, 244–45, 249
Form 1040EZ, 239, 243, 246, 247, 249
Form 1099-INT, 239, 248
fractions: addition of, 66–72, 75, 76; conversions among types of, 47, 48, 52–53, 54, 56, 57, 110–12, 114; decimal, 52–53, 54, 57; definition of, 46, 56; division of, 86, 91–94, 96; higher terms, raising to, 51–52, 57; improper, 47–49, 56; introduction to, 46–55; like, 66–67, 70–71; lowest terms, reducing to, 49–51, 56–57; and mixed numbers, 47; multiplication of, 86–91, 95; percentage shown with, 124, 129; proper, 47, 56; subtraction of, 66, 72–73, 76; terms used in, 46, 66; unlike, 66, 67, 72

G

greatest common factor, 50–51, 56, 57
gross earnings: calculating, 205–9, 220; terms used in computing, 204–5
gross profit (markon), 189, 192

H

health insurance, 256–58
Health Management Organizations (HMOs), 257
higher terms, raising fractions to, 51–52, 57
Hindu-Arabic (decimal) number system, 2
hospitalization insurance, 256–58, 266
hourly paid Personnel, 205, 220

I

improper fractions: converting mixed numbers to, 48–49, 56; converting to mixed numbers, 48, 56; converting to whole numbers, 47, 56

income tax: federal, 210, 211–14, 219, 221, 237, 239–45, 249; state, 214
increase: actual, finding, 138–39, 140–41, 146; and decrease, percent of, 138–44, 146; definition of, 138, 146; percent of, finding, 139
inspection method or adding and subtracting fractions, 66, 68, 75
installment (consumer) loans: calculations used with, 316–17, 326, 327; cash purchases, comparison with, 320; final payment, calculating, 319–20, 327; mortgage loans, 321–23, 325, 328; rebates, 318–19, 325, 327; total installment price and finance charge amount, determining, 317–18
insurance: automobile, 258–60, 267; coinsurance coverage, 262–64, 267, 268; health, 256–58, 267; life, 260–61, 268; property, 261–64, 268; terms used in, 256
insured, 256, 266
interest: annual percentage rate (APR), 311, 312, 316; Compound, 278, 286–95, 297, 300; exact, 280, 281; ordinary, 280, 281; principal, rate, or time, finding when interest is known, 283, 299; simple, 278, 279, 297, 298–99
interest-bearing promissory note, 284, 297
interest income form (Form 1099-INT), 239, 248
interest period, 286, 297
interest rate, 279, 283, 284, 297, 299
invoice total, determining, 188, 193
invoices, 182–88, 193
itemized deductions, 240, 248

L

length, measurement using metrics, 340–42
liability insurance (automobile), 259–60, 266
life insurance, 260–61
like fractions, 66–67, 70–71, 75
limited payment life insurance, 260, 266
list price, 182, 192
loans: face value of, 279; installment, 315–23; maturity date of, 281–82; maturity value

of, 280; mortgage, 321–23, 325, 328; payment schedule, preparing, 322–23

lowest common denominator, 66, 75

lowest terms, reducing fractions to, 49–51, 56–57

M

manufacturer, 182, 192

markdown, 189, 190, 192, 194

market value, 232, 248

markon (markup), 189, 192, 194

maturity date, 281–82, 284, 285

maturity value, 280, 297, 298

Medicare, 210, 219, 221

merchandise: pricing of, 189–90; Purchasing of, 182–88; returned, 184–85, 193

Metric Conversion Act of 1975, 338

metric system of measurement: area, 342–43, 350; base units, 339; facts about, 338; length, 340–42, 350; prefixes, 340; temperature, 346, 350; terms used in, 339; volume, 344–46, 350; weight, 343–44, 350

mill, 233

minuend, 17, 32, 33

mixed numbers: addition of, 70–72, 76; converting to decimal fractions, 54, 57; converting improper fractions to, 48, 56; converting to improper fractions, 48, 56; division of, 93–94, 96; fractions and, 47; multiplication of, 90–91, 95

monthly installment loan payments, computing, 316–17

monthly mortgage payments, computing, 321–22

mortgage loans, 321–23, 325, 328

mortgagee, 321, 325

mortgagor, 321, 325

multiplicand, 21–22, 32, 33

multiplication: checking of, 22, 33; definition of, 21, 32; of fractions, 86–91, 95; of mixed numbers, 90–91; products, accumulation of, 25, 33; terms used in, 21–22; of whole numbers and decimals, 22–24, 33: with zeros, 23–24, 33

multiplier, 21–22, 32, 33

N

n/EOM, 183, 192

n/30EOM, 183, 192

n/30ROG, 183, 192

negative numbers: combining addition and subtraction, 20–21; and positive numbers, subtracting, 33; in subtraction, 18, 32

net amount due on invoice, 183, 184, 185

net cost, 182

net pay, 210, 219

net profit, 189, 192

new charge account balance, calculating, 314–15

non-interest-bearing promissory note, 284, 297

numbers: decimal system, 2, 7; mixed, 47, 48, 54, 70–72, 76, 90–91, 93–94, 95; negative, 18, 20–21, 32, 33; prime, 66, 68–70, 75; rounding, 5, 8; whole, 16–28, 47, 89–90, 92, 95; and word forms, 3–4, 7–8

numerator, 46, 56

O

Omnibus Trade and Competitiveness Act of 1988, 338

open-ended credit, 311, 325

operating expenses, 189, 192

option account, 311, 325

ordinary interest, 280, 281

outstanding checks, 166–68, 172

outstanding deposits, 166–68

overdraft protection, 159

overtime, 205–6, 219, 220

P

partial dividend, 26, 32, 34

payroll deductions: calculating, 210–15, 221; computing net pay, 214–15, 221; federal income tax, 211–14, 221; FICA, 210–11, 221; other deductions, 214–15; terms used in computing, 209–10

payroll register, 210, 216–17, 219

percentage, base, and rate: base, finding, 126–27, 129; percent problems, identifying elements of, 127, 129; percentage, finding, 122–24; percentage shown with decimals, 123–24, 129; percentage shown with fractions, 124, 129; rate, finding, 125–26, 129; terms used in, 122

percents: converting decimals to, 109–10, 114; converting to decimals, 108–9, 114;

converting fractions to, 112, 114; converting to fractions, 110–11, 114; introduction to, 108–13; percent of increase and decrease, 138–44, 146; percent problems, identifying elements of, 127, 129; percentage distribution, 138, 13–4, 146; terms used in, 108, 138

personal check writer, 163

personal identification number (PIN), 159

Piecework wage system, 205, 209, 219, 220

places (positions) in decimal system, 2, 5

policy, insurance, 256, 266

policyholder, 256

positions (places) in decimal system, 2, 5

Preferred Provider Organization (PPO), 257

premium, insurance, 256, 266

present value, 286, 294, 295, 297, 300

previous balance method of calculating finance charges, 312, 325, 326

pricing: equations for, basic, 189, 194; markdown sale price, determining, 190, 194; terms use in, 189

prime factorization of a number, 68

prime numbers, 66, 68–70, 75

principal, 279, 283, 297, 299

proceeds of a promissory note, 284, 297

product, 22, 32

products, accumulation of, 25, 33

promissory notes, 284–85, 297, 299

proper fractions, 47, 56

property insurance, 261–64, 267, 268

property tax, 232, 233–34, 248

purchasing: cash discounts, calculating, 183–87; complement method of determining net amount due, 184; invoice total, determining, 188; returned merchandise, 184–85; sales tax, calculating, 187–88; shipping and insurance charges, 186–87; terms used in, 182, 183

Q

quarterly compounding, 287, 290, 293, 300

quotient, 25, 26, 32, 34

R

rate of increase or decrease, 138
rates of percentages, 122, 125–26, 129
rebates on installment loans, 318–19, 325, 327
reconciliation of bank statement, 167–71, 172
regrouping (borrowing), 18–20, 32
remainder: in division, 25, 26, 27–28, 32, 34; in subtraction, 17, 32, 33
restrictive endorsement, 161, 172
retail installment contract, 315
retail price, 189
retailer, 182, 192
returned checks, 166
returned merchandise, 184–85, 193
revolving charge account, 311, 325
rounding, 4, 5, 8
Rule of 78ths, 318, 325
running balance, 162

S

salaried personnel, 205, 207–8, 219, 220
sales tax, 187–88, 192, 193
selling price, 189, 192
semiannual compounding, 289–90, 293, 300
service charges, 159, 166–67, 172
shipping and insurance charges, 186–87, 193
simple interest, 278, 279, 297, 298–99
Social Security, 210–11, 219, 221, 237
standard deduction, 240, 248
state unemployment insurance tax (SUTA), 234–35, 249
stated interest rate, 287, 316
statement of bank account, 165–67, 172
statement reconciliation, 167–71, 172
straight time, 205, 219, 220
subtraction: checking of, 18, 33; definition of, 17, 32; of fractions, 66, 72–73, 76;

negative numbers, combining addition and subtraction, 20–21; of negative and positive numbers, 20–21, 33; regrouping in, 18–20; terms used in, 17–18, 66; of whole numbers and decimals, 18, 33
subtrahend, 17, 32, 33
Sum-of-Digits method of computing interest rebates, 318, 325
sum, total, or amount, 16, 32

T

tax rate, 233, 248
tax tables, 240–43, 248
taxable income, 240, 248
taxes: income, federal, 210, 211–14, 219, 221, 237, 239–45, 249; income, state, 214; property, 232, 233–34; sales, 187–88, 192, 193; tax rate, 233, 248; terms used in, 232–33; unemployment insurance, federal, 236; unemployment insurance, state, 234–35
temperature, measurement using metrics, 346, 350
term insurance, 260, 266
term of insurance policy, 256, 266
terms of cash discounts, 183, 192, 193
thirty-day charge account, 311, 325
time-and-a-half, 204, 219, 220
time to repay a loan or earn interest on an investment, 279, 283–84, 297, 299
total income, 239
total installment price, 317, 325, 327
total, sum, or amount, 16, 32
trial-and-error method of reducing fractions to lowest terms, 49–50, 56
Truth in Lending Act, 310

U

unemployment insurance tax: federal, 236, 249; state, 234–35, 249

universal life insurance, 260, 267
unlike fractions, 66, 67, 72, 75

V

variable life insurance, 261, 267
variable universal life insurance, 261, 267
volume, measurement using metrics, 344–46, 350

W

W-2 form. *See* Wage and Tax Statement
W-4 form. *See* Employee's Withholding Allowance Certificate
Wage and Hour Law, 204, 219
Wage and Tax Statement (W-2 form), 238, 248
weight, measurement using metrics, 343–44, 350
whole life insurance, 260, 266
whole-number part of a decimal number, 2
whole numbers: converting improper fractions to, 47, 56; in decimal system, 2, 7; division of fractions and, 92, 96; division of mixed numbers and, 93–94, 96; multiplication of fractions and, 89–90, 95; multiplication of mixed numbers and, 91, 95; operations with decimals and, 16–28
wholesaler, 182, 192
withholding tax table, 213
word forms of numbers, 3–4, 7–8

Z

zeros: in multiplication, annexing/appending, 24, 32; in multiplication, difficulty with, 23–24, 33; in regrouping, 19–20

Answers to Assignments (Odd-Numbered)

The answers shown have been calculated by hand according to the instructions in the text. If you have used a calculator, your answers may vary slightly for those questions involving fractions, cents, and decimals.

Chapter 1–Assignment 1

A.
1. 4 tens
3. 2 hundredths
5. 2 millions
7. 4 units
9. 4 hundreds
11. 5 hundreds
13. 1 hundred thousandths
15. 3 hundred thousands
17. 9 hundred thousandths
19. 7 thousands

B.
21. 130
23. 73.65
25. 5,600.034
27. 1,002
29. 4,000,000,000

C.
31. five hundred ninety-five

Chapter 1– Assignment 2

1. four hundred sixty-five thousand, six hundred two
3. two and four hundred sixty-nine thousandths
5. three million, four hundred fifty-two thousand and forty-two thousandths
7. two and forty-two hundredths
9. four hundred sixty-two thousandths
11. twelve thousand, five hundred ninety-two and seven tenths
13. five billion, four hundred twenty-nine million, one thousand, nineteen
15. eight hundred eighty-nine thousandths
17. two thousand three hundred thirty-five
19. forty-seven and nine tenths
21. two hundred fifty-five
23. six thousand, eight hundred eighty-one and one tenth

Chapter 1–Assignment 3

A.
1. 5 units
3. 9 millions
5. 4 hundred thousandths
7. 5 thousands
9. 4 hundredths

B.
11. two hundred sixty-nine
13. thirty and nine tenths
15. four thousand, nine hundred ninety-nine and two hundred sixty-two thousandths

C.
17. 2,000,000.11   19. 964.12

Chapter 1–Assignment 4

A.
1. eight and fourteen hundredths
3. one million four hundred fifty-seven thousand nine hundred ninety-nine
5. six and one hundred sixty-eight thousandths
7. eight and nine thousand forty-seven ten thousandths
9. twenty-three and two hundred twenty-two thousandths

B.
11. 68,022
13. 80,000,001,100.1
15. 708

C.
17. 9 thousandths
19. 4 ten thousands
21. 1 ten millions
23. 8 thousands
25. 1 unit

Chapter 1–Assignment 5

A.	1. 950	3. 5,320	5. $5,270
B.	7. 642,500	9. $400	
C.	11. 423	13. 42,963	15. 5,472
D.	17. 59,012,000	19. 36,000	
E.	21. 1.0	23. 8,494.2	25. 10,000.0
F.	27. 469.25	29. 29,465.25	
G.	31. 249.957	33. 5.068	35. 7.649
H.	37. 0.3612	39. 0.0037	
I.	41. 50	43. 9	45. 5,001
J.	47. 41.9625	49. 1.0000	
K.	51. 47.9	53. 1.5	55. 8.1
L.	57. 100	59. 800	

Chapter 1–Assignment 6

A.	1. 651	3. 7,909	5. 1
B.	7. 4,480	9. 99,830	
C.	11. 0.3	13. 237.9	15. 7.2
D.	17. 600	19. 400	
E.	21. 0.12	23. 35.35	25. 1.46
F.	27. 28,000	29. 23,000	
G.	31. 0.487	33. 69.178	35. 0.838
H	37. 0.1212	39. 258.5588	
I.	41. 1,800,000	43. 150,000	45. 10,000
K.	47. 300,000	49. 900,000	
L.	51. 408	53. 76	55. 10,696
M.	57. 10,800	59. 400	

Chapter 2–Assignment 1

A. **1.** 4,764 **3.** 88,589 **5.** 60,461
7. 2,686 **9.** 202.62 **11.** 8.987
13. $128.55 **15.** $223.26
B. **17.** 3,934 **19.** 4,909,232 **21.** 2,037
23. 990,184 **25.** $20.30 **27.** $758,680
C. **29.** $54.43 **31.** 4,110 **33.** 1,457,903

Chapter 2–Assignment 2

A. **1.** 20,674 **3.** $1,102.34 **5.** $2,384.01
7. 32,737 **9.** 1,187.12
B. **11.** 6,015 **13.** 933 **15.** 19.752
C. **17.** 558 **19.** 14.849 **21.** 1.26
23. 22.07 **25.** 161.14
D. **27.** $212.09 **29.** 247 **31.** 43.181
E. **33.** $62.92 **35.** −99.77

Chapter 2–Assignment 3

1. 0.19 **3.** 1.44 **5.** $0.01
7. 0.63 **9.** 20.73 **11.** 8.28
13. 0.31 **15.** $10.51 **17.** 154.22
19. 8.94

Chapter 2–Assignment 4

A. **1.** 2.88 **3.** 28.45 **5.** 240.86
7. 10.60 **9.** 2,368.00 **11.** 369.98
13. 686.80 **15.** 273.00 **17.** 9.23
19. 21.33 **21.** 5.65 **23.** 97
25. 83
B. **27.** 29.21 **29.** 0.61 **31.** $53.86
33. $3.97 **35.** 6.84 **37.** 20.00
39. 0.17 **41.** 28.57 **43.** $0.04

Chapter 2–Assignment 5

1. $3,230 **3.** $40,000 **5.** $17.35
7. $265.00 **9.** $161.70 **11.** $7.37

Chapter 2–Assignment 6

1. $127.25 **3.** $409.75 **5.** $16,100.40
7. 312 **9.** $39.77 **11.** $258.00
13. 12.25 in.

Chapter 3–Assignment 1

A. **1.** I **3.** M **5.** I **7.** P
B. **9.** $9\frac{3}{8}$ **11.** $1\frac{2}{5}$ **13.** $12\frac{3}{8}$ **15.** 8
C. **17.** $\frac{25}{4}$ **19.** $\frac{16}{3}$ **21.** $\frac{603}{8}$ **23.** $\frac{19}{4}$
D. **25.** 4 **27.** 5 **29.** 6 **31.** 21

Chapter 3– Assignment 2

A. **1.** $\frac{1}{2}$ **3.** $\frac{1}{6}$ **5.** $\frac{3}{7}$ **7.** $\frac{1}{2}$
B. **9.** $\frac{50}{125}$ **11.** $\frac{20}{96}$ **13.** $\frac{12}{57}$ **15.** $\frac{35}{50}$
C. **17.** 0.16 **19.** 9.42 **21.** 0.44 **23.** 7.14
D. **25.** $\frac{3}{4}$ **27.** $\frac{3}{100}$

Chapter 3–Assignment 3

A. **1.** $4\frac{11}{12}$ **3.** 11 **5.** $6\frac{1}{4}$ **7.** $4\frac{1}{3}$
B. **9.** $\frac{135}{30}$ **11.** $\frac{73}{16}$ **13.** $\frac{53}{16}$ **15.** $\frac{34}{7}$
C. **17.** $3\frac{3}{5}$ **19.** $8\frac{39}{1,000}$ **21.** $9\frac{469}{1,000}$ **23.** $3\frac{33}{50}$
D. **25.** 0.80 **27.** 2.29 **29.** 4.38 **31.** 3.43

Chapter 3–Assignment 4

1a. 0.60 **b.** $\frac{20}{12} = 1\frac{2}{3}$
3a. $28\frac{2}{5}$ **b.** $36\frac{4}{5}$ **c.** $64\frac{1}{2}$
d. $32\frac{3}{5}$ **e.** $8\frac{3}{25}$
5a. 0.67 **b.** 0.33
7a. $\frac{27}{50}$ **b.** $\frac{491}{1,000}$ **c.** $1\frac{17}{50}$
d. $\frac{41}{125}$ **e.** $32\frac{163}{1,000}$

Chapter 3–Assignment 5

1a. $34.50 **b.** $36.13 **c.** $38.38
d. $37.40 **e.** $35.75
3a. $7\frac{3}{20}$ **b.** $23\frac{2}{5}$ **c.** $2\frac{3}{4}$
d. $48\frac{4}{5}$ **e.** $69\frac{3}{5}$
5a. 0.17 **b.** 0.33 **c.** 0.50
7. $\frac{59}{246}$

Chapter 3–Assignment 6

1a. $79\frac{31}{250}$ **b.** $\frac{2,069}{2,500}$ **c.** $\frac{11}{20}$
d. $3\frac{17}{20}$ **e.** $\frac{1}{4}$
3a. 0.25 **b.** 0.50 **c.** 0.13 **d.** 0.13
5. $\frac{35}{80}$ or $\frac{7}{16}$, 0.44
7a. 12, 12.00 **b.** $12\frac{11}{13}$, 12.85
c. $49\frac{1}{6}$, 49.17 **d.** $166\frac{1}{3}$, 166.33
e. $3\frac{1}{2}$, 3.50

Chapter 4–Assignment 1

A. 1. 75   3. 20   5. 12   7. 75

B. 9. $1\frac{3}{4}$ 11. $\frac{19}{25}$ 13. $1\frac{2}{5}$ 15. $1\frac{1}{3}$

C. 17. $79\frac{1}{4}$ 19. $21\frac{1}{2}$ 21. $12\frac{14}{35}$ 23. $12\frac{1}{13}$

D. 25. $\frac{7}{8}$ 27. $1\frac{1}{84}$ 29. $1\frac{1}{6}$ 31. $2\frac{1}{16}$

E. 33. $48\frac{1}{24}$ 35. $24\frac{59}{72}$ 37. $23\frac{11}{56}$ 39. $122\frac{3}{10}$

F. 41. $1\frac{29}{48}$ 43. $1\frac{3}{8}$ 45. $1\frac{11}{20}$

Chapter 4–Assignment 2

A. 1. $\frac{1}{2}$ 3. $\frac{1}{5}$ 5. $\frac{1}{2}$ 7. $\frac{1}{4}$

B. 9. $6\frac{1}{8}$ 11. $4\frac{11}{21}$ 13. $9\frac{1}{2}$ 15. $102\frac{1}{4}$

C. 17. $\frac{5}{12}$ 19. $\frac{3}{8}$ 21. $\frac{1}{10}$ 23. $\frac{2}{15}$

D. 25. $12\frac{11}{35}$ 27. $5\frac{8}{21}$ 29. $1\frac{1}{12}$ 31. $6\frac{5}{18}$

E. 33. $\frac{7}{48}$ 35. $\frac{1}{6}$ 37. $\frac{1}{63}$

F. 39. $1\frac{1}{2}$ 41. $3\frac{1}{12}$ 43. $9\frac{7}{15}$ 45. $2\frac{1}{4}$

Chapter 4–Assignment 3

A. 1. $4\frac{3}{4}$ 3. $14\frac{5}{6}$ 5. $96\frac{2}{3}$ 7. $1\frac{1}{2}$

B. 9. $13\frac{1}{8}$ 11. $11\frac{15}{28}$ 13. $4\frac{37}{70}$ 15. $6\frac{9}{20}$

C. 17. $2\frac{1}{4}$ 19. $6\frac{7}{9}$ 21. $8\frac{11}{12}$ 23. $2\frac{6}{11}$

Chapter 4–Assignment 4

1. $2\frac{3}{8}$ yards   3. $38\frac{1}{6}$ gallons

5. $2,820\frac{6}{10}$ miles

7a. $-\$4\frac{5}{8}$ b. $-\$1\frac{7}{8}$ c. $-\$\frac{2}{3}$

d. $-\$2\frac{3}{8}$ e. $-\$1\frac{7}{8}$

Chapter 4–Assignment 5

1. $9\frac{1}{20}$ pounds   3. $23\frac{11}{24}$ yards sand

5. $1\frac{11}{12}$ reams   7. $4\frac{3}{4}$ hours   9. $\$14\frac{1}{2}$

Chapter 4–Assignment 6

1. $7\frac{3}{4}$ hours   3. $3\frac{1}{7}$ feet long

5. $1\frac{37}{168}$ gross   7. $\$64\frac{19}{24}$ 9. $16\frac{1}{4}$ feet

Chapter 5–Assignment 1

1. $\frac{5}{36}$ 3. $\frac{1}{3}$ 5. $\frac{5}{12}$ 7. $1\frac{2}{3}$

9. $\frac{25}{27}$ 11. $2\frac{7}{9}$ 13. $12\frac{11}{24}$ 15. $7\frac{3}{16}$

17. $29\frac{1}{6}$ 19. $\frac{2}{3}$ 21. $\frac{1}{6}$ 23. $7\frac{5}{16}$

25. $19\frac{29}{32}$ 27. $\frac{3}{14}$ 29. $\frac{5}{16}$ 31. $\frac{1}{12}$

33. $34\frac{8}{25}$ 35. $3\frac{3}{35}$ 37. $\frac{7}{48}$ 39. $\frac{28}{45}$

41. $14\frac{1}{6}$ 43. $\frac{3}{25}$ 45. $\frac{35}{48}$ 47. $26\frac{11}{20}$

49. $5\frac{1}{4}$ 51. $4\frac{1}{8}$

Chapter 5–Assignment 2

1. $2\frac{2}{3}$ 3. $\frac{7}{8}$ 5. $1\frac{7}{9}$ 7. $1\frac{11}{19}$

9. $2\frac{2}{5}$ 11. $2\frac{22}{39}$ 13. $\frac{56}{75}$ 15. $\frac{7}{9}$

17. $1\frac{1}{3}$ 19. $2\frac{4}{7}$ 21. $1\frac{2}{3}$ 23. 2

25. 9   27. $5\frac{10}{11}$ 29. $\frac{196}{513}$ 31. $41\frac{1}{4}$

33. $\frac{75}{76}$ 35. $1\frac{1}{3}$ 37. $1\frac{1}{8}$ 39. $\frac{1}{3}$

41. $\frac{3}{5}$ 43. $\frac{2}{3}$ 45. 2   47. $\frac{4}{5}$

49. $\frac{9}{112}$ 51. $2\frac{2}{3}$

Chapter 5–Assignment 3

A. 1. $\frac{7}{16}$ 3. $\frac{7}{18}$ 5. $\frac{88}{135}$ 7. 15

9. $\frac{5}{12}$ 11. $\frac{1}{24}$ 13. $\frac{9}{49}$ 15. $\frac{11}{15}$

17. $\frac{1}{8}$ 19. $3\frac{3}{8}$ 21. $1\frac{53}{275}$ 23. $2\frac{2}{3}$

25. $2\frac{2}{9}$ 27. $\frac{6}{49}$ 29. $\frac{1}{5}$

Chapter 5–Assignment 4

1. $102\frac{9}{10}$ 3. $7\frac{1}{2}$ feet   5. $7\frac{7}{32}$

7. $\$25,000$ 9. $329\frac{1}{16}$ 11. $\$450$

13. John: 18; David: 6

Chapter 5–Assignment 5

1. $\frac{8}{15}$ yards   3. 120   5. 70

7. $\$31,500$ 9. 244   11. $\$8,333.33$

Chapter 5–Assignment 6

1. $62\frac{1}{2}$ yards   3. 48   5. $\$2,842.50$

7. $\$51.43$ 9. $5\frac{1}{15}$ hour   11. $\$3.23$/yd.

Chapter 6–Assignment 1
A. **1.** 0.03 **3.** 0.152 **5.** 0.312
7. 0.0254 **9.** 1.28 **11.** 0.15
13. 0.07 **15.** 3.55 **17.** 0.44
19. 2.12
B. **21.** 630% **23.** 7% **25.** 43%
27. 30% **29.** 89.7% **31.** 5%
33. 12.5% **35.** 2.5% **37.** 60%
39. 40%

Chapter 6–Assignment 2

1. $\frac{2}{25}$ **3.** $\frac{1}{1,000}$ **5.** $1\frac{3}{4}$

7. $\frac{53}{500}$ **9.** $\frac{3}{50}$ **11.** $\frac{22}{25}$

13. $\frac{13}{40}$ **15.** $2\frac{3}{5}$ **17.** $\frac{3}{32}$

19. $\frac{429}{1,000}$

B. **21.** 275.0% **23.** 37.5% **25.** 83.3%
27. 62.5% **29.** 75.0% **31.** 1,283.3%
33. 377.8% **35.** 566.7% **37.** 14.0%
39. 87.5%

Chapter 6–Assignment 3
A. **1.** 66.7% **3.** 47% **5.** 145%
7. 41.7% **9.** 0.3% **11.** 9.1%
13. 47% **15.** 561% **17.** 0.8%
19. 31.7%
B. **21.** 0.155 **23.** 0.30 **25.** 0.72
27. 0.00625 **29.** 0.007 **31.** 1.19
33. 1.00 **35.** 0.255 **37.** 0.09
39. 0.32125

Chapter 6–Assignment 4
1. −3.6%; 1.2%; 4.32%; −0.98%
3. 0.087
5. 0.019; 0.009; 0.011; 0.007; 0.003; 0.015; 0.016

Chapter 6–Assignment 5
1. 75%; 23.4%; 0.85%; 1.1%; 36.7%;
3.3%; 4.25%; 101%; 0.25% 87.5%
3. 0.0325; 0.0302; 0.0445; 0.027; 0.0125; 0.0567
5. 0.23; −0.11; 0.673; −0.08; 0.01

Chapter 6–Assignment 6
1. 0.008; 0.0075; 0.0125; 0.021;
0.028; 0.013; 11.66

3. $\frac{9}{20}$; $\frac{37}{100}$; $\frac{1}{2}$; $\frac{1}{4}$; $\frac{3}{10}$

5. 40.0%; 85.7%; 66.7%;
28.6%; 80.0%; 37.5%

Chapter 7–Assignment 1
A. **1.** ?; 34%; 65
3. 108.3; 100%; ?
5. 85.7; 56%; ?
7. ?; 19%; 133
9. 6; ?; 9
B. **11.** 0.75 **13.** 20.82 **15.** 2.88
C. **17.** 47.50% **19.** 17.33% **21.** 10.00%
D. **23.** 49.60 **25.** 75

Chapter 7–Assignment 2
A. **1.** 15.36 **3.** 270.00 **5.** 21.92
7. 6.03 **9.** 89.27
B. **11.** 11.2% **13.** 13.2% **15.** 58.3%
17. 80.9% **19.** 14.7%
C. **21.** 28.44 **23.** 370.37 **25.** 933.33
27. 200.00 **29.** 1,600.00

Chapter 7–Assignment 3
1. $1,396.08 **3.** 15.1% **5.** $200.00
7. $175.80 **9.** $82.05 **11.** 27.8%
13. $13.36 **15.** 49.0% **17.** $30.00
19. 38.88 **21.** $8,333.33

Chapter 7–Assignment 4
1. $116.85 **3.** $14.58 **5.** $2,250
7. $6,303.00 **9.** $296.49 **11.** $223.75

Chapter 7–Assignment 5
1. $192.92 **3.** 70.0% **5.** yes, 49.6%
7. $2,999.48 **9.** 252 people

Chapter 7–Assignment 6
1. $338.75 **3.** $1,890.00
5. 22,500 bottles **7.** 3,480 people
9. 11.5% **11.** 45%

Chapter 8–Assignment 1
A. **1.** $18,599; 28.6% **3.** $9,076; 31.4%
B. **5.** $5,078; 6.9% **7.** $16,734; 19.7%
C. **9.** 22.3% **11.** 12.0%

Chapter 8–Assignment 2
A. **1.** 15.2% **3.** 11.7% **5.** 28.1%
7. 28.3% **9.** 16.6%
B. **11.** 35.0% **13.** 17.5% **15.** 9.2%

Chapter 8–Assignment 3
A. **1.** $1,570; 6.1% **3.** $1,820; 5%
5. $1,150; 2.8%
B. **7.** $0.07; 6.3% **9.** $0.05; 4.5%
C. **11.** +16.2% **13.** −4.8% **15.** +7.7%

Chapter 8–Assignment 4

1. $56,895.82   3. $5,054
5. $210;   8.0%   7. 20%
9. 3,734 books;   11.3%
11. 87 employees;   493 employees
13. 59.4%

Chapter 8–Assignment 5

1. $24,801.88   3. $500,000
5. $54.56   7. 18.1%
9. $449,108.48   11. $43.27
13. 71.4%

Chapter 8–Assignment 6

1. $792,792   3. 22.0%
5. 49,900.0%   7. $312.17
9. $407.14   11. 273 pounds
13. $1,395,000

Chapter 9–Assignment 1

1. $950.00   3. $527.60   5. $460.40
7. $545.95   9. $523.65   11. $329.99
13. $595.73   15. $438.33   17. $713.87
19. $673.57

Chapter 9–Assignment 2

1. $800.90   3. $767.80   5. $767.80
7. $761.25   9. $761.25   11. $729.75
13. $729.75   15. $716.30   17. $716.30
19. $693.60   21. $1,443.60   23. $1,346.40
25. $1,346.40   27. $1,273.80   29. $1,273.80
31. $1,160.80   33. $1,260.80   35. $1,232.21

Chapter 9–Assignment 3

1. $1,167.90   3. $957.87   5. $1,257.87
7. $7.00   9. $785.37   11. $472.50

Chapter 9–Assignment 4

1. $1,693.50   3. $5,685.50   5. $686.80
7. more; $8   9. $1,911.90   11. $3,142.10

Chapter 9–Assignment 5

1. $1,171.11;   yes   3. $545.40
5. $27.53   7. $801.47; $801.47
9. $995.42   11. $63.00

Chapter 9–Assignment 6

1. no; $5.93   3. $693.14   5. $1,647.80
7. $540;   yes;   decreased $540 too much
9. $350.97   11. $1,112.48

Chapter 10–Assignment 1

A.   1. $12.47   3. $139.07   5. $9.60
    7. $3.85   9. $12.47   11. $82.30
   13. $62.51   15. $40.88   17. $101.24
   19. $90.02
B.   21. $10.70   23. $6.05   25. – 0 –
   27. – 0 –   29. $25.13   31. – 0 –
   33. $2.07

Chapter 10–Assignment 2

A.   1. $224.83   3. $120.02   5. $20.95
    7. $17.59   9. $52.07   11. $97.22
   13. $90.02
B.   15. $72.16   17. $38.51   19. $62.02
   21. $24.23   23. $630.00
C.   25. $37.98   27. $7   29. $111.70

Chapter 10–Assignment 3

1. $35.70   3. $989.85   5. $77.70
7. $499.75   9. $642.50   11. $19,917.15
13. $597.51   15. $1,159.18   17. $20,567.82

Chapter 10–Assignment 4

1. $7.88   3. $237.93   5. $258.93
7. $264.67   9. $117.04   11. $495
13. $1,259.39

Chapter 10–Assignment 5

1. $4,213.40   3. $889.91   5. $644.81
7. $6,249; $6,311.7   9. $1,098.02   11. $44

Chapter 10–Assignment 6

1. $14.64; $717.36
3. $194.77   5. $757.44; $773.22; $789.00
7. $321.01   9. $4,231.11

Chapter 11–Assignment 1

A.   1. $4,733   3. $1,350   5. $1,050
    7. $3,508   9. $1,625
B.   11. $174.00   13. $252.30   15. $90
C.   17. $364.00   19. $568.75   21. $315

Chapter 11–Assignment 2

B.   1. $278.00; $62.55; $340.55
    3. $380.00; $57.00; $437.00
    5. $350.00; $196.88; $546.88

Chapter 11–Assignment 3

A.   1. $25.30   3. $32.00   5. $20.64
    7. $43.91   9. $5.62   11. $15.91
   13. $7.55   15. $54.62   17. $37.93
   19. $24.63
B.   21. $25   23. – 0 –   25. $50
   27. $36   29. $19   31. – 0 –
   33. $26   35. – 0 –   37. $28
   39. $5

Chapter 11–Assignment 4
A. **1.** $463.16   **3.** $568.33   **5.** $651.50
B. **7.** $638   **9.** $487.50

Chapter 11–Assignment 5
1. $936.10   **3.** $312.50   **5.** $631
7. $393.75   **9.** $368.82   **11.** $674.00
13. $120.00

Chapter 11–Assignment 6
1. $4,090   **3.** $43.71   **5.** $295.37
7. $44.67   **9.** $3,582.25; $3,582
11. $1,165.50   **13.** $63

Chapter 12–Assignment 1
A. **1.** $61.20   **3.** $29.63   **5.** $3,600.00
7. $21.60   **9.** $156.00
B. **11.** 8%   **13.** 6%   **15.** 4%
17. 1%   **19.** 7%

Chapter 12–Assignment 2
A. **1.** $156.80   **3.** $251.72   **5.** $143.98
7. $252.00   **9.** $252.00
B. **11.** $47.92   **13.** $56.00   **15.** $56.00
17. $33.80   **19.** $16.47
C. **21.** $1,016   **23.** $3,311   **25.** $1,384
27. $1,699   **29.** $2,779

Chapter 12–Assignment 3
1. $19,750   **3.** $14,550   **5.** $13,600
7. $6,910   **9.** $13,030   **11.** $12,250
13. $14,200   **15.** $7,350

Chapter 12–Assignment 4
1. $11,300   **3.** $11,540   **5.** $13,900
7. $12,400   **9.** $7,500

Chapter 13–Assignment 1
1. $25   **3.** $4.17   **5.** – 0 –
7. $58.33   **9.** $100.00   **11.** $56.25
13. $0.83   **15.** – 0 –

Chapter 13–Assignment 2
A. **1.** $50,000   **3.** $60,000   **5.** $47,000
7. $57,000
B. **9.** $893.75   **11.** $2,500.00
C. **13.** $660.00   **15.** $49.50   **17.** $215.00

Chapter 13–Assignment 3
A. **1.** $60,000   **3.** $39,200   **5.** $280,000
7. $88,000   **9.** $20,000
B. **11.** $90,000   **13.** $20,000   **15.** $500,000
17. $210,000   **19.** $40,000

Chapter 13–Assignment 4
1. – 0 –   **3.** $2,301.00   **5.** $12,000
7. $8,300   **9.** $11,500   **11.** – 0 –
13. $356   **15.** $2,136

Chapter 13–Assignment 5
1. $120   **3.** $325.00   **5.** $50,922.16
7. $192.50   **9.** $706   **11.** $525

Chapter 13–Assignment 6
1. $561.00   **3.** $100,720   **5.** $125,000
7. $180,000   **9.** $143.75   **11.** $92,000

Chapter 14–Assignment 1
A. **1.** $418   **3.** $371.25
B. **5.** $26.65; $846.65   **7.** $222.00; $3,922.00
9. $22.05; $1,002.05
C. **11.** $52.50; $51.78   **13.** $117.00; $115.40
D. **15.** June 8   **17.** December 12
19. 86

Chapter 14–Assignment 2
A. **1.** $1,890.00   **3.** 18%   **5.** 16%
7. 60 days   **9.** 120 days
B. **11.** $110.64; $5,421.10
13. $46.56; $2,281.53
15. $24.72; $1,211.28
17. $144.55; $7,082.95
19. $73.10; $3,581.90

Chapter 14–Assignment 3
1. $45.13   **3.** $182.50   **5.** 13%   **7.** 106
9. $177.33; $3,977.33; $122.63; $3,854.73

Chapter 14–Assignment 4
A. **1.** $1,347.84   **3.** $1,167.39
B. **5.** $5,858.30; $858.30 **7.** $1,360.49; $360.49
C. **9.** $6,691.10   **11.** $6,963.28
D. **13.** $225.92   **15.** $2,000.77

Chapter 14–Assignment 5
A. **1.** $2,342.56; $742.56 **3.** $3,376.53; $376.53
5. $541.22; 41.22
B. **7.** $2,422.70   **9.** $4,619.23
11. $15,769.86

Chapter 14–Assignment 6
1. $6,348.67; $1,348.67
3. $251.02   **5.** $39,123.63
7. $18,267.25   **9.** $6.14

Chapter 15–Assignment 1
A. **1.** $6.58   **3.** $5.17
B. **5.** $468.06   **7.** $502.02
C. **9.** $23.23; $1,021.63 **11.** $23.98; $1,120.28
13. $25.22; $1,074.00

Chapter 15–Assignment 2

1. $10.52; $585.17   **3.** $28.11
5. $6.44

Chapter 15–Assignment 3

1. $8.51   **3.** $286.53
5. $5.45; $411.14   **7.** $296.83
9. $359.58

Chapter 15–Assignment 4

A.   **1.** $649.94; $164.94   **3.** $403.70; $104.70

B.   **5.** $\frac{78}{171}\left(\frac{26}{57}\right)$ **7.** $\frac{120}{300}\left(\frac{2}{5}\right)$

C.   **9.** $28.81   **11.** $60.31
D.   **13.** $422.91   **15.** $458.17

Chapter 15–Assignment 5

1. $485.06   **3.** $3,309   **5.** $252.30
7. $617.92
9. interest payment = $450.00
principal payment = $89.84
balance on principal = $59,910.16

Chapter 15–Assignment 6

1. $133.84   **3.** $351   **5.** $55.77
7. $126   **9.** $28.87
11. $157.21; $3,082.79

Chapter 16–Assignment 1

A.   **1.** 39.6   **3.** 424.1   **5.** 4,100.0
     7. 8,014.0   **9.** 60.2
B.   **11.** 5,770.0   **13.** 6,000.0   **15.** 31,110.0
     17. 152,090.0   **19.** 4,110.0
C.   **21.** 274.805   **23.** 136.765   **25.** 73.660
D.   **27.** 30.666   **29.** 7.625
E.   **31.** 250.800   **33.** 25,242.140 **35.** 1,090.388

Chapter 16–Assignment 2

A.   **1.** 47.3   **3.** 110.0   **5.** 21.3
B.   **7.** 612,160.0   **9.** 7,410,000.0
C.   **11.** 343.678   **13.** 1,814.000   **15.** 1,956.150
D.   **17.** 512.0   **19.** 21.2
E.   **21.** 9.180   **23.** 1,926.565
F.   **25.** 8

Chapter 16–Assignment 3

A.   **1.** 1,400   **3.** 6,280.0   **5.** 4,000.0
     7. 68,790.0   **9.** 19.3
B.   **11.** 47.2   **13.** 9,410.0   **15.** 2.9
     17. 9,990.0   **19.** 3,860.0
C.   **21.** 378.0   **23.** 144.7   **25.** 100.5
     27. 1,940.0   **29.** 570.0

Chapter 16–Assignment 4

1. 11.355   **3.** 1,680   **5.** 20
7. 188.100   **9.** 20°

Chapter 16–Assignment 5

1. 96.8°F   **3.** 10 ft, 0.499 in.
5. 39.246 m²   **7.** 11,700.3 t
9. 31.7°C; 24.5°C; 0°C; 12.8°C; 8.3°C; −5.6°C

Chapter 16–Assignment 6

1. 5.149; 49.885; 343.319; 5.488
3. 50.747 ha   **5.** 4.113 m
7. 10.131 mi   **9.** 302.250 m²

Answers to Math Alerts

Chapter 1–Math Alert

1. $1.42; $1.20; $1.46; $1.60

2.

Your Name Address City, State Zip	No. *101* $\frac{11-71}{690}$
	Nov. 1 19 —
PAY TO THE ORDER OF *Chase Visa*	$ *455 12*
Four Hundred Fifty-Five and $\frac{12}{100}$	DOLLARS
First Texas Bank San Antonio, Texas 78223-6031	For Classroom Use Only *Your Name*

⑆069000712⑆ ⑈016172711⑈

Chapter 2–Math Alert

1. 49,649;   68,052;   58,089;   175,790

3. 1,942;   80.917 rounded to 81

2. 26,930;   8,904;   18,026

4. see below

5a. $58 **b.** $142 **c.** $40 **d.** $160 **e.** $827

Lucky Western Wear

No. 292001

2866 Hines Boulevard
Dallas, TX 75201-6328

Phone: (214) 555-2229

Customer order no.: *A1109* **Date:** *10-20-19--*

Sold to: *J. R. Barnes Corp.*

Address: *1406 56th Street*

Dallas, TX 75241-2201

Terms: *N/A*   **Sales representative:** *R.L.*

Qty.	Stock No.	Description	Price	Amount
18	B2245	Boot, Leather	79.99	1. $1,439.82
6	H2933	Belt, Leather	28.95	2. $173.70
20	H2935	Belt Buckle, Silver	34.95	3. $699.00
15	I9986	Shirt, Long Sleeve	34.95	4. $524.25
6	K2916	Hat	75.00	5. $450.00
3	K1468	Hat	82.50	6. $247.50
24	L1456	Jeans	54.00	7. $1,296.00
15	M129	Hatband	67.50	8. $1,012.50
12	N1698	Bootjack	12.95	9. $155.40
144	P1698	Hatpin, Assorted	3.95	10. $568.80
		Total		11. $6,566.97

Chapter 3–Math Alert

1a. $\dfrac{593}{1,000}$ **b.** $\dfrac{491}{1,000}$ **c.** $\dfrac{62}{125}$ **d.** $\dfrac{41}{125}$ **e.** $\dfrac{461}{1,000}$

2. 3.50

Chapter 4–Math Alert

1a. $166\dfrac{3}{4}$ **b.** 166.75

2a. $14\dfrac{7}{12}$ lb **b.** 14.583 lb

3a. $61\dfrac{7}{8}$ **b.** 61.875 **c.** $58\dfrac{1}{2}$

Chapter 5–Math Alert

Solution: $50\dfrac{1}{4} \div 6 = 8\dfrac{3}{8}$

Chapter 6–Math Alert

1. 0.0506 **2.** 0.04 **3.** −0.0397 **4.** 0.1379
5. 0.0000 **6.** −0.1039 **7.** 0.2051 **8.** −0.0602
9. 0.0000 **10.** 0.0161

Chapter 7–Math Alert

1. 57.0% **2.** 56.7% **3.** 50.7%
4. 48.9% **5.** 45.4% **6.** 40.4%

Chapter 8–Math Alert

1. 1,197 registrations **2.** 2,477 units
3. 150 units **4.** 1,017 units

Chapter 9–Math Alert

Solution: Solutions will vary

Chapter 10–Math Alert

A Invoice totals are correct.

B **1a.** Original $19.95 − 16 = 3.95 ÷ 19.95 = 19.8% (almost 20%)

1b. Original $45.95 − 28 = 17.95 ÷ 45.95 = 39.1% (almost 40%)

2. Yes, $40 × 0.50 = $20 (only a penny off.)

Chapter 11–Math Alert

1. see below

Employee	Hours Worked			Hourly Rate	Earnings		
	Straight	Overtime	Total		Straight	Overtime	Total
Holley, J.	40	6	46	$6.60	264.00	59.40	323.40
Sanchez, L.	40	8	48	$5.50	220.00	66.00	286.00
Stroner, B.	40	-0-	40	$4.25	170.00	-0-	170.00
Barrett, G.	40	10.5	50.5	$6.00	240.00	94.50	334.50

2. see below

Payroll Register

Employee No.	Ex.	Hrs. Wkd	Pay Rate	Reg.	O.T.	Gross Earnings	Federal In. Tax	Soc. Sec.	Medi-care	Med. Ins.	Union Dues	Total Ded.	Net Pay
							Deductions						
1	0	44	4.70	188	28.20	216.20	14.00	13.40	3.14	21.00	12.00	63.54	152.66
2	2	40	5.60	224	-0-	224.00	1.00	13.89	3.25	21.00	12.00	51.14	172.86
3	5	48	4.40	176	52.80	228.80	-0-	14.19	3.22	21.00	12.00	50.51	178.29
4	2	42	5.60	224	16.80	240.80	4.00	14.93	3.49	21.00	12.00	55.42	185.38
5	3	40	6.30	252	-0-	252.00	-0-	15.62	3.65	21.00	12.00	52.27	199.73
6	0	40	4.75	310	-0-	310.00	29.00	19.22	4.50	21.00	12.00	85.72	224.28
7	4	42	4.25	170	12.75	182.75	-0-	11.33	2.65	21.00	12.00	46.98	135.77
8	1	41	6.95	278	10.43	288.43	17.00	17.88	4.18	21.00	12.00	72.06	216.37
9	0	43	5.95	238	26.78	264.75	21.00	16.42	3.84	21.00	12.00	74.26	190.52
10	2	44	4.50	180	27.00	207.00	-0-	12.93	3.00	21.00	12.00	48.83	158.17
Totals				2,240	174.76	2,414.76	86.00	149.71	35.02	210.00	120.00	600.73	1,814.00

3a. $2,333.33 **b.** understated **c.** $200 understated

Chapter 12–Math Alert

Adjusted Gross Income = $13,225.00
Taxable Income = $6,975.00
Tax = $1,046.00
Refund = $303.00

Chaper 13–Math Alert

1. $29,750 (remember the $250 deductible)
2. $500 (The deductible applies only to Section I.)
3. $544
4. $2,750 ($3,000 less the $250 deductible)

Chapter 14–Math Alert

1. Solution:
   $I = 1,800 \times 0.025 \times 1 = \$45$;
   $MV = \$1,800 + \$45 = \$1,845.00$
2. Compound periods = $10 \times 4 = 40$
3. Rate = $6\% \div 4 = 1.5\%$
4. Table amount = 0.55126232
5. $\$40,000 \times 0.55126232 = \$22,050.49$ will grow to $40,000 in 10 years.

Chapter 15–Math Alert

95 monthly payments × $55
= $5,225 total installment price
$5,225 installment price − $1,994 cash price
= $3,231 finance charge

Proof of monthly charge: $\dfrac{\$3,231 + \$1,994}{95 \text{ months}}$
= $55/month

2a. 1.65% **b.** 1.49% **c.** 1.58% **d.** 1.24%

Chapter 16–Math Alert

1. See below

Temperature: January 1 to January 14

Day	Fahrenheit	Celsius
1	20°	−6.672 or −7° Celsius
2	18°	−7.784 or −8° Celsius
3	22°	−5.560 or −6° Celsius
4	32°	0° Celsius
5	38°	3.336 or 3° Celsius
6	39°	3.892 or 4° Celsius
7	45°	7.228 or 7° Celsius

2. See below

Town	Kilometers	Miles
Coutts to Lithbridge	117.477	73 miles
Lithbridge to Fort Macleod	56.324	35 miles
Fort Macleod to Claresholm	37.013	23 miles
Claresholm to High River	65.980	41 miles
High River to Calgary	61.152	38 miles
Total miles		210 miles

3. $235.71

Answers to Proficiency Quiz Reviews

Chapters 1–2–Proficiency Quiz

A.
1. tens 2. 93.11 3. $90.00 4. 0.7
5. 0.9836 6. thousandths
7. 9.037 8. 50,000 9. 3,894.8 10. 466

B.
11. 10,867 12. 546.712 13. $888.58
14. 66.734 15. 131,280 16. 25.38
17. 115.92 18. 2,560.91 19. 256 20. 8.461
21. $36.64 22. 18 hours

C.
23. 18,083 24. 542 25. $662.09
26. −0.731 27. 83 28. $189.84
29. $886.16 30. $16.01 31. $1,266
32. 67 33. 601 34. 819
35. 2,917 36. −2,970

D.
37. 1,632 38. 104,328 39. 3,664
40. 51,200 41. 147,288 42. 0.2793
43. 8.9122 44. 27.450192 45. $219.45
46. $0.8325 47. $11.88; $6.45; $5.95; $24.28
48. 990 49. 4,950 50. 6,222
51. 10,020 52. 93,104 53. 1,951,026
54. $14,400 55. $4,366

E.
56. 12.2632 57. 1,002.0000 58. 2,028.0000
59. 103.4314 60. 304.6400 61. 3,672.7273
62. 5.2478 63. 2,000.0000 64. 1,397.6522
65. 25.4824 66. 309.8750 67. 1,188.0667
68. 515.8000 69. 22.2343 70. 95.5000

Chapters 3–5–Proficiency Quiz

A.
1. $6\frac{1}{5}$ 2. $34\frac{1}{2}$ 3. $20\frac{1}{9}$ 4. $5\frac{3}{8}$
5. $18\frac{1}{2}$ 6. $5\frac{1}{4}$ 7. $13\frac{2}{3}$ 8. 13
9. $2\frac{2}{7}$ 10. 12

B.
11. $\frac{87}{4}$ 12. $\frac{58}{9}$ 13. $\frac{21}{8}$ 14. $\frac{95}{7}$
15. $\frac{15}{4}$ 16. $\frac{3}{2}$ 17. $\frac{19}{4}$ 18. $\frac{17}{2}$
19. $\frac{20}{3}$ 20. $\frac{29}{5}$

C.
21. $\frac{1}{4}$ 22. $\frac{1}{2}$ 23. $\frac{12}{13}$ 24. $\frac{8}{49}$
25. $\frac{5}{9}$ 26. $\frac{1}{2}$

D.
27. $\frac{60}{180}$ 28. $\frac{24}{52}$ 29. $\frac{64}{72}$ 30. $\frac{32}{96}$

E.
31. $\frac{61}{100}$ 32. $\frac{9}{200}$ 33. $\frac{3}{10}$
34. $\frac{1,493}{5,000}$ 35. $\frac{463}{1,000}$ 36. $\frac{3}{1,000}$
37. $\frac{193}{10,000}$ 38. $\frac{1}{5}$ 39. $\frac{2,453}{5,000}$ 40. $\frac{8}{25}$

[Chapters 1–2 continued — right column]

F.
41. 0.300 42. 3.250 43. 6.500 44. 0.060
45. 0.100 46. 0.012 47. 0.700 48. 0.640

G.
49. 120 50. 32 51. 36 52. 120

H.
53. $1\frac{3}{4}$ 54. $2\frac{3}{7}$ 55. $1\frac{7}{12}$
56. $14\frac{1}{45}$ 57. $6\frac{5}{8}$ 58. $1\frac{29}{63}$

I.
59. $\frac{1}{2}$ 60. $\frac{3}{10}$ 61. $\frac{1}{15}$
62. $4\frac{1}{3}$ 63. $2\frac{7}{12}$ 64. $1\frac{1}{9}$
65. $1\frac{3}{8}$ 66. $\frac{16}{63}$

J.
67. $\frac{1}{18}$ 68. $\frac{16}{21}$ 69. $10\frac{2}{3}$ 70. $7\frac{1}{8}$
71. $\frac{2}{25}$ 72. $\frac{1}{12}$ 73. $\frac{1}{6}$ 74. $5\frac{1}{3}$
75. $13\frac{2}{9}$ 76. $\frac{9}{28}$

K.
77. 1 78. 16 79. 2 80. $3\frac{4}{7}$
81. 5 82. $9\frac{1}{3}$ 83. $\frac{4}{5}$ 84. $1\frac{1}{39}$
85. 6 86. $\frac{2}{5}$

L.
87. $29\frac{1}{4}$ hours 88. $28\frac{5}{24}$ yards
89. $23.00 90. $\frac{13}{24}$ cake
91. 4,335 ft^2 92. $7,066.05
93. $58.88

Chapters 6–8–Proficiency Quiz

A.
1. 0.000235 2. 0.63 3. 0.085
4. 0.07 5. 0.1384 6. 0.214
7. 0.0045 8. 0.734 9. 1.5
10. 0.001459 11. 0.017936 12. 0.2495
13. 3.231 14. 0.85

B.
15. 41% 16. 6.51% 17. 11.15%
18. 463% 19. 72.80% 20. 5.38%
21. 0.01% 22. 2% 23. 365%
24. 9.99% 25. 0.045% 26. 615.04%
27. 19.65% 28. 148.9%

C.
29. $\frac{13}{20}$ 30. $\frac{1}{2}$ 31. $1\frac{11}{100}$
32. $\frac{1}{20}$ 33. $\frac{9}{10}$ 34. $1\frac{1}{2}$
35. $\frac{4}{25}$ 36. $\frac{9}{20}$ 37. $2\frac{1}{2}$
38. $\frac{3}{20}$

D.

39. 20%	**40.** 37.5%	**41.** 50%
42. 66.67%	**43.** 62.5%	**44.** 12.5%
45. 90%	**46.** 87.5%	**47.** 33.3%
48. 40%		

E.

49. 240	**50.** 0.12	**51.** 7.5
52. 44.4	**53.** 25	**54.** $10.20
55. $9.15	**56.** $318.99	

F.

57. 20%	**58.** 21.4%	**59.** 51.4%
60. 50%		

G.

61. $90.00	**62.** 3.44%	**63.** $386.67
64. $121.80	**65.** $2,500	**66.** 14%

67. 20%	**68.** $433.33	**69.** $136.00
70. 8%	**71.** $36.00	**72.** $760.00
73. $750.00	**74.** $250.00	**75.** $34.65
76. 22.5%	**77.** $1.76	**78.** 50%

H.

79. 20.12%	**80.** 87.6%	**81.** −14.5%
82. 42.76%	**83.** −38.47%	**84.** −7.78%
85. 44.22%	**86.** −3.02%	**87.** 76.26%
88. −22.82%	**89.** 33.9%	**90.** 44.28%

I.

91. 9.98%	**92.** 14.89%	**93.** 29.35%
94. 45.77%		

Chapter 9–Proficiency Quiz

A.

DATE	CHECK NUMBER	DESCRIPTION OF TRANSACTION	√	AMOUNT OF DEPOSIT		AMOUNT OF CHECK		BALANCE		
								$ 5,952	80	
4/01/–	1329	Segal Mfg. Co.				310	15	310	15	
		Accts. Payable						5,642	65	1.
4/05/–		Deposit		3,671	50			3,671	50	
		Professional fees						9,314	15	2.
4/10/–	1330	Dayton Eq. Co.				75	90	75	90	
		Dental drills						9,238	25	3.
4/11/–	1331	Moore Dental Supplies				180	75	180	75	
		Dental supplies						9,057	50	4.
4/12/–		Deposit		2,190	90			2,190	90	
		Professional fees						11,248	40	5.

B.

No. _1_ $ **65.39**
August 4, 19 --
TO *ABC Offfice*
FOR *Office supplies*

	Dollars	Cents	
Bal. Bro't. For'd.	620	50	
Amt Deposited	200	00	
Total	820	50	**6.**
Amt. This Check	65	39	**7.**
Bal. Car'd For'd.	755	11	**8.**

No. _2_ $ **25.00**
August 4, 19 --
TO *Right-Way Page*
FOR *Phone Expense*

	Dollars	Cents	
Bal. Bro't. For'd.	755	11	**9.**
Amt Deposited	—	—	
Total	755	11	**10.**
Amt. This Check	25	00	**11.**
Bal. Car'd For'd.	730	11	**12.**

No. _3_ $ **39.45**
August 5, 19 --
TO *Big Deals Furn.*
FOR *Office Furniture*

	Dollars	Cents	
Bal. Bro't. For'd.	730	11	**13.**
Amt Deposited			
Total	730	11	**14.**
Amt. This Check	39	45	**15.**
Bal. Car'd For'd.	690	66	**16.**

No. _4_ $ **6.21**
August 6, 19 --
TO *US Post Office*
FOR *Postage Expense*

	Dollars	Cents	
Bal. Bro't. For'd.	690	66	**17.**
Amt Deposited	—	—	
Total	690	66	**18.**
Amt. This Check	6	21	**19.**
Bal. Car'd For'd.	684	45	**20.**

No. _5_ $ **88.89**
August 7, 19 --
TO *Mkt Auto Rental*
FOR *Leasing Expense*

	Dollars	Cents	
Bal. Bro't. For'd.	684	45	**21.**
Amt Deposited	451	39	
Total	1135	84	**22.**
Amt. This Check	88	89	**23.**
Bal. Car'd For'd.	1046	95	**24.**

No. _6_ $ **60.00**
August 7, 19 --
TO *Miller Clinic*
FOR *Medical Exp.*

	Dollars	Cents	
Bal. Bro't. For'd.	1046	95	**25.**
Amt Deposited	—	—	
Total	1046	95	**26.**
Amt. This Check	60	00	**27.**
Bal. Car'd For'd.	986	95	**28.**

C.

Reconciliation of Bank Statement

Bank balance on statement		*$3,811.42* **29.**
Less outstanding checks	*32.11*	
	10.95	
	98.10	
	300.50	
	211.05	*652.71* **30.**
		$3,158.71 **31.**
Plus outstanding deposits	*700.00*	
		700.00 **32.**
Adjusted bank balance		*$3,858.71* **33.**
Checkbook balance		*$3,265.21* **34.**
Less unrecorded debits	*12.50*	
		12.50 **35.**
Plus unrecorded credits	*606.00*	
		606.00 **36.**
Adjusted checkbook balance		*$3,858.71* **37.**

D. **38.** $672.90   **39.** $215.40   **40.** $454.30

Chapter 10–Proficiency Quiz

1. $51.32   **2.** $2,514.68   **3.** $63.00
4. $3,087.00   **5.** $49.56   **6.** $1,602.44
7. $59.78   **8.** $2,929.22   **9.** $57.84
10. $1,870.16 **11.** $60.00   **12.** $2,940.00
13. $50.34   **14.** $1,627.66 **15.** $66.69
16. $2,156.31 **17.** $36.18   **18.** $1,772.82
19. $61.50   **20.** $1,988.50

B. **21.** $12.00   **22.** $598.30   **23.** $1.87
24. $185.08   **25.** $3.38   **26.** $130.79
27. $15.04   **28.** $751.69   **29.** $11.29
30. $381.59

C. **31.** $12.74   **32.** $287.61   **33.** $12.32
34. $365.76   **35.** $139.30   **36.** $2,160.79
37. $78.12   **38.** $1,070.51 **39.** $12.69
40. $211.39   **41.** $40.69   **42.** $582.29

D. **43.** $106.45   **44.** $44.95   **45.** $363.09
46. $405.40   **47.** $150.50

E. **48.** $39.25   **49.** $501.10   **50.** $2,328.00
51. $2,868.35 **52.** $2,868.35 **53.** $–86.05
54. $2,782.30 **55.** $222.58   **56.** $3,004.88
57. $65.00   **58.** $3,069.88

Chapter 11—Proficiency Quiz

A. **1.** $3,292   **2.** $1,775   **3.** $1,158
4. $1,538   **5.** $4,583   **6.** $2,463

B. **7.** $340   **8.** $127.50   **9.** $467.50

C. **10.** $370   **11.** $124.88   **12.** $222
13. $716.88

D. **14.** $274.82   **15.** $674.82   **16.** $1,298.18
17. $1,623.18

E. **18.** $295.75

F. **19.** $54.25   **20.** $12.69   **21.** $38.47
22. $9.00   **23.** $22.64   **24.** $5.29
25. $17.47   **26.** $4.09   **27.** $25.61
28. $5.99   **29.** $86.21   **30.** $20.16
31. $60.32   **32.** $14.11

G.

	Hrs. Worked					Deductions							
Date	Reg.	O.T.	Regular	Overtime	Gross Earnings	Federal In. Tax	Soc. Sec.	Medi-care	Med. Ins.	Union Dues	Total Ded.	Net Pay	Accum. Earnings
4/1	40	12	**33.** 380.00	**34.** 171.00	**35.** 551.00	**36.** 58.00	**37.** 34.16	**38.** 7.99	**39.** 13.75	**40.** 3.00	**41.** 116.90	**42.** 434.10	**43.** 4,701.67
4/8	40	1	**44.** 380.00	**45.** 14.25	**46.** 394.25	**47.** 34.00	**48.** 24.44	**49.** 5.72	**50.** 13.75	**51.** 3.00	**52.** 80.91	**53.** 313.34	**54.** 5,095.92

H.

Payroll Register												
Name	Allow.	Hrs. Pay		Earnings		Gross Earnings	Deductions				Total Ded.	Net Pay
		Wkd.	Rate	Reg.	O.T.		Federal In. Tax	Soc. Sec.	Medi-care	Med. Ins.		
Bates, N	0	29	6.50	**55.** 188.50	**56.** -------	**57.** 188.50	**58.** 10	**59.** 11.69	**60.** 2.73	**61.** 12.50	**62.** 36.92	**63.** 151.58
Davis, B	2	45	5.75	**64.** 230.00	**65.** 43.13	**66.** 273.13	**67.** 8	**68.** 16.93	**69.** 3.96	**70.** 12.50	**71.** 41.39	**72.** 231.74
Stewart, R	1	42	8.00	**73.** 320.00	**74.** 24.00	**75.** 344.00	**76.** 26	**77.** 21.33	**78.** 4.99	**79.** 12.50	**80.** 64.82	**81.** 279.18

Chapter 12-13–Proficiency Quiz

A. **1.** $91.16 **2.** $221.49 **3.** $241.20
B. **4.** 4% **5.** 16% **6.** 4%
C. **7.** $270.00 **8.** $196.65 **9.** $71.97
D. **10.** $48.98 **11.** $23.89 **12.** $56.00
E. **13.** $1,894 **14.** $2,269
F. **15.** $2,756 **16.** $1,706
G. **17.** $59 **18.** $119
   19. $37.17 **20.** $77.17
H. **21.** $36,000 **22.** $10,000
   23. $46,000 **24.** $4,050
I. **25.** $424 **26.** $178
J. **27.** $31,500

Chapter 14–Proficiency Quiz

A. **1.** $156.67 **2.** $450.00 **3.** $37.50
   4. $540.00
B. **5.** $135.00 **6.** $1,635.00 **7.** $245.00
   8. $2,695.00 **9.** $754.00 **10.** $6,554.00
C. **11.** $93.00 **12.** $91.73 **13.** $60.67
   14. $59.84 **15.** $120.38 **16.** $118.73
D. **17.** 240 days **18.** 12% **19.** 3,000
E. **20.** $78.38 **21.** $3,538.20 **22.** $271.32
   23. $12,248.60
F. **24.** $860.60 **25.** $1,949.62 **26.** $285.48
   27. $3,380.48 **28.** $1,657.34
G. **29.** $395.78 **30.** $253.58 **31.** $1,460.99
   32. $183.24 **33.** $807.86
H. **34.** $113.75 **35.** 12%

Chapter 15–Proficiency Quiz

A. **1.** 1.54% **2.** 18% **3.** 20% **4.** 2%
B. **5.** $2.87 **6.** $290.31 **7.** $8.36 **8.** $550.53
C. **9.** $327.98 **10.** $5.05
D. **11.** $64.55
E. **12.** $17.71 **13.** $607.71
   14. $20.23 **15.** $570.23
F. **16.** $478.80 **17.** $3,138.80 **18.** $261.57
   19. $179.82 **20.** $1,178.82 **21.** $98.24
   22. $243.00 **23.** $1,593.00 **24.** $132.75
   25. $121.50 **26.** $796.50 **27.** $66.38
   28. $651.60 **29.** $4,271.60 **30.** $355.97
G. **31.** $5,117.68 **32.** $1,059.68
H. **33.** $\frac{21}{136}$ **34.** $\frac{45}{171} = \frac{5}{19}$
I. **35.** $5.77 **36.** $58.24
J. **42.** $702.40

Chapter 16–Proficiency Quiz

A. **1.** 5,000.0 **2.** 45.9 **3.** 235.9
   4. 21.1 **5.** 2.9 **6.** 3,659
   7. 86.9 **8.** 40 **9.** 39.5
   10. 25.4
B. **11.** 86,110 **12.** 78,960 **13.** 8,800
   14. 5,390 **15.** 990 **16.** 43,390
   17. 1,230 **18.** 4,890
C. **19.** 78.69 **20.** 44.79 **21.** 72.41
   22. 223.52 **23.** 272.67 **24.** 159.29
D. **25.** 1,170 **26.** 2.1900
E. **27.** 4.5 **28.** 1,056.7 **29.** 5,729.4
   30. 250.8
F. **31.** 0.088 **32.** 440 **33.** 59.800
   34. 88.700 **35.** 33.300 **36.** 30
G. **37.** 5,960 **38.** 2,430 **39.** 458,690
   40. 8,574,650
H. **41.** 1,587.6 **42.** 251.97 **43.** 1,814
I. **44.** 78.6 **45.** 34.2 **46.** 589.7
   47. 2,854.6
J. **48.** 226.734 **49.** 23.463 **50.** 87.055
   51. 192.231
K. **52.** 31° **53.** 6° **54.** 18°
   55. 2°
L. **56.** 315.0 **57.** 396.0 **58.** 67.2
   59. 93,555.0 **60.** 194.3 **61.** 2,376.5
   62. 139.7 **63.** 57,472.0

Student Progress Record   Name _____

Chapter	Assignment	Date Assigned	Date Completed	Score/ Grade	Comments
Chapter 1					
Chapter 2					
Proficiency Quiz Ch. 1-2 Review					
Chapter 3					
Chapter 4					
Chapter 5					
Proficiency Quiz Ch. 3-5 Review					
Chapter 6					
Chapter 7					
Chapter 8					
Proficiency Quiz Ch. 6-8 Review					
Chapter 9					

Chapter	Assignment	Date Assigned	Date Completed	Score/ Grade	Comments
Proficiency Quiz Ch. 9 Review					
Chapter 10					
Proficiency Quiz Ch. 10 Review					
Chapter 11					
Proficiency Quiz Ch. 11 Review					
Chapter 12					
Chapter 13					
Proficiency Quiz Ch.12-13 Review					
Chapter 14					
Proficiency Quiz Ch. 14 Review					
Chapter 15					
Proficiency Quiz Ch. 15 Review					
Chapter 16					
Proficiency Quiz Ch.16 Review					